Acoustics
for
Audiologists

Acoustics for Audiologists

Peter Haughton

Department of Medical Physics
Hull Royal Infirmary
Hull, UK

ACADEMIC PRESS

An imprint of Elsevier Science

Amsterdam   Boston   London   New York   Oxford   Paris   San Diego
San Francisco   Singapore   Sydney   Tokyo

This book is printed on acid-free paper. ∞

Copyright 2002, Elsevier Science (USA)

Academic Press
An Imprint of Elsevier Science
525 B Street, Suite 1900, San Diego, California 92101-4495, USA
http://www.academicpress.com

Academic Press
An Imprint of Elsevier Science
Harcourt Place, 32 Jamestown Road, London NW1 7BY, UK
http://www.academicpress.com

Library of Congress Catalog Card Number: 00-2001098440

International Standard Book Number: 0-12-332922-1

PRINTED IN THE UNITED STATES OF AMERICA
02 03 04 05 06 MB 9 8 7 6 5 4 3 2 1

Contents

Foreword

Audiology is a relatively new discipline and acoustics a well-established branch of physics. Audiology and acoustics have widened their applications over the years, and while acoustics has remained firmly based in physics, audiology has, by its nature, become a paramedical discipline. One of the effects of this is that a knowledge of acoustics is not seen as of primary importance to many professionals in the field of audiology. The value of this book is that it brings together the two areas of knowledge and demonstrates how acoustics can be seen as a very important underpinning of audiology.

If one goes back to the early days of audiology in the United Kingdom, which are well within the living memory of our more senior professionals, there were no audiologists. There were physicists, engineers, medically qualified people, and technical staff. The only people who had 'audiology' in their titles were audiology technicians. Many science and engineering graduates in those days would have undergone, in some part of their training, lectures on acoustics. Today it is interesting to note that the number of courses available at the graduate level in acoustics is very small and the subject is not taught in most physics courses. My personal experience, that confirms this situation to me, in teaching a module on audiology in a medical physics course over a period of five years, is that not one of the hundred or more honours graduates in physics had undertaken a course containing any element of acoustics. In fact the majority of them did not even know what a decibel was. It is therefore against this background that one must welcome a book that can be read at levels ranging from a nonmathematical introductory level to that of the science graduate dealing with both the theory and practice of acoustics related to audiology.

A further underpinning of audiology, and indeed any science that involves measurement, is the availability of standards. Standards take many forms and range from codes of practice to dealing with fine technical details. One of the main uses of standards is to ensure that measurements taken by an individual in one place at a certain time can be replicated within given tolerances in a different place at a different time. In audiology the relevant standards relate to procedures for measurement, to the hardware required to undertake the measurements, and to the environment in which the measurements are taken. To complicate the issue, many of the standards involved have to be based on subjective measurements relating to defining thresholds of hearing, as well as to the acoustical characteristics of the ear, head, and torso. Standards are therefore vital to the good practice of audiology and involve both acoustics and audiology to obtain meaningful results. However, possibly because standards today are developed in an international arena, namely, the ISO and IEC, and are contained within committees that are largely concerned with acoustical matters, they are not well known to many who practise audiology in spite of the fact that many of their daily clinical measurements rely on the validity of these standards. It is therefore very encouraging to see this rectified in this book, with a detailed reference being made to appropriate standards and their most relevant details described.

It might be a forlorn hope that the presence in the book of so many references to standards would awaken an interest in this area vital to the good practice of audiology. The current situation is that there is almost no support for standardization work in the United Kingdom and elsewhere, and the future of standards rests with a very small group, worldwide, of dedicated people who produce the data required to upgrade existing standards and to develop new ones. Peter Haughton rightly expresses in the book his frustration with the inadequacies of existing standards. However, the answer lies in the audiological profession and their employers giving a much higher priority to standards development and ensuring that appropriate resources are provided to ensure that these developments take place. The failure to develop standards does not lie with the standards organizations but with the audiological profession for not ensuring that appropriate support is made available.

Many advances in audiology today are related to the application of acoustic technology. A knowledge of the fundamentals of acoustics is essential if practitioners in audiology are going to understand the new applications and maximize their use.

I hope that this book will become required reading for training courses in audiology and related professions, as it brings together the relevance of acoustics to audiology, which is not emphasized enough today.

Mike Martin, OBE

Preface

This book is written for audiologists. The name has come to mean someone who applies the methods of audiological science to the diagnosis of hearing disorders and their treatment. An audiologist is, then, someone who works directly with patients or clients. The book is therefore for audiologists in hospitals, for those in hearing aid practice, and for clinicians who, though well acquainted with the medical aspects of hearing, may wish to learn more about the underlying physics. The text will also provide preparatory material for those undertaking courses in audiological science.

Many who come to work in the medical or paramedical professions have backgrounds in the biological rather than the physical sciences and, understandably, sometimes feel uncomfortable with physics itself and with the mathematical formalism it involves. But like it or not, acoustics is thoroughly rooted in classical physics, and to understand the science of sound it is necessary to have some appreciation of the physics. Fortunately, a little learning goes a long way and most of this text will be understandable to anyone who is familiar with those basic principles, particularly of mechanics, that were taught in school but that may, perhaps, have been forgotten.

The same is true for much of the mathematics which is essential to an appreciation of any of the physical sciences. It is amazing what can be done with just elementary algebra and trigonometry, but the difficulty is that any facility with mathematics requires not just logical thinking but also memory, experience, and recent practice. For many readers the memory may be dim, the experience forgotten, and the practice none too recent. For this reason mathematical treatments have been kept to a reasonable minimum. What is required, though — and here the reader must make due effort — is a full understanding of mathematical notation, that is, of the symbolism and rules of basic algebra. Mathematical notation provides, without ambiguity, the most concise expression of the relationships between physical quantities, and without it descriptions would become impossibly long-winded. With the foregoing in mind, the first chapter starts with a review of the basic physics. Some essential mathematics is included in Appendix A. Readers who have purchased this book but who find the mathematical expressions unwelcome should give particular attention to the introductory remarks in Chapter 1 if they are not to be disappointed.

There is, of course, far more to acoustics than the physics. The sciences of anatomy and physiology describe the structure of the ear and attempt to explain how it is that we hear, psychology or psychoacoustics deals with the perception of sound, and medical science considers disorders of hearing and their alleviation. The audiologist (as defined earlier) may have to contend with all these aspects, or at least have some knowledge of them. This book, however, will not stray far from the physics. There is some justification for this because there are many good texts that deal with the nonphysical parts of hearing science but few that treat physical acoustics at an appropriate level. It is the author's hope, therefore, that the reader will find in these chapters something helpful which is not readily obtainable elsewhere. It may seem curious that a book written for audiologists has only a short chapter on hearing aids. However, a recent Internet search of books about hearing aids provided a list of 104 titles. Many of these works were written by experts in the hearing aid industry, and an attempt to add to this literature would seem almost presumptuous. The reader will be better served by consulting current publications, particularly because the technology is changing almost daily as we enter the digital era.

Questions and exercises have been included at the end of each chapter (except the first). Their order corresponds approximately to that of the subjects as they appear in the main text. Students will find that working through the questions will impart a familiarity with the physics and generally make it easier to understand the ground that is being covered. Many of the questions include numerical exercises. Although they may seem challenging at first sight, the exercises usually require little more than substitution of numerical information into the formulae given in the text. Attention to the exercises will enhance the student's arithmetical skill and help him or her to remember the meaning of the symbols and the technical terms and their relationship to each other. Diligence is essential — very little can be learnt without practice.

Unless they are fortunate enough to have worked abroad, authors of technical works tend to write for their own countrymen. The choice of subjects treated in this book has to some extent been determined by the syllabuses of courses run in Britain for audiology technicians, for industrial audiometricians, and for hearing aid dispensers whose practice is regulated by statute under the Hearing Aid Council Act. The range is, however, wider than such courses require, and the treatment is for the most part more thorough. The physics is, of course, universal. The text should therefore be as relevant to audiological practice beyond these shores as it is to those on home ground.

P. M. Haughton

Acknowledgments

The help and encouragement I received from colleagues while writing this book is gratefully acknowledged. I particularly wish to thank Michael Spicer, Head of Bioengineering, for his painstaking reading of the text and for the many helpful suggestions that he offered along the way. Michael spent many hours in this task and but for his patient help there would be many more errors than those which inevitably remain. I also thank Richard Stubbs for the many lively discussions of technical issues and for his assistance in producing the computer-generated facsimile of Newton's rings in Figure 3.8.

Most of the illustrations in this book are original, but some have been taken directly from other publications or adapted to suit the needs of this text. I wish to record my gratitude to those who have generously allowed me to use their work. I also thank my colleague Tracy Kemp for creating the free-hand drawings in Figures 3.1, 3.20, and 3.21; and Rob Walker, Rachel Quaranta, and Colin Parker of the mechanical workshop and medical illustration department for their help with technical drawings, photographs, and various other illustrations.

1
The Basics

This book is about the physics of sound: how it is produced, how it is propagated, how it can be measured, and so on. We should therefore start by reminding ourselves of some of the basic principles of physics, making special reference to those principles which are directly relevant to the science of acoustics. This first chapter provides a brief, highly selective review of elementary science and, in particular, elementary mechanics and properties of matter. Although these things are of course taught at school, they are often forgotten or only partially remembered after our school days are over. This chapter moves quite rapidly from one topic to the next and does not attempt to provide much in the way of explanation. It should be treated rather like a glossary except that the material is presented in a logical rather than an alphabetical order. The intention is not so much to teach the physics as to explain terms that occur later in the text. Readers who have difficulty with elementary physics are advised that further reading should be considered.

It will be apparent from the outset that mathematical notation is used more widely in this book than in the introductory material found in other basic audiology textbooks. It is worth repeating the statement in the preface that mathematical notation provides a concise and unambiguous expression of the relationships between physical quantities and, providing it is understood, it can be a valuable complement to ordinary language. The great difficulty here is that many aspiring audiologists have little knowledge of mathematics (they probably detested it in their school days) and no ability to manipulate algebraic expressions. For them mathematics is an anathema. Their interest in audiology is not so much in the physics or even the acoustics as perhaps in the medical aspects and in the rewards of providing a personal service to another human being in need of help. But most courses in audiology and most examinations essential to the acquisition of professional qualifications do require at least some understanding of acoustics, in which physics and mathematics are virtually indispensable. So what is to be done? There are two things that should encourage the mathematically diffident reader. The first is that although mathematical expressions will be used freely throughout the book, it will not be a disaster if they are not all understood. Wherever possible the ordinary text stands on its own so that

1

it is intelligible even if equations and other expressions are not meaningful. The second thing is that the mathematics is used only for its illustrative value and not as a means to derive one expression from another. Statements such as 'it can be shown . . .' may be taken to mean just that. There are inevitably some areas in which mathematical notation and mathematical ideas do have to be understood. The sine function, for example, is so important to acoustics that it is essential to know what this function is, to fully understand its principal characteristics, and to appreciate its relationship to other trigonometic functions. Appendix A should help, and there are many textbooks that can supply this information.

Physics is a difficult subject, and acoustics, which is part of physics, is not particularly easy. It is important to recognize where the difficulty lies because it is not only in the mathematical content. Physics is not all about the discovery of 'scientific facts'; it is about ideas, about the explanation of one observation in terms of another, and about a quest for ever greater generalization. Many of the ideas — the so-called laws — have taken centuries to develop and have exercised the minds of some of the most able thinkers of all time. We should not be surprised if they are at times difficult to comprehend. But for the most part the physics in this book is straightforward and should be within the grasp of any intelligent person. There are really two difficulties. The first is that physics relies on very precise statements in which ordinary words are given narrowly defined, special meanings which they do not generally have in an everyday context. Familiarity with the special meanings is essential. The other difficulty is that almost any discourse requires a knowledge of many facts that the reader may be unaware of, and introduces as the argument proceeds many statements that have to be recalled. These are difficulties that can be overcome only through perseverance.

Moving Bodies

The title of this section may sound like an advertisement for the mortician, so it would be as well to explain what is meant here by a *body*. In dynamics, the term denotes any distinct object which moves or is capable of being moved as a whole. There is no restriction on size: a body can be anything from a star to a grain of sand and can comprise any material substance, real or imaginary. The ordinary (nonrotational) motion of such a body can be described by that of a specified point within the body, its centre of mass, for example, and in elementary dynamics bodies are often treated as though they existed at a single point. The orbit of Jupiter, for example, can be described in terms of the relative motion of two point-masses, the sun and the planet. A point is, of course, a mathematical abstraction, but when we draw diagrams to represent the movement of some object we are usually content with this abstract representation and perhaps even unaware of it. It is an obvious and necessary simplification.

The rate at which a body moves is called its *velocity*. To describe velocity completely, it is necessary to know both its magnitude (how many metres per second) and its direction. Quantities whose description includes a magnitude and a direction are called *vectors*. Velocity and its cousin, acceleration, are vectors. The term *speed* is sometimes used to denote the magnitude part of a velocity, but there is no need to be pedantic about this. It is usually obvious from the context whether a complete or partial specification is needed, and it would not be a cardinal sin to describe the velocity of sound as, say, 340 m/s while omitting to specify the direction of propagation.

If a body is moving with constant velocity in a straight line, the distance moved in a given time is obviously the product of the velocity and the time. So if a body has a constant velocity u, the distance moved in time t is ut, which is the area of the rectangle in Figure 1.1a. The rate of change of velocity is called *acceleration*. Note that acceleration may involve a change in speed or direction or both. When a body is forced to move in a circle, for example, it has an acceleration towards the centre of the circle by virtue of the changing direction needed to maintain the circular path even though its speed along the circumference may be constant. If a body moves in a straight line with constant acceleration, its velocity increases at a steady rate. If a body, initially at rest, undergoes an acceleration a, its velocity v at time t will be at. The distance it has moved in this time is $\frac{1}{2}vt$, which equals $\frac{1}{2}at^2$, the area of the triangle in Figure 1.1b. One way of looking at this is to say that the average velocity during the interval 0 to t is $\frac{1}{2}v$ and that the distance moved must be the average velocity multiplied by the time. This is fair enough when the acceleration is constant, but more generally it may be said that whether the acceleration is constant or not, the distance moved in the interval from time t_1 to t_2 is the area beneath the velocity-time curve during the interval. This is illustrated in Figure 1.1c. Finally, we see from the foregoing that the following formulae apply if a body starts at rest and moves in a straight line with constant acceleration:

$$v = at$$

and

$$s = \tfrac{1}{2}vt$$

where a is the acceleration, v is the velocity, and s is the distance moved after time t. Therefore,

$$s = \tfrac{1}{2}at^2$$

and

$$s = \tfrac{1}{2}v^2/a \qquad\qquad 1.1$$

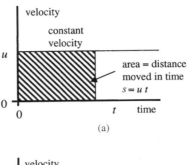

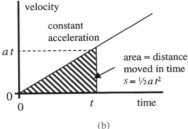

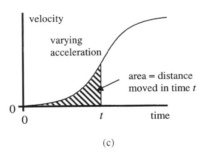

FIG. 1.1 (a) Unaccelerated motion. (b) Constant acceleration. (c) Varying acceleration.

Mass

The *mass* of a body is often said to be 'the quantity of matter' it contains, but this begs the question as to how such a quantity might be recognized. Mass shows itself in two ways: *inertia* and *gravity*. Inertia is apparent in the effort needed to change the velocity of a body, and gravity is a universal attraction which exists between all material objects. The *weight* of a body is the gravitational force between it and the earth. For any two bodies, regardless of substance, the ratio of their masses m_1/m_2 is identical to the ratio of their weights w_1/w_2. This equivalence of inertial and gravitational mass has been verified experimentally with great precision and is one of the fundamental tenets of the general theory of relativity. The unit of mass is that of the standard kilogram, and the mass of other objects can be determined by comparing their weights with that of the standard.

The gravitational force F between two bodies is described by Newton's formula:

$$F = G\frac{m_1 m_2}{d^2}$$

1.2

where m_1 and m_2 are the masses of the bodies and d is the distance between them. The gravitational constant G is a universal constant that depends only on the units of measurement and not on the material of which the bodies may be composed. Though surpassed by Einstein's theory of relativity, this beautiful and simple law holds true on almost all scales from the near-atomic to the cosmic. Gravity is not the only interaction between matter, however; over very short distances, intermolecular and interatomic forces dominate. The d in Equation 1.2 is never quite zero, and F is never infinite.

Centre of Mass (Centre of Gravity)

Matter is not, of course, concentrated at a point but is distributed within and among the material bodies in any chosen region. For any aggregation of matter there is a point called the *centre of mass*, and in a uniform gravitational field this is also the *centre of gravity*. The simplest definition of the centre of gravity of a body is to say that it is the point at which the weight of the body acts. This means that if the body is freely suspended in any way whatsoever, its centre of gravity will always be directly below the point of suspension. The centre of gravity is also the point about which the weight of the body is perfectly balanced.

We can, however, think about the centre of mass in a more fundamental way. Any collection of matter can be imagined to consist of an indefinite number of particles having masses m_1, m_2, m_3,... at distances r_1, r_2, r_3,... from some arbitrary fixed point. Suppose we square these distances and multiply each by its respective mass and then add the results to form the sum S, that is,

$$S = m_1 r_1^2 + m_2 r_2^2 + m_3 r_3^2 + \cdots$$

The sum will depend on the position of the point from which the distances have been measured. The centre of mass is that position for which S has the least possible value. In statistics we are used to the idea that the sum of the squares of a set of values is a minimum when the values are measured relative to the mean. One can think of the centre of mass as a point which represents the mean position of the masses under consideration where the mean is statistically weighted (no pun intended) according to mass. The positions of the individual masses can be described by their coordinates x, y, and z in some arbitrary frame of reference. The coordinates of the *centre of mass* are therefore the following weighted means:

$$\bar{x} = \frac{m_1 x_1 + m_2 x_2 + m_3 x_3 \cdots}{m_1 + m_2 + m_3 + \cdots} \qquad \bar{y} = \frac{m_1 y_1 + m_2 y_2 + m_3 y_3 \cdots}{m_1 + m_2 + m_3 + \cdots}$$

$$\bar{z} = \frac{m_1 z_1 + m_2 z_2 + m_3 z_3 \cdots}{m_1 + m_2 + m_3 + \cdots}$$

Momentum

The product of mass and velocity, mv, is called *momentum*. Because velocity is involved (and because mass is obviously nondirectional, that is, scalar), momentum is a vector whose direction is the same as that of the velocity. The truly important thing about momentum is that it is conserved; that is, within any closed system the total momentum in any specified direction is constant. This principle is a valuable aid to calculations in dynamics. Here is an example. Consider a 'system' comprising just two bodies of masses m_1 and m_2 and suppose that they are moving towards each other along the same straight path with velocities v_1 and $-v_2$, respectively. They will collide. Suppose the bodies are made of soft inelastic material such as putty so that on collision they coalesce to become a single body of mass $(m_1 + m_2)$ moving at velocity v. We can calculate v by applying the principle that the momentum is conserved. Before the collision it was $m_1 v_1 - m_2 v_2$, and afterwards it became $(m_1 + m_2)v$, and since these must be equal so as to keep the momentum unchanged, we can write

$$m_1 v_1 - m_2 v_2 = (m_1 + m_2)v$$

Therefore

$$v = \frac{m_1 v_1 - m_2 v_2}{(m_1 + m_2)}$$

Force and the Laws of Motion

Newton's law of gravitation, his theory of dynamics, and his explanation of the motion of celestial bodies together constitute one of the greatest achievements of the human mind. We are nowadays so familiar with these ideas that we often overlook their wonderful simplicity and near-universal application. Newton established three laws — the so-called laws of motion — which are the bedrock of classical mechanics. They may be stated as follows:

1. A body remains at rest or in uniform motion in a straight line unless acted on by a force.
2. The rate of change of momentum of a body is proportional to the force acting on it and takes place in the direction of the force.
3. For every action there is an equal and opposite reaction.

The first law is a definition of *force*; the corollary is that a force is that which changes, or tends to change, the state of rest of a body or its uniform motion in a straight line. When the mass of a body is constant (as it usually is), the rate of change of momentum is simply the mass times rate of change of velocity, that is, mass times acceleration. The second law can therefore be paraphrased as 'The force is proportional to the mass of the body multiplied by its acceleration'. In the third law, 'action' means 'force' or the act of applying a force. It is never possible to create a force without also creating a force of equal strength acting in the opposite direction. When bodies collide, each is subject to the same force acting in opposing directions; if you fire a gun, the force that drives the bullet is equal and opposite to the recoil. It is impossible to move the centre of mass of any closed system by any action occurring entirely within it.

The first law gives a qualitative definition of force, and the second law provides a way of defining the unit of measurement in terms of the acceleration imparted to a body of unit mass. A force of 1 newton is that force needed to produce an acceleration of 1 ms^{-2} when acting on a mass of 1 kg. Within any consistent set of units, we can say that 'force equals mass times acceleration' and write this symbolically as

$$F = ma$$

Philosophers have objected to the concept of force as an agent which brings about a change in the motion of a body or which, in gravitation, facilitates action at a distance. Matter accelerates towards other matter at a rate inversely proportional to the square of the separation, and to say that this is the action of a force does not, it is claimed, add new knowledge or provide a meaningful explanation. The idea of a force is superfluous, and Newton's equations can be written without involving it. But force is an extremely helpful intermediary whose use pervades the whole of science and engineering. It gives us a sense of what is happening and helps our understanding of everyday events.

Work and Energy

It may be the curse of the drinking classes, but in physics *work* means the product of force and the distance an object moves during the action of that force. As already stated, force is a vector and distance is also, for it clearly has direction as part of its identity. Thus, work is the product of two vectors. When vectors are multiplied the result can be another vector or it can be scalar, according to which definition is chosen (see Appendix A). Work is defined as scalar, that is, without direction. This means that the force and the distance moved must be in the same direction. If the force is in some other direction, then its component in the direction of the movement must be considered in reckoning the work done.

Work is measured in joules, 1 joule being the work done when a body moves 1 metre during the action of a force of 1 newton.

Work is said to be done *by* some entity *on* some other entity, and as a result the former is said to lose energy while the latter is said to gain it. Within any isolated system the losses and gains are always equal so that the total energy[1] of the system remains constant whatever is happening within it. The principle of the conservation of energy is one of the most important and enduring in science and has never been shown to fail. It is often said that energy can take many forms — mechanical, electrical, chemical, thermal, internal, and so on. This is certainly true, but all forms are in one or other of two categories: potential and kinetic. *Potential energy* is energy that a body or system has as a consequence of its state or configuration; *kinetic energy* is energy due to movement. We will consider kinetic energy first.

A body can move in one or both of two ways: it can move along a line (translational motion) or it can rotate. In translational motion under the action of a constant force F, a body of mass m will have an acceleration equal to F/m. If it starts at rest, the distance s that it moves before acquiring a velocity v is given by Equation 1.1 as

$$s = \tfrac{1}{2}v^2/a = \tfrac{1}{2}v^2 m/F$$

so that

$$Fs = \tfrac{1}{2}mv^2$$

But Fs is the product of the force responsible for the acceleration and the distance moved during the action of that force, that is to say, the work done on the body in order to give it the velocity, v. The kinetic energy E_k acquired by the body is equal to the work done on it. Therefore the kinetic energy of

[1]The energy includes that associated with mass-energy transformations in accordance with the theory of relativity.

a body moving with velocity v is given by

$$E_k = \tfrac{1}{2}mv^2 \qquad\qquad 1.3$$

The kinetic energy associated with rotation will not be considered here except to say that it can be expressed in a similar way, but instead of velocity and mass we find angular velocity and a quantity called the moment of inertia.

Potential energy has many varieties, but in all these it is energy that is stored after work has been done to cause some change in whatever it is that holds the stored energy. Two examples follow.

Example 1: Potential Energy due to Gravity in a Uniform Gravitational Field

The force on a body due to the earth's gravity is its weight; we will digress briefly to say what this means. There are two ways of thinking about weight. If a body is allowed to fall, it does so with an acceleration g (approximately $9 \cdot 81$ ms^{-2}). We can say that gravity exerts a force W equal to the mass of the body times its acceleration; that is, $W = mg$. We should therefore call g the 'acceleration due to gravity'. Alternatively, we can think of the earth as the cause of a gravitational field (which indeed it is). The *intensity* of a field is the force it exerts on unit 'something', so for a gravitational field it is unit mass, for an electric field it is unit electric charge, and so on. The constant g can therefore be called the 'gravitational intensity'. So we measure g in ms^{-2} or in Nkg^{-1}; it makes no difference. In most everyday events, objects when they move do so close to the earth's surface; that is, they are usually at distances from the surface that are minute compared with the earth's radius. Accordingly, their weight remains almost constant. Indeed, unless objects are taken into space the loss of weight with increasing height is difficult to detect. This means that when something is lifted, the work done on it is equal to a constant force (its weight) multiplied by the distance raised. The potential energy E_p of a body raised to a height h above the ground is therefore

$$E_p = mgh$$

The potential energy E_p is equal to the kinetic energy that the body has the potential to acquire if it falls freely through a distance h.

It may be noticed that the earth's surface is a quite arbitrary datum from which to measure the height. There is, in fact, no absolute reference, and the potential energy in the preceding expression is the potential relative to that at the ground. Potential energy can only be expressed in relative terms, and in applying the concept it is the *change* in potential that matters. The same is true for kinetic energy because velocity is also relative, but for most purposes we think of velocity as being relative to that of the earth; that is, we regard objects as stationary if they are motionless relative to our ordinary terrestrial reference.

Example 2: Potential Energy and Elasticity

Elasticity will be considered later in this chapter because it is particularly important in the context of mechanical vibrations and acoustic phenomena. The point to be made here is that if work is done in changing the shape of an object, the change, if it is reversible, is a source of potential energy. The difference between this potential and the gravitational potential described in the previous example is that the force which causes the change in shape is not constant. The potential energy is the area under the force-displacement curve, where *displacement* means the distance moved at the point where the force is applied. In a simple linear elastic system, the force increases (or decreases) in proportion to the displacement. If additionally the force is zero when the displacement is zero, we can express this symbolically as $F = sx$, where s is a constant called *stiffness* and x is the displacement (see Figure 1.2). The work done in establishing the displacement is therefore $\frac{1}{2}Fx$, and the corresponding potential energy is

$$E_p = \tfrac{1}{2}Fx = \tfrac{1}{2}sx^2 \qquad\qquad 1.4a$$

or

$$E_p = \tfrac{1}{2}F^2/s \qquad\qquad 1.4b$$

The quantity $1/s$ is called *compliance*.

Transformation of Energy Between Various Forms

Nearly all physical events lead to energy transformations from one form to another. Imagine, for example, a stone falling into a pond. Its initial gravitational energy becomes kinetic energy, but this is soon transferred to the water, where it is transported in the ripples as kinetic energy of the moving water and potential energy associated with changes in the height of the water surface (gravitational) and the shape of the surface (elastic because

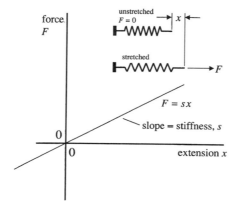

FIG. 1.2 Linear elastic system: the variation of force with extension of a simple spring.

of surface tension). But the ripples spread out and eventually disappear, so where has the energy gone? The answer is that it has been converted into heat, which ultimately contributes to the internal energy of the pond, its surroundings, and so on indefinitely. Internal energy can be thought of at a microscopic level as the kinetic and potential energies of the particles which make up matter, that is, the energies of the molecules, atoms, and subatomic particles themselves. The movement of these particles is haphazard. The local order associated with the presence of the stone at a known point above the pond has become disorder; the total energy of the universe has not changed, but its state has become more chaotic.

Power

The rate at which work is done is called *power*. Power is measured in watts, 1 watt being a rate of working of 1 joule per second. Although the origin of this definition is mechanical, the idea applies in situations where the source of energy is not directly mechanical. For example, a chemical reaction generates energy in the form of heat and light. The rate at which this energy is produced is of course measurable in watts. Similarly, the electromagnetic radiation from, say, a radio transmitter can be expressed in watts. Power is particularly important in acoustics as an expression of the acoustic energy emanating from a source, and many acoustic quantities, such as sound level, are directly related to it.

Electrical Current, Potential Difference, Resistance, and Power

Electricity is, in its own right, an important part of physics, but it is also relevant to acoustics for several reasons. As concerns theory, some methods in acoustics have their direct counterparts or even their origins in the theory of electrical circuits and transmission systems. Acoustic equipment often includes electroacoustic transducers, such as microphones and speakers, which have electrical components; and electronic amplifiers are an integral part of acoustic instruments, sound reproduction equipment, and acoustic amplifiers, including, of course, hearing aids. Although the terms used in describing electrical processes are well known, students may not be familiar with their precise definitions which, accordingly, are given in this section.

The unit of electric current, the *ampere*, is a fundamental unit (a 'base' unit) in the International System of Units (SI). The ampere is defined through its magnetic effect. The official definition is given below without explanation:

> The ampere is that constant current which, if maintained in two straight parallel conductors of infinite length, of negligible circular cross-section, and placed 1 metre apart in vacuum, would produce between these conductors a force equal to 2×10^{-7} MKS unit of force (newton) per metre of length.

The unit of electric charge (quantity of electricity) is the *coulomb*. It is the charge carried by a current of 1 ampere flowing for 1 second. If electric charge is taken from one point in a circuit to another, work must be done to overcome the resistance of the conductors and the influence of any electric or magnetic fields. The electric potential difference between the points is said to be 1 *volt* if 1 joule of work is required for each coulomb carried. When a steady current flows in a conductor, the potential difference between two points on the conductor is directly proportional to the current. This is Ohm's law. The constant of proportionality is called *resistance*, and the unit of resistance, the *ohm*, is defined such that a potential difference of 1 volt leads to a current of 1 ampere if the resistance is 1 ohm. Ohm's law can be expressed symbolically as

$$V = IR$$

where I is the current, V the potential difference, and R the resistance.

It follows from these definitions that the power W dissipated (as heat) in an electrical resistance is given by the following:

$$W = VI \qquad\qquad 1.5a$$

or

$$W = V^2/R \qquad\qquad 1.5b$$

or

$$W = I^2 R \qquad\qquad 1.5c$$

It should be noted that electrical potential is not potential energy as such, but the potential energy per unit charge, that is, energy in joules per coulomb. Notice also that power involves the *square* of the voltage or the current. The correspondence with expressions of acoustic power will become apparent later.

Electrical Capacitance and Inductance

Circuit components capable of storing electric charge are called *condensers* or *capacitors*. Their electrical *capacitance* is a measure of their ability to store charge. A capacitor usually has two terminals. The stored charge is proportional to the capacitance and to the potential difference (the voltage) between the terminals. The unit of capacitance is the farad (symbol F). It is equal to the number of coulombs of charge stored for the application of 1 volt. The stored charge Q coulombs is therefore given by

$$Q = CV$$

where C is the capacitance in farads and V is the voltage difference between the terminals.

The potential energy stored in a capacitor is equal to the work that has to be done to charge it. This energy is equal to $\frac{1}{2}CV^2$. Note the similarity between this expression and the corresponding expression for the mechanical energy stored in an elastic system (Equation 1.4b). Voltage corresponds to force, and capacitance to $1/s$, the reciprocal of mechanical stiffness. We shall refer to this similarity again in Chapter 7 when discussing the various forms of impedance.

When an electric current flows in a conductor, it creates a magnetic field. If a loop or coil of wire is placed in a changing magnetic field, it is found that a voltage appears that causes current to flow in the wire. If, therefore, a coil of wire is connected to a nonsteady electrical supply that causes the current in the coil to change, an electrical potential difference will be developed across the ends of the coil by virtue of the changing magnetic field that it is itself producing. The phenomenon is called *self-inductance*, and the induced potential difference e opposes the change in current. For this reason the potential difference is known as a *back emf* (emf means electromotive force). The emf is proportional to the rate of change of the current, so that we can write

$$e = -L \times \text{rate of change of current}$$

where L is a constant called the *inductance* of the coil. The minus sign reminds us that the emf opposes the change in the current. Inductance is measured in henries (symbol H). A back emf of 1 volt is generated by a current changing at 1 ampere per second when the inductance is 1 henry.

If a current I is established in a coil of wire, work has to be done to create the associated magnetic field. The energy stored in the magnetic field is equal to $\frac{1}{2}LI^2$. Note the similarity between this expression and the one for kinetic energy (Equation 1.3). Inductance L is analogous to mass, and current is analogous to velocity.

Density and Specific Gravity

The *density* of a substance tells us how much of it, in terms of its mass, is contained in a given volume. Density is therefore defined as mass per unit volume and expressed in kg m^{-3}. A lowercase Greek rho is often used to denote density, so if the mass m of a substance occupies a volume V, we can write the density as

$$\rho = m/V$$

It is sometimes convenient to express density relative to the density of pure water (1000 kg m^{-3}). Density expressed in this way is called *specific gravity*. For example, the specific gravity of the acid in a car battery is about 1·25, meaning that the density of the acid is 1·25 times the density of water, that is, 1·25 × 1000 kg m^{-3}). A brewer may give the 'original gravity' of his wort as '1038', omitting the decimal point and meaning that the specific gravity of the liquor is 1·038.

Pressure

For a force distributed over a surface, *pressure* is defined as force per unit area of the surface. In the SI system, pressure is measured in newtons per square metre (N m^{-2}) or pascals (Pa). The pascal is the name given to a unit of pressure equal to 1 newton per square metre.

Force is a vector, and in defining pressure, the force is that which is directed at right angles (normal) to the surface. Pressure might therefore be regarded as a vector, but at any point in a fluid (liquid and gas) it is found to have the same value in all directions, so that it is, in effect, scalar. This does not, of course, mean that pressure has the same value (magnitude) at all places in a fluid. In general, pressure does vary from one place to another; the variation with distance in a given direction is called the *pressure gradient*, which clearly is a vector.

Variation of Pressure with Depth in a Liquid

Suppose a liquid of density ρ is contained in a cylinder having a cross-sectional area S. Consider the liquid above a horizontal plane at depth h (Figure 1.3). The volume of the liquid is hS and its weight is therefore $\rho g h S$. This weight is the force distributed over the cross section of the container, and since this section has an area S, the pressure in the liquid at the depth h is given by

$$p = \rho g h \qquad\qquad 1.6$$

This formula gives the pressure at the base of a column of liquid of height h. Pressure is often expressed in terms of the height of an equivalent column of water or mercury. It may be noticed that Equation 1.6 makes no reference to atmospheric pressure and therefore gives pressure relative to atmospheric pressure; that is, it gives the amount by which the pressure at depth h exceeds the pressure in the surrounding air. There are many circumstances when this is appropriate. For example, a patient's blood pressure is important, not in absolute terms, but as a pressure relative to atmospheric pressure. Similarly, the pressure in the middle ear as measured by tympanometry is best expressed in relative terms because this gives directly the pressure difference across the tympanic membrane. On the other hand, if air

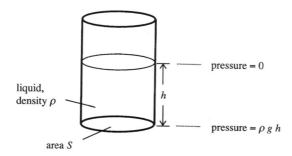

FIG. 1.3 Hydrostatic pressure within a column of liquid.

pressure were needed in order to determine the density of the air, the absolute rather than the relative measure would be required.

Referring again to the measurement of middle ear pressure, modern equipment usually expresses the result in decapascals (daPa), whereas older equipment used millimetres of water. The density of water is 1000 kg m^{-3}, and the gravitational intensity has already been given as 9·81 ms^{-2}. Therefore the pressure in a column of water whose height is 1 mm is $1000 \times 9·81 \times 10^{-3} = 9·81$ Pa or 0·981 daPa. So for practical (audiological) purposes the two units are equivalent.

Elasticity

The ability of a body to recover its size and shape after being deformed is called *elasticity*; a perfectly elastic body is one in which the recovery is complete. Failure to recover size and shape is called *plasticity*, and a perfectly plastic body is one in which the deformation is permanent. Most materials are neither perfectly elastic nor perfectly plastic, but many approach these limits quite closely. Spring steel, for example, is almost perfectly elastic if the deformation is small; the elastic compression of air, essential to the transmission of airborne sound, is perfectly reversible, but liquid/solid composites such as putty can be almost perfectly plastic.

The changes in shape and size of a body when forces are applied to it depend on the nature of the forces and the material of which the body is composed. It will be assumed that the material is homogeneous (same composition throughout) and isotropic (same elastic properties in all directions). Its properties can be specified in terms of a number of constants, namely, the elastic moduli and Poisson's ratio. A *modulus* is the ratio of stress to strain. *Stress* (like pressure) is a force per unit area, and *strain* is a measure of the relative change in size or shape that the stress produces. Strain is always dimensionless because it is the ratio of two like quantities, such as change in length or volume divided by the original length or volume. A modulus therefore has the same units as those of stress (Nm^{-2}). The type of modulus depends on the way that the stress is applied and the nature of the strain to be accounted for.

Any deformation can be accomplished in two steps: a change in size (volume) without a change in shape, followed or preceded by a change in shape without a change in size. The two elastic moduli related to these changes are therefore particularly important. We will, however, start with the tensile modulus because this is easily visualized and it provides a useful introduction to the terminology.

Tensile Modulus (Young's Modulus)

A *tensile* force is one which stretches a body in a specified direction; *tensile stress* is tensile force per unit area, where the area is in a surface at right angles to the direction of the force. *Tensile strain* is the increase in length per unit length of the body in the direction of the stress. Figure 1.4 shows a wire whose unstretched length is l and whose cross-sectional area is S. The application of the force F causes the length to increase to l'. The increase in length is therefore $l' - l$, and the tensile strain is $(l' - l)/l$. The tensile modulus (also known as Young's modulus) is defined as

$$Y = \frac{\text{tensile force per unit area}}{\text{increase in length per unit length}} = \frac{F/S}{(l' - l)/l}$$

This definition can be expanded to include the case where the tensile force is reversed, making it compressive and making the strain a relative decrease in length. Nevertheless, Young's modulus is applicable only to the analysis of the elastic behaviour of solids because liquids and gases need containment, which creates lateral forces in addition to the axial force F.

As the wire is stretched, the increase in volume due to the elongation is partly compensated by a small reduction in diameter—the wire becomes longer and thinner. The reduction in diameter is a strain in a radial direction. The ratio of this strain to the axial strain is called *Poisson's ratio* (symbol σ). For hard materials such as steel its value is typically between 0·2 and 0·3; for soft materials such as rubber it approaches, but is always less than, 0·5.

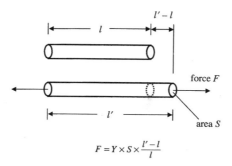

$$F = Y \times S \times \frac{l' - l}{l}$$

FIG. 1.4 Young's modulus: the ratio of tensile stress to tensile strain.

Bulk Modulus

The *bulk modulus* is used to specify the decrease in volume that occurs under the action of a uniform pressure. The *volume strain* is the decrease in volume of the body, divided by the original volume. It expresses a change in size without a change in shape. The bulk modulus, κ, is the ratio of the stress to the strain. Accordingly,

$$\kappa = \frac{\text{compressive force per unit area}}{\text{decrease in volume per unit volume}} \qquad 1.7$$

The bulk modulus exists for solids, liquids, and gases. In solids the stress can be directed outwards (in effect a negative pressure) with a corresponding reversal of the strain, but liquids and gases cannot normally exist unless contained by a positive pressure. For these states the compressive stress that defines the modulus is a stress added to the preexisting pressure. The bulk modulus is particularly important in acoustics because it is one of the factors that determine the velocity of sound and other important acoustic quantities. (See also the section Bulk Modulus of a Gas, later in this chapter.)

Modulus of Rigidity

Rigidity means, somewhat loosely, the tendency of an elastic body to resist deformation under the action of external forces. In the context of the *modulus of rigidity*, it refers specifically to a type of strain called a *shear*. This strain is a change in shape without a change in volume. Shear is produced by a force acting tangentially (parallel) to a surface; the *shear stress* is the tangential force per unit area of the surface. Deformation in shear is illustrated in Figure 1.5. It may be imagined as the progressive displacement of successive slices of the material parallel to the direction of the applied force. Suppose two planes ('slices') are separated by a distance h and that the shear produces a relative movement of material in these planes through a distance x. The shear strain is defined as x/h. If x is small, this ratio is equal to the angle θ shown in the diagram.[2] The modulus of rigidity n is defined as

$$n = \frac{\text{tangential force per unit area}}{\text{shear strain, } \theta}$$

In solids the modulus of rigidity is typically 0·3 to 0·5 times the bulk modulus; in fluids (liquids and gases) *it is zero*.

---

[2]The angle θ is measured in radians. See Appendix A.

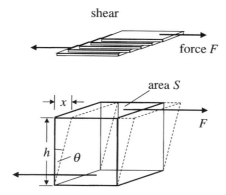

shear

force F

area S

F

FIG. 1.5 Modulus of rigidity: the ratio of shear stress to shear strain. Stress = F/S; strain = $x/h = \theta$; modulus = n; $n = (F/S) \div (x/h)$.

Because any deformation can be accomplished by combining a change in size and change in shape, the elastic constants are related. It can be shown, for example, that

$$\sigma = \frac{3\kappa - 2n}{6\kappa + 2n} \quad \text{and} \quad Y = \frac{9n\kappa}{3\kappa + n}$$

Potential Energy

Expressed in terms of stress and strain, the elastic potential energy is the elastic energy per unit volume of the deformed material (J m^{-3}).

$$\text{energy per unit volume} = \tfrac{1}{2} \times \text{stress} \times \text{strain}$$

$$= \tfrac{1}{2} \times \text{modulus} \times (\text{strain})^2$$

$$= \tfrac{1}{2} \times (\text{stress})^2 \div \text{modulus} \qquad 1.8$$

(Compare Equation 1.4.)

It may be noted that potential energy is proportional to the square of the stress or the strain; we will see later that acoustic energies are similarly proportional to the squares of such terms. Because the shear modulus is zero in liquids and gases, the elastic energy in these materials comes only from compressive (or expansive) strains and, anticipating later parts of this text, it may be expected that vibrations can be propagated as sound waves only if they are associated with compression or expansion, that is, if the vibrations are in the direction of propagation. In solids, however, transverse modes involving shear are also possible. For example, the vibrations of chime bars and tuning forks are in effect transverse waves associated with bending and governed principally by Young's modulus, which is related to both shear and bulk elasticity.

Viscosity

Although fluids are not elastic in shear, the movement of one layer of a fluid relative to an adjacent layer is not unhindered. The tangential force created by such movement is proportional, not to the shear strain itself, but to the rate at which the shear is occurring. The rate of shear is the rate of change of velocity with distance, that is, the velocity gradient in a direction perpendicular to the stress (see Figure 1.6).

The shear elasticity in solids is due to permanent interactions between the constituent molecules of the material — it is these interactions that bind the solid together and give it its strength. The atoms and molecules of a solid move about fixed positions within the material but are generally unable to migrate to other localities. In gases, on the other hand, the molecules are widely separated, so that their movement is unhindered almost all of the time. Interactions between molecules occur only briefly during collisions. The liquid state is intermediate between these extremes. Molecular interaction is important. It is manifest in properties such as surface tension and the fact that liquids are quite difficult to

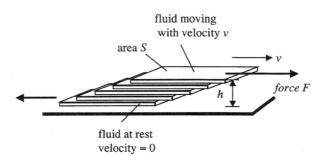

FIG. 1.6 Fluid viscosity: the ratio of shear stress to rate of shear strain. Compare with Figure 1.5. Stress = F/S; velocity gradient = v/h; viscosity = η; $\eta = (F/S) \div (v/h)$.

compress, but the molecules do not occupy fixed sites and they are not linked on any large scale to other molecules in their neighbourhood. The free movement of molecules in liquids and gases allows the momentum in one part of a fluid to be transferred to other parts, and the transfer of momentum is the origin of the rate-dependent shear force. The resistance to flow that it creates is called *viscosity*. It is characterized by the *coefficient of viscosity*, η, defined as

$$\eta = \frac{\text{tangential force per unit area}}{\text{velocity gradient}}$$

In acoustics and fluid mechanics, the coefficient of viscosity often appears in association with the density ρ. The quantity η/ρ is called the *kinematic viscosity*. It should be understood that the velocity gradient in these definitions applies to streamline (nonturbulent) flow. The air resistance in turbulent flow that we notice on a windy day, for example, is not produced by viscosity but by an aerodynamic phenomenon called *drag*. Curiously, although drag is independent of the magnitude of the viscosity, it would theoretically disappear if the viscosity were truly zero.

Viscoelasticity

In the foregoing discussion of elasticity no consideration was given to the rate at which elastic changes occur following the application of a stress. No deformation can be instantaneous because of inertia, but even when this is negligible there may be internal molecular processes which delay the elastic response. If the relationship between stress and strain is manifestly time dependent, the material is said to be *viscoelastic*. Viscoelasticity is caused by, for example, the breaking of intermolecular bonds and the realignment or disentanglement of large molecules as deformation proceeds. Polymers often show marked viscoelastic behaviour. The plastic materials used in food packaging provide a good demonstration: if the wrapper from a package of crisps is crumpled and released, the delayed recovery of its shape is obvious. Biological tissue is viscoelastic; in an audiological context, this property is relevant to the acoustic performance of the middle ear. When work is done to change the size or shape of a viscoelastic body, the rate-dependent part of the work is associated with an irreversible conversion to heat. The part which does not depend on time is stored as elastic potential energy as previously described.

Properties of Gases

The analysis of airborne sound is naturally the aspect of acoustics most relevant to audiology; consequently, the physical properties of air, and gases in general, that relate to acoustics are of special interest. At the temperatures and pressures that generally prevail in our environment, air is almost an ideal gas. An *ideal gas* is one for which it is always exactly true that

$$\frac{\text{pressure} \times \text{volume}}{\text{temperature}} = \text{constant}$$

The relationship between pressure, volume, and temperature is called the *equation of state*. It can be written as

$$PV = RT \qquad\qquad 1.9$$

where V is the volume of 1 gram-molecule of the gas, R is the *universal gas constant*, and T is the absolute temperature of the gas. These terms need explanation, and we digress slightly to provide it.

Molecular Weight: The Mole

Atomic and molecular weights express the masses of atoms and molecules relative to that of a standard. In the SI system the standard is an atom of the isotope carbon 12, which is deemed to have an atomic mass of exactly 12. A mole (abbreviation mol) is defined as a mass which always contains the same number of 'elementary entities'. These entities, which must be specified, are, for example, atoms, molecules, electrons, and so forth, and a mole is the mass containing the same number of the named particles as there are atoms in 0·012 kg of carbon 12. A mole in the sense of a gram-molecule of substance therefore contains the same number of molecules as there are atoms in 12 grams of carbon 12. This is, of course, an enormous number. It is known as *Avogadro's number* and is approximately $6·0 \times 10^{23}$. One of the wonderful things about gases is that, at a given temperature and pressure, equal volumes of gases of any type always contain the same number of molecules. If, in Equation 1.9, the volume V is chosen to be the volume containing 1 mole of the gas, the constant R has the same value for any kind of gas (hence it is a universal constant). It is approximately $8·31$ J K^{-1} mol^{-1}.

Absolute Temperature

According to the laws of thermodynamics, there is a theoretical limit to the lowest temperature that can be attained. This temperature is called *absolute zero* and it defines the starting point of the thermodynamic temperature scale. The unit of temperature on this scale is called the kelvin (K). It has the same magnitude as 1 degree on the Celsius scale; that is, the temperature range from the freezing point to the boiling point of water is both 100 K and 100°C. Absolute zero is approximately -273°C, so to convert degrees Celsius to absolute temperature, one should add 273. The freezing temperature of water is therefore 273 K and the boiling point is 373 K. In Equation 1.9, T is the absolute temperature and absolute zero is the temperature at which the pressure of the hypothetical ideal gas would vanish.

Isothermal and Adiabatic Processes

Suppose that a gas is contained in a cylinder that is closed at one end by a moveable piston. If the piston is allowed to move outwards the gas will expand, and as it does so work is done on the piston. This work is done at the expense of the internal energy of the gas, and as a consequence its temperature falls unless heat is supplied from an external source. The opposite changes occur when the gas is compressed. In an *isothermal* process, the heat supplied or removed exactly compensates the change in internal energy so that the temperature remains constant. In an *adiabatic* process, no heat is supplied or removed and temperature falls with expansion and rises with compression. Volume changes that occur rapidly, including those associated with the presence of sound, are usually adiabatic because there is too little time for the exchange of heat; slow changes are more likely to be isothermal.

The gas equation (Equation 1.9) tells us that in isothermal changes the product of the pressure and volume of a gas is constant, that is,

$$PV = \text{a constant}$$

This relationship is known as Boyle's law. The law for an adiabatic process is more complicated because changes in temperature have to be taken into consideration. It can be shown for an adiabatic change that

$$PV^{\gamma} = \text{a constant} \qquad\qquad 1.10$$

where γ is the ratio of the principal specific heats as explained in the next subsection. Equation 1.10 can be combined with Equation 1.9 to give corresponding relationships between other pairs of the variables P, V, and T.

Specific Heats

The *specific heat* of a substance is the heat needed to raise the temperature of a specified mass of it by 1 degree. Specific heat depends on the molecular composition of the substance and on its state: specific heats of gases are usually greater than those of liquids, and specific heats of liquids are usually greater than those of solids. The specific heats of gases are important in acoustics because they are related to the elastic bulk modulus and therefore to acoustic properties such as the speed of sound.

The specific heat of a gas depends on whether or not it is allowed to expand when heat is added. If expansion occurs, work has to be done to move the boundaries of the container, and accordingly, the specific heat will be greater than it would have been had the volume remained constant. The *principal specific heats* C_p and C_v are those defined for the conditions of constant pressure and constant volume, respectively. The difference between C_p and C_v is the same for all gases so long as their equation of state is approximately that of an ideal gas. It can be shown that $C_p - C_v = R$ when a mass of 1 gram-molecule is specified. The ratio of the principal specific heats is denoted by γ; that is, $\gamma = C_p/C_v$. This ratio does depend on the type of gas and particularly on its molecular form. For monatomic gases it is approximately 5/3, for diatomic gases 7/5, and for more complicated molecules it is closer to unity.

Bulk Modulus of a Gas

In defining the bulk modulus (Equation 1.7), the volume strain was said to be the relative decrease in volume caused by the application of a pressure. A gas has to be contained in some way to prevent it expanding indefinitely, and this means that a finite pressure is always present. In acoustics, this pressure is often atmospheric pressure and it may be regarded as constant because atmospheric changes take place very slowly compared with the pressure changes due to sound. So far as the modulus is concerned, the stress should be thought of as a very small increase, δP, in the prevailing pressure P. The volume strain is similarly defined as a very small (negative) increase,

$-\delta V$, in the preexisting volume. Therefore,

$$\kappa = \frac{\text{increase in pressure}}{\text{increase in volume} \div \text{original volume}} = \frac{\delta P}{-\delta V/V} = -V\frac{\delta P}{\delta V}$$

From this definition and Equations 1.9 and 1.10 it can be shown that

$$\kappa = P \text{ for an isothermal change} \qquad\qquad 1.11a$$

and

$$\kappa = \gamma P \text{ for an adiabatic change} \qquad\qquad 1.11b$$

The Atmosphere

Air contains 78% nitrogen (by volume) and 21% oxygen, together with small quantities of other substances, notably water vapour and carbon dioxide. At sea level, atmospheric pressure is approximately 1×10^5 Nm^{-2}, although small fluctuations are always occurring with changing meteorological conditions. It is customary to express atmospheric pressure in millibars (mbar). One bar is 10^5 Nm^{-2}, so the pressure at sea level is approximately 1000 mbar. This is also 1000 hectopascals (hPa), the prefix *hecto* meaning '100'. But for changes in temperature, air pressure would diminish in an almost perfectly exponential fashion with increasing altitude. Near sea level the decrease is approximately 130 mbar for every 1000 metres. If the air is dry its temperature falls steadily by about 6·5°C for every 1000 metres increase in height up to 11,100 metres, which, in the middle latitudes, is the base of a region called the stratosphere. Here the temperature remains almost constant at $-56\cdot5$°C up to an altitude of 20,000 metres, beyond which it increases slightly.

In order to provide a general description of the atmosphere and, in particular, to calibrate and use aviation flight instruments, a *Standard Atmosphere* has been specified by the International Civil Aviation Organization (Table 1.1). At sea level the standard temperature is 15°C (288·15 K) and the standard pressure is 1·01325 mbar. (The reason for choosing this rather curious pressure as the standard is that it is a conversion to millibars of a former standard, namely 760 torr, where 1 torr is the pressure equal to 760 mm Hg). The standard density of air at sea level is 1·2250 kg m^{-3}.

TABLE 1.1   Some Properties of the ICAO Standard Atmosphere

Altitude		Temperature (°C)	Pressure (mbar)	Density $(kg\,m^{-3})$	Speed of sound (ms^{-1})
Metres	Feet				
Sea level	0	15	1013·25	1·2250	340·3
1,000	3,281	8·5	899	1·112	336·4
2,000	6,562	2·0	795	1·006	332·5
3,000	9,843	−4·5	701	0·909	328·6
5,000	16,404	−17·5	540	0·736	320·5
7,000	22,966	−30·5	411	0·590	312·3
10,000	32,808	−49·9	265	0·414	299·5
15,000	49,213	−56·5	121	0·195	295·1

2

Vibrations

Terminology

An object is said to *vibrate* if it undergoes a rapid to-and-fro movement about a more or less fixed position. The term *vibration* is a general one by which such movement may be described, and there is no requirement for the movement to be regular or cyclic — it could in fact be completely haphazard. We usually think of vibrations in the mechanical sense, that is, as a description of the movement of physical bodies or structures or parts of such things, but the term and the analysis of vibrations extend to a description of motion in elastic solids and fluids and to descriptions of nonmaterial objects, including, for example, the vibrations in an electromagnetic wave. An *oscillation* is a particular kind of vibration in which the movement is periodic. This means that the pattern of the displacement relative to the fixed position repeats itself at regular intervals. The interval is called the *period* of the oscillation It is usually represented by the symbol T and measured in seconds. The sequence of events that describes the motion during the period is called a *cycle*. The rate at which cycles are repeated is called the *frequency*. Frequency is the reciprocal of the period. It is measured in hertz (Hz). Therefore,

$$f = 1/T$$

These definitions of *period*, *cycle*, and *frequency* can be extended to include oscillations for which successive cycles are not identical. For example, the vibrations of a tuning fork or a plucked string gradually diminish, but their periodicity, and therefore their frequency, remain constant.

Vibrations that follow an initial disturbance of a system but which depend only on the properties of the system itself are called *free vibrations*. The period of a free vibration is called the *natural period*, and the corresponding frequency is called the *natural frequency*. A vibration that is maintained by the application of a periodic force is called a *forced vibration*.

Sinusoidal Oscillations: Simple Harmonic Motion

Sinusoidal oscillation (also called *harmonic* oscillation) is of first importance in acoustics and in the science of vibrations generally. The reason for this is not that sinusoids occur frequently in nature (they do occur, but not frequently), but that all periodic vibrations — and indeed most nonperiodic ones — can be produced, at least theoretically, by the addition of a large number of sinusoids. This will be considered further in Chapter 5, but the point to be made here is that the analysis of complex vibrations in terms of sinusoids is a great help to understanding many acoustic phenomena, and whether or not the analysis has been made formally, the fact that it is possible is often tacitly and perhaps unconsciously assumed. In audiology, for example, one may talk of the various frequency components of speech sounds. Although the speech signal itself does contain regions which are quasi-periodic, it would be impossible to ascribe a frequency to the signal overall. It is the concept of its synthesis from sinusoidal elements that makes the frequency-specific description possible. It should be noted that many terms apply only to sinusoidal motion, and care should be taken to avoid using them in other contexts.[1]

In learning about vibrations, a good starting point is to consider the properties of an elementary oscillator consisting of just a mass and a spring. An oscillator of this type is shown in Figure 2.1. In the configuration depicted here, one end of the spring is fixed and the other end of the spring supports a mass lying vertically below the fixture. We know that if the mass is pushed upwards slightly to compress the spring and then released, it will oscillate in the vertical direction. It is not easy to think of practical examples of oscillators made in exactly this way, but the working principle is found in many applications. For example, the balance wheel of a mechanical clock or watch is a similar form of oscillator, though it differs from the one described here because its motion is rotary rather than linear. An ordinary pendulum swinging through a small arc also oscillates in a similar way, again with a rotary instead of a linear motion and with gravity instead of a spring as the source of the restoring force. The important thing to recognize is that many oscillators are essentially the combination of a mass and a spring. Moreover, these components have their counterparts in many electrical circuits and acoustic systems. Providing that the displacement is not too great, the spring behaves in a linear fashion — it supplies a force that is directly proportional to the displacement. An oscillator comprising a mass and linear spring is called a *simple harmonic oscillator*.

[1]Examples are *angular frequency*, *amplitude*, *phase*, *wavelength*, and *impedance*.

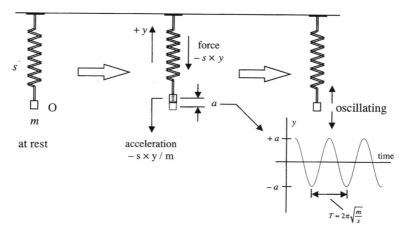

FIG. 2.1 Simple harmonic oscillator.

Let us consider the mass-spring system shown in the figure and describe exactly how the mass moves under the action of the force provided by the spring. The influence of gravity on the attached mass is constant and need not be considered, and the spring itself is assumed to have negligible mass. We will suppose that the mass moves only in the direction of the axis of the spring, that is, vertically in the configuration shown in the figure. The spring is characterized by its *stiffness*, defined as the ratio of the force produced by the spring divided by the amount the spring is extended or compressed. To start with, suppose that the system is completely at rest with the mass m at a position O. Now let the mass be raised through a distance a (in the positive y-direction) and released so that the spring, which was compressed by raising the mass, begins to expand, driving the mass downwards. In general, if the compression of the spring is y, and, if its stiffness is s, it exerts a force of magnitude sy upon the mass. This force is directed towards O, that is, in the opposite direction to the displacement y and it is opposed by the inertia of the mass. The mass must therefore move in accordance with the following statement:

$$\text{mass} \times \text{acceleration} = -(\text{stiffness} \times \text{displacement})$$

The minus sign is needed because the force provided by the spring is always in the opposite direction to the displacement of the mass. We write this equation symbolically as

$$m\ddot{y} = -sy \qquad \therefore \ m\ddot{y} + sy = 0 \qquad\qquad 2.1$$

where the symbol $\ddot{y}$ denotes acceleration in the y-direction.

Equation 2.1 is called the *equation of motion* of the oscillator. We see that it describes movement of a mass about a fixed point such that the acceleration is always directed towards that point and is always proportional to the distance from it. This is called *simple harmonic motion*. Now, the equation of motion is only a general statement about how the mass will move in keeping with Newton's laws once oscillation has been started. In order to understand the oscillator more thoroughly, we need to ask where the mass will be at any given time. An answer to this question is called a *solution* of the equation of motion. It depends, of course, on when timing is started (how time $t = 0$ is chosen) and the conditions prevailing at that time. By 'conditions' we mean the initial state of the oscillator, for example, the displacement and velocity of the mass at $t = 0$. It can be shown that simple harmonic motion is strictly sinusoidal and that the equation of motion has the solution

$$y = a \sin(\omega t + \phi) \qquad\qquad 2.2$$

The symbols a, ω, and ϕ represent constants whose meaning will now be explained.

Amplitude

The sine function in Equation 2.2 is dimensionless (without units) and its value ranges from -1 to $+1$. The displacement y, on the other hand, has units of length (metres), and the range of values it can take depends on the amount by which the oscillator was disturbed from its initial state of rest, just as the vibration of a tuning fork or a guitar string depends on how hard the fork was struck or how firmly the string was plucked to start with. The constant a therefore has to incorporate the units of measurement and ensure that the displacement is given over the correct range of values. This constant is called the *amplitude* of the vibration. It is equal to the greatest displacement in the cycle. The term *amplitude* should be used only to denote the greatest value as just defined; the tendency, particularly prevalent in sciences other than mathematics and physics, to use it as a general term is to be resisted.

It can be shown that if the displacement varies sinusoidally then the velocity and the acceleration also vary sinusoidally. The same is true according to context of other quantities, such as voltage, current, force, and sound pressure, when sinusoidal oscillation is involved. The definition of amplitude can therefore be extended to include velocity amplitude, pressure amplitude, and so on, as the names given to the greatest value in the cycle.

Angular Frequency

A sine function is a function of an angle. In mathematics this angle is called the *argument* of the function. The rate at which the angle is changing in Equation 2.2 is equal to ω radians per second. If the angle increases continuously, its sine is repeated for each increase of 2π radians so that the number of cycles completed in every second is $\omega/2\pi$. The *angular frequency* ω is therefore defined as 2π times the frequency of the sinusoidal oscillation, that is,

$$\omega = 2\pi f$$

We have been describing the oscillation of a body that is moving in a straight line, and it may seem strange that an angle—moreover, one that is constantly changing—has now appeared in the description. It should be understood that this is in a sense an analogy because the rotation of an actual physical object is not involved. The analogy, which comes of course from geometry, is a convenient and universally recognized description of harmonic oscillation, in which the movement of the oscillating object is likened to the movement of a point around the circumference of a circle. We should notice that the word *cycle*—so necessary for the portrayal of oscillatory behaviour—has its origins aı a Greek word meaning a circle. The radius to the moving point on the circumference is a line that rotates about the centre of the circle like the spoke of a wheel. Its projection in a fixed direction generates the sine function used here to describe the linear oscillation (see Appendix A). Angular frequency has a meaning only in the context of sinusoidal motion, whereas ordinary frequency (the reciprocal of the period) can be used in the description of any repetitive event.

For the simple harmonic oscillator it can be shown that

$$\omega^2 = s/m \qquad\qquad 2.3$$

and therefore

$$f = \frac{1}{2\pi}\sqrt{\frac{s}{m}}$$

These equations make quantitative what one would expect intuitively, namely, that a stiff system, with a large value of s, vibrates more rapidly than a compliant one, and that a system with a small mass m vibrates more rapidly than one with a large mass.

Phase

The word *phase* means the stage reached in a periodically recurring sequence. When the sequence can be described by a sinusoid, the stage in the sequence can obviously be described in terms of the angle for which the sine function is calculated. In Equation 2.2 the whole quantity in the parentheses is an angle, and the constant part, ϕ, which forms part of the total angle is called the *phase angle*. This term usually denotes the *difference* in the phases of two sinusoids, for example, the one given by the equation and a corresponding sinusoid for which $\phi = 0$. The phase angle has been included in Equation 2.2 to allow for the possibility that time may be reckoned from any instant in the cycle. If time zero is chosen to be a moment when the displacement is zero and about to increase, then ϕ is zero (or a multiple of 2π) and it can be omitted. If, however, time zero is chosen as the moment when the displacement attains its greatest positive value, then $\phi = \pi/2$.

It often happens that we want to compare two sinusoids, for example, the displacements of two oscillators running at the same time, or the sinusoids representing the force on the mass and the velocity of the mass. When the frequencies are the same, we can write these sinusoids as

$$y_1 = a_1 \sin \omega t$$

and

$$y_2 = a_2 \sin(\omega t + \phi)$$

Here it is clear that the phase angle expresses the difference in phase between the vibrations y_1 and y_2. If ϕ is positive a given epoch occurs earlier in y_2 than in y_1. When this happens we say that y_2 *leads* y_1. Conversely, if the phase difference is negative, we say that y_2 *lags* y_1.

Displacement, Velocity, and Acceleration

Velocity, the rate of change of displacement, and acceleration, the rate of change of velocity, are important attributes of a vibrating system. For any sinusoidal variation, the displacement, velocity, and acceleration are related to each other by the angular frequency. The following important relationships should be noted:

velocity amplitude = angular frequency × displacement amplitude

acceleration amplitude = angular frequency × velocity amplitude

In the foregoing text we have used the symbol a to denote the amplitude of a displacement y and have represented velocity and acceleration by $\dot{y}$ and $\ddot{y}$, but in other parts of this book we will often use the symbols ξ, u, and a to denote, respectively, displacement, velocity, and acceleration. Let us add the subscript 'p' when an amplitude (rather than the general value) is required. Then, for example, a sinusoidal oscillation with displacement amplitude ξ_p can be represented by

$$\xi = \xi_p \sin(\omega t + \phi)$$

The previously shown relationships between the amplitudes of displacement, velocity, and acceleration can then be written algebraically as

$$u_p = \omega \xi_p \qquad a_p = \omega u_p$$

These are important relationships which should be memorized if they are not intuitively obvious. They apply equally to root mean square (rms) values.

Energy of Free Vibrations

The simple harmonic oscillator depicted in the previous section is a hypothetical construction because once set in motion it continues indefinitely without loss of energy. The energy it possesses is at all times the energy given to it during the disturbance which started the oscillation. This energy is shared between the kinetic energy of the moving mass and the potential energy stored in the spring, and the two energies are constantly interchanged during the cycle. Motion ceases momentarily when the displacement is greatest, and at these instants the energy is entirely potential. When, on the other hand, the displacement passes through zero, the energy is all kinetic. Since the total energy is the same at any point in the cycle, this energy must be equal to the potential energy when the displacement is greatest, that is, when y equals a. Recalling that elastic potential energy is equal to the stiffness multiplied by half the square of the displacement (Chapter 1), we see that the energy E of the oscillator is given by

$$E = \tfrac{1}{2}a^2 s$$

and, by substituting for s from Equation 2.3,

$$E = \tfrac{1}{2}m\omega^2 a^2$$

The important thing to notice here is that the energy is proportional to the square of the amplitude. A similar relationship exists between the energy in a sound wave and the vibration of the medium in which the sound is carried.

Damped Vibrations

Perpetual motion is the stuff of science fiction; any real vibration is associated with a progressive loss of energy that will eventually bring it to an end. An oscillator can lose energy in several ways depending on how it interacts with its surroundings, but one of the simplest and most general ways is through a force that opposes the motion in proportion to the velocity at any given moment. Fluid viscosity can provide just such a force, though in acoustics viscosity is often unimportant and instead energy dissipation may come from the transfer of vibrations themselves to other vibrating systems or regions of the sound field through the creation of sound waves. This will be considered later, but it may be remarked here that in many circumstances the acoustic radiation from a vibrating object does exert a reaction on the object that is, to a fair approximation, proportional to its velocity.

The process whereby the amplitude of oscillation of a system is progressively diminished through an irreversible loss of energy is called *damping*. The force that resists the motion divided by the velocity is called the *damping coefficient*, and if this ratio is constant then the damping is described as *viscous damping*. We can express this symbolically by adding the damping term to the equation of motion (Equation 2.1), which now becomes

$$m\ddot{y} + c\dot{y} + sy = 0 \qquad\qquad 2.4$$

where c, a constant, is the damping coefficient and $\dot{y}$ is the velocity in the y-direction.

The solutions to this equation fall into two categories depending on the magnitude of the damping coefficient, that is, on how much damping is present. If the damping is heavy, the system does not oscillate but instead makes a gradual return to equilibrium following the initial disturbance. On the other hand, if the damping is light, the system does oscillate but the oscillations diminish exponentially. At the boundary between these conditions the damping is said to be *critical*. This happens when $c = 2\omega m$, where ω is the angular frequency for undamped vibrations (Equation 2.3). The influence of viscous damping on free vibrations is illustrated in (Figure 2.2). The oscillation shown at the top of the diagram is undamped. The oscillations shown below this are subject to progressively more damping, as indicated by the value of the parameter $c^2/4sm$. Critical damping occurs when this is equal to 1.

Damping is an important consideration in the design of systems that are able to undergo free oscillations. In a moving-coil meter (of the type often found in sound-level meters and other measuring equipment), the movement

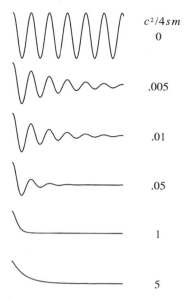

FIG. 2.2 Damped harmonic motion.

has to be damped to prevent the pointer oscillating every time there is a change in the input being measured. Similarly, microphones and sound sources may have to be damped to prevent oscillation ('ringing'). Ideally, the damping should be close to the critical value to give as rapid a response as possible without unwanted oscillation.

Forced Vibrations

Many vibrating systems normally have their motion maintained by an external input that provides a driving force. Relevant examples include loudspeakers, earphones, and bone vibrators maintained by an electrical input, as well as the electronic oscillators and filters found in most electronic equipment. The force needed to maintain a vibration must of course vary with time, and we can represent it symbolically as $f(t)$. The equation of motion for the simple mass-spring-damper system described earlier becomes, with the addition of this force,

$$m\ddot{y} + c\dot{y} + sy = f(t)$$

If $f(t)$ is periodic (a sinusoid, for example), the motion of the system is described as a *forced vibration* or *forced oscillation*. Its frequency is the same as that of the driving force.

In general, the ideas presented here can be applied to complex systems that comprise one or more parts in which free oscillation could occur. The driving force can be thought of, or described as, a sinusoidal 'stimulus' or 'excitation' at the input to the system to which the system as a whole responds. The *response* is a defined measure of the output that accompanies a defined excitation. It is influenced by the damping and natural frequency of any oscillatory components. A response can be expressed, for example, in terms of quantities such as displacement, velocity, acceleration, sound pressure, current, voltage, and so on according to context. The term *frequency response* denotes the way in which a given form of response depends on the excitation frequency; *phase response* denotes the way in which the phase of a given response varies with frequency relative to the phase at the input. Readers will be familiar with frequency response in the context of hearing aids, where the term describes the variation in the output of the aid as a function of the frequency of the sound at its input. Phase response is important in the design of filters and amplifiers, where it is often desirable that the input and output should have the same phase regardless of frequency.

Resonance

The *Oxford English Dictionary* explains resonance as 'the reinforcement or prolongation of sound by reflection or synchronous vibration'. It then defines resonance in the mechanical sense as 'a condition in which an object or system is subjected to an oscillating force having a frequency close to its own natural frequency'. These definitions certainly accord with our everyday understanding of resonance, and indeed the word is often used figuratively to convey the sense of reinforcement. In the acoustical sense, resonance is easily understood in the context of simple vibrations. Hold down a key on a piano and sound the same note with a tuning fork. The piano string will be set in motion as its natural vibration is reinforced by the synchronous vibration of the tuning fork. The concept of synchronous reinforcement is also found in areas that have nothing to do with acoustics. The following quotation is from Lord Rayleigh's treatise, *The Theory of Sound*:

> Illustration of the powerful effects of isochronism must be within the experience of every one. They are often of importance in very different fields from any with which acoustics is concerned. For example, few things are more dangerous to a ship than to lie in the trough of the sea under the influence of waves whose period is nearly that of its own natural rolling.

There is an alternative definition[2] of resonance in terms of frequency response that avoids direct reference to natural vibrations. In general, resonance is a condition associated with forced sinusoidal oscillation. If a system is driven sinusoidally by a force of constant magnitude (that is, if the amplitude of the sinusoid describing the force is constant), then there may exist one or more frequencies for which the response of the system is a maximum. In other words, the frequency-response curve may contain peaks. Resonance is said to occur if the system is driven at a frequency where the response is a maximum. This frequency is called the *resonance frequency*. Any small change in frequency relative to the resonance frequency will cause a decrease in the response. Correspondingly, *antiresonance* is said to occur at minima that may exist in the frequency response. The frequency-response definition of resonance has the advantage that it can be applied to quite complex systems, including, of course, the human ear. The difficulty is that peaks in the frequency response may occur for reasons unconnected with the existence of natural vibrations. It would therefore be wise to speak of resonance only when it is associated with some clearly identifiable natural vibration. The maxima and minima in parts of the frequency-response function of a hearing aid may well be attributable to resonance and antiresonance in the output tube and earmould, in which free vibration is clearly possible. On the other hand, a peak in the response of a sound-level meter might come from the combination of diffraction at the microphone giving a rising output with increasing frequency, and a weighting circuit with a falling frequency characteristic in the amplifier. The peak in the overall response would then not be resonance in the sense of the first of the previous definitions.

The peak in the frequency response around a resonance frequency may be quite broad or it may be sharply defined. Whether a particular response is desirable depends on circumstances. In hearing aids, for example, we would usually consider resonances to be undesirable, and a broad peak would be preferable to a narrow one. On the other hand, a sharply tuned circuit might be desirable in a filter designed to accept (or exclude) only a narrow band of frequencies. The sharpness of a peak in the frequency response, that is, the ratio of the height of the peak to its width, is called the *quality factor* (*Q-factor*). This factor depends on the type of response, but unless otherwise specified it may be assumed when dealing with mechanical or acoustic systems that the required response is the velocity response. We therefore think of the *Q*-factor as a measure of the sharpness of a resonance peak in the function that describes how the velocity of a system varies with

---

[2]See, for example, International Electrotechnical Commission (IEC) 60050-801 (1994), *International electrotechnical vocabulary — Chapter 801: Acoustics and electroacoustics.*

frequency when it is driven by fixed-amplitude sinusoidal force or pressure. In the electrical case, we speak of voltage rather than force and of current rather than velocity. The Q-factor is defined as 2π times the mean stored energy in an oscillator divided by the energy that has to be supplied during each cycle to maintain the oscillation at resonance.

An alternative definition gives Q as the resonance frequency divided by the 'half-power bandwidth'. This definition is often used when describing the performance of electrical filters. To understand it, suppose that an alternating voltage supply is connected to a load (such as a resistance) through a filter. The power dissipated in the load is proportional to the square of the current passing through it, and we can measure this power at the resonance frequency. There are two frequencies, one above and one below the resonance frequency, at which the power is half that at resonance. The *half-power bandwidth* is the difference between these frequencies. The half-power frequencies are sometimes called the *cut-off frequencies* of the filter. They are frequencies at which the attenuation produced by the filter is 3 dB greater than the attenuation at resonance. The velocity resonance of a simple oscillator and the definition of the Q-factor are illustrated in Figure 2.3.

Root Mean Square Values

When describing free vibrations, it was noted that the energy stored in a free harmonic oscillator is proportional to the square of the amplitude of the vibration. This is true when the amplitude is a measure of peak displacement, peak velocity, peak acceleration, or indeed the peak value of any quantity that is proportional to these measures.

The kinetic and potential energies of an oscillator vary throughout the cycle. At a given instant, the kinetic energy is proportional to the square of the velocity at that instant, and the potential energy is proportional to the square of the displacement. The average kinetic energy during a complete cycle is therefore proportional to the average value of the square of the velocity (the mean squared velocity), and the average potential energy is proportional to the average of the square of the displacement. The square root of the average squared value is called the *root mean square* or *rms* value. Therefore the mean kinetic energy in the cycle is proportional to (rms velocity)^2, and the mean potential energy in the cycle is proportional to (rms displacement)^2. Because, at a given frequency, the rms displacement, rms velocity, and rms acceleration are proportional to one another, the mean kinetic energy and the mean potential energy are proportional to the squares of the rms values of any of these quantities. It is therefore true to

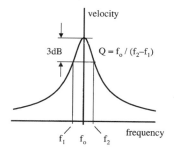

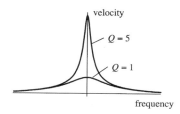

FIG. 2.3 Resonance and *Q*-factor.

say that the total energy of an oscillator (kinetic plus potential) is proportional to the square of the rms displacement or rms velocity, and so forth.

The force needed to maintain an oscillation also varies sinusoidally throughout the cycle. It too can be represented by an rms value. It can be shown that the average power dissipated in viscous damping is given by

$$W = F_{rms} u_{rms} \cos \phi$$

where W is the average power dissipated in the cycle, F_{rms} and u_{rms} are the root mean square force and velocity, and ϕ is the phase difference between the sinusoids representing the force and the velocity. Readers who have some knowledge of electronics will see that this expression is the same as that for the power dissipated in a circuit carrying an alternating current.

Many observations in acoustics are fundamentally related to acoustic power, often as power radiated by a source or carried in a sound wave. The root mean square expression of quantities related to power is therefore especially important. Sound-measuring equipment includes circuits that automatically square and average the incoming signal and present the result in terms of sound power, expressed usually in decibels. This will be described later.

Finally, it may be noted that in any sinusoid the rms value is equal to the amplitude divided by $\sqrt{2}$. If the rms values are replaced by the corresponding amplitudes, the previous equation becomes

$$W = \frac{F_p}{\sqrt{2}} \frac{u_p}{\sqrt{2}} \cos \phi = \frac{F_p u_p}{2} \cos \phi$$

where F_p and u_p are the force amplitude and velocity amplitude, respectively.

Questions and Exercises

Practical exercises are marked with an asterisk.

2.1   What is meant by *angular frequency*? A sinusoidal vibration and a vibration with a triangular waveform each have a period of 2·3 ms. What is the angular frequency of the sinusoid? Why is it not possible to assign the same angular frequency to the other vibration?

2.2   Find the angular frequencies for sinusoidal oscillation of frequencies of 32·5, 630, 500, 2000, and 8000 Hz.

2.3   Find the frequencies in Hz corresponding to angular frequencies of 16, 157, 1500, 6283, and 33,000 radian s^{-1}.

2.4   Sketch a graph to show sinusoidal motion as a function of time during at least one complete cycle. On the same axes draw a sinusoid to represent sinusoidal motion with the same frequency but with a phase that leads the first vibration by 90°. Express this angle in radians.

2.5   The displacement of an oscillating particle from its equilibrium position varies sinusoidally with time. Sketch a graph to show displacement during one half-cycle starting at an instant when the displacement is greatest. On the same graph show a sinusoidal displacement of the same frequency, that is twice the amplitude and lagging by one-eighth of a cycle behind the first vibration. Express the phase lag in radians.

2.6   What is meant by an *harmonic* oscillation? If a particle is in simple harmonic motion, at what stage in the cycle is its kinetic energy greatest?

2.7   In the cycle of a simple harmonic oscillator, when is the oscillator at rest? What form of energy, if any, does it possess at such times?

2.8   If a particle undergoes harmonic motion, when in the cycle is its acceleration greatest? What is its kinetic energy at these times?

2.9   Distinguish between *displacement amplitude* and *velocity amplitude* of an harmonic oscillator. Two oscillators comprising mass and spring elements have the same mass, but one runs at twice the frequency of the other. Explain why they each have the same average kinetic energy when their velocity amplitudes are equal, but different kinetic energies when their displacement amplitudes are equal. For the latter condition, what is the ratio of the energy of the higher-frequency oscillator to that of the lower-frequency oscillator?

2.10   Make a sketch to show the motion of a damped oscillator following an initial disturbance when the oscillator is (a) underdamped, (b) overdamped, and (c) critically damped.

2.11   Explain the terms *frequency response* and *phase response* in a system undergoing forced harmonic vibration.

2.12   What is meant by the stiffness s of a spring? If a spring has a stiffness of 220 Nm^{-1}, what force is required to compress it by 3 mm? If the spring is attached to a mass of 2·5 kg, what is the frequency of free vibration assuming that the mass of the spring itself can be neglected? If the mass of the spring were taken into account, would the result be greater or less?

2.13   A lightly damped oscillator is in free vibration not sustained by an alternating force. In each cycle, when is the force on the damper greatest and when is it least?

2.14   What does the Q-factor tell us about the frequency response of a resonant system? The peak voltage response of an electrical filter occurs at 1 kHz. The voltage falls to $1/\sqrt{2}$ times the peak value at frequencies of 910 and 1100 Hz. What is the value of the Q-factor?

2.15   Sketch a graph to show the sinusoidal waveform of an alternating voltage during two complete cycles. On the same time axis draw a graph to show the square of the voltage.

2.16   What is meant by the *root mean square value* of an alternating quantity? 'Sample' a sinusoid at 10 points throughout one cycle by finding the values of $\sin x$ where $x = 18, 54, 90, \ldots 342$ degrees. Square the values of $\sin x$, add them, and divide by 10 to find the average. Find the square root of the average. The answer may be a surprise!

2.17   For a sinusoidal displacement, the rms velocity is ω times the rms displacement, and the rms acceleration is ω times the rms velocity, where ω is the angular frequency. Calculate the rms velocity and the

peak acceleration for a 2-kHz oscillation with a displacement amplitude of 0·5 μm.

2.18   An hydraulic damper is driven sinusoidally at a frequency of 7 Hz.
The stroke of the damper is 3 mm. A force transducer shows that the
rms force applied to the damper is 27 N. Calculate the rms velocity
of the driving rod at the point where the force is applied and,
assuming that the force is proportional to the velocity of the driving
rod throughout the cycle, calculate the average absorbed power.

2.19*   The period T of a simple pendulum is given by the formula

$$T = 2\pi \sqrt{\frac{L}{g}}$$

where L is the length of the pendulum and g is the gravitational acceleration
9·81 ms^{-2}. Make a simple pendulum by suspending a weight from a
fixed object with some suitable thread. Use a stopwatch or other
timer to measure the period of oscillation for various different lengths
measured from the point of suspension to a fixed mark on the weight.
The swing of the pendulum should be small—not more than 30°. Plot
a graph of T^2 against L. It should be a straight line. Verify that the
slope of the line is equal to $4\pi^2/g$.

2.20*   Make a simple pendulum as described in the previous question. Use
a weight of about 1 kg and make the thread reasonably long—several
metres if possible. With a piece of card or other suitable material,
construct a scale that will enable you to estimate the amplitude of the
oscillation. It may help to attach a pointer to the weight. Set the
pendulum in motion and record the amplitudes of oscillation at
suitable intervals. For example, record the amplitudes of every third
or every fifth oscillation. Expect the amplitude to diminish exponentially, that is, $A_n = A_0 e^{-kn}$, where A_n is the amplitude of the nth
oscillation counting from any arbitrary starting point, and where A_0
and k are constants. Plot a graph of $\log A_n$ against n. Expect the
graph to be a straight line. If logarithms to base 10 are used, the slope
of the line will be $-k/2\cdot3$, from which the decay constant k can be
calculated.

3
Sound Waves

What is Sound?

Sound is both the name given to the sensation of hearing caused by vibrations of the air and the name that describes the vibrations themselves. Much of audiology is concerned with the relationship between these perceptual and physical attributes of sound. Audiometry, for example, relates the onset of sensation to frequency and intensity; hearing aid technology considers speech recognition in relation to the many physical characteristics that define the speech signal itself and the acoustic background that accompanies it. Generally when we talk of sound, we do so in the context of vibrations at frequencies that are audible to the human ear. The frequency range for hearing is often given as 20 Hz to 20 kHz, but in fact the limits are not so clearly defined. The highest audible frequency depends considerably on the listener, and it declines quite rapidly with increasing age. Tones whose frequencies are above the upper limit are completely inaudible, but at the other extreme tones may in fact be audible down to frequencies of a few hertz. Below about 15 Hz, however, sound loses its tonal quality. Very low frequency sound is called *infrasound*. It is created by gunfire, for example, and has been used in the location of enemy artillery. At the other end of the spectrum, sound at frequencies from 10 kHz to frequencies well beyond the range of human hearing is called *ultrasound*. It has numerous medical and industrial applications.

When an object vibrates, it disturbs the air around it and the disturbance is propagated as a sound wave. The vibrating object is called a *source*, and the region in which sound waves exist is called a *sound field*. The air in direct contact with a vibrating surface of a source moves with the surface, but as the movement is imparted to the surrounding air, the inertia of the air together with its capacity for elastic compression and expansion results in a motion that is delayed relative to that of the source. The compression and expansion is accompanied by a local increase and decrease in pressure relative to the pressure that normally exists in the absence of sound. The latter, called *static pressure*, is the normal atmospheric pressure. The fluctuating pressure is known as *sound pressure*. It is perhaps the most important manifestation of sound. It provides the physical input that is the basis of human hearing and indeed the hearing of all animal species that possess this sense, and it is the sound property detected by most (though not all) microphones. Sound pressure is a characteristic of the sound that creates it. Its alternation has the same frequency as the sound, and its variation with time is the pressure waveform of the sound which we would see, for example, with an oscilloscope connected to a microphone via a suitable amplifier. We should think of sound pressure at any instant as a small quantity, positive or negative, added to the prevailing static pressure. It is important to understand just how small sound pressure is in comparison with atmospheric pressure. For example, the peak sound pressure in a sinusoidal tone at 100 dB SPL is about 2·8 Pa; atmospheric pressure is approximately 100,000 Pa.

Sound Particles

To enquire further into the nature of sound, it is necessary to have some way of describing vibrations in the air caused by the presence of a sound field. At first this may seem something of a problem because not only is air invisible, but so far as sound is concerned, we know that the movement we are interested in varies from place to place. We are not interested in the movement of air in bulk but in the local vibrations that constitute sound. The problem of description is solved by imagining that the air consists of particles that vibrate in response to the passage of sound. Such particles are often called molecules, but this can be misleading because in science the word *molecule* denotes a special object, namely, an atom or cluster of atoms with specific chemical and physical properties. The *particle* in acoustics is merely a very small 'piece' of air that shares the vibration of the air as sound is carried. It is not a real object in the sense that a molecule is, but a mental construction to aid description. Its size and shape are undefined. The only requirement is that it should be very small compared with the wavelength of the prevailing sound. Sound particles do not move haphazardly with thermal energy as do physical molecules. Their movement is determined entirely by the sound vibrations.

If we fix our attention on any chosen region within a sound field, we can imagine the particles to move about fixed positions in time with the sound vibrations. The movement of a particle can then be described in terms of its displacement from the fixed point, its velocity, its acceleration, and so on. We therefore have, as a description of the movement of the air, terms such as *particle displacement*, *particle velocity*, and *particle acceleration*.

Sound Waves

Most of us have some idea of what a wave is. We have seen waves on the seashore or ripples on the surface of a pond. Hold one end of a rope and shake it, and a wave is seen to travel towards the other end. Although the idea seems clear enough, a precise definition of what is meant by a wave is not so easy to formulate. Yet the description and analysis of physical phenomena in terms of wave theory is one of the great unifying themes in science. It is found in the theory of optics and of electromagnetic radiation generally, in mechanics and acoustics, in the conduction of heat, and, through quantum theory, as part of the very nature of the elementary particles that make up the universe. So far as acoustics is concerned, we seek a mental picture of exactly what sound waves are. We will find that acoustic waves share many of the properties of waves of all types and that the general principles of wave theory provide a useful insight into the way in which sound is propagated and the way it interacts with objects in the environment.

A sound wave is a mental or mathematical construction that describes the sound particle displacement in terms of both time and position within a sound field. The waves seen on the surface of a pond provide an analogy, and although there are several important differences between acoustic waves and surface waves in water, the visibility of the latter is helpful in providing the appropriate mental picture. If we look at such waves two features are immediately obvious. The first is that at any instant the vertical displacement of the surface forms an extended pattern consisting of a rise and fall of the surface relative to its undisturbed level. The second is that the pattern as a whole appears to move outward from the source of the disturbance with a steady speed. However, an object floating on the surface would move only up and down as waves pass — it would not move outwards with the wave. The wave system is produced by vertical movements in adjacent areas of the surface so timed as to create the moving pattern (Figure 3.1). Waves that move, that advance from regions where vibrations exist to regions hitherto undisturbed, are called *progressive* waves. This distinguishes them from *standing waves*, which will be described later.

A defining feature of wave motion is that the displacement depends on time and position — the pattern exists both temporally and spatially, and providing that the speed of the wave is independent of frequency, the profile of the wave remains unchanged. To illustrate this we will consider waves travelling in a specified direction, which, for the sake of discussion, will be the x-axis, as shown in Figure 3.2. Suppose we have a source located at the origin, that is, at $x = 0$. Any source would do, but for the sake of this

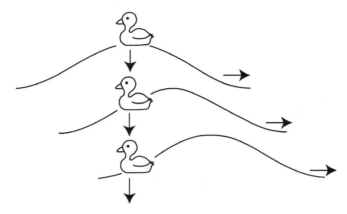

FIG. 3.1 Wave motion: a floating object moves only up and down as the wave passes. It is not caried forward with the wave.

explanation it is helpful to choose one that produces an asymmetrical waveform, a periodic triangular (sawtooth) wave for example, with a gradual displacement in one direction and a rapid displacement in the other. Imagine that the source is first turned on at time $t = 0$. As time passes, the wave train, emerging from the source, moves outwards as a whole into the undisturbed medium.

Figure 3.3 shows the wave in two ways. In the upper diagram it is shown as it was in the previous figure, as a function of distance from the source at the arbitrary time t'. The medium is disturbed over the distance ct' that the wave has advanced since the source was turned on, but beyond this point it is still at rest. In the lower diagram the wave is shown as a function of time, that is, we see the time waveform of the vibration at a fixed distance x' from the source. The wave, travelling outwards from the source with velocity c, takes a time x'/c to reach this point. Up to this time the medium is undisturbed, and afterwards its movement is the same as that of the source from the moment it was first turned on. In the water surface analogy, it is the movement of a cork floating at a distance x' from the origin of the waves. Note that the displacement-time pattern is the same as the displacement-distance pattern except that it is reversed from left to right.

The ideas just discussed can be expressed mathematically in a very neat way by observing that the particle displacement ξ depends not on time alone or on position alone but on the *difference* between the distance that the wave travels in time t and the distance x from the source. We can write this

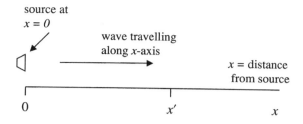

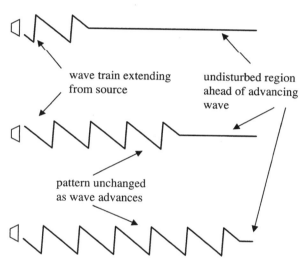

FIG. 3.2 Wave motion illustrated in this example by a wave with a triangular waveform. Particle displacement is shown on the vertical axis; distance from the source is shown on the horizontal axis.

formally[1] as

$$\xi = f(ct - x) \qquad 3.1$$

Similar expressions apply, of course, to waves travelling along the *y*- or *z*-axes.

Waves on the surface of water are called *transverse* waves, meaning that the displacement of the surface is at right angles to the direction in which

[1]The equation should be read as 'displacment is a *function of ct − x*', meaning that displacement depends on the difference between *ct* and *x*. For *function* see Appendix A.

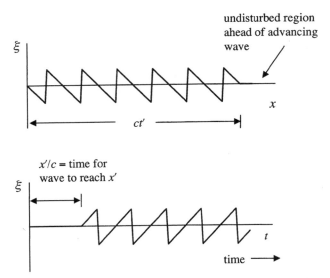

FIG. 3.3 Wave motion, showing how the waveform is represented as a variation in particle displacement with distance from the source at a given instant (top), and as a variation with time at a given location (bottom).

the wave is advancing. The vibrations of a string, in a musical instrument for example, can also be treated as waves (standing waves). Again, these are transverse waves. Sound waves in gases and liquids, however, are always *longitudinal waves*, in which the particles move in the same direction as the wave. For the type of disturbance normally described as a sound wave, it is essential that neighbouring parts of the medium should be coupled elastically. The forces on a particle in one small volume have to be communicated to neighbouring particles by what can be imagined as an elastic link. For sound waves in air, the elasticity is associated with compression or expansion as previously stated. Transverse acoustic waves are possible in solids, where they produce a shear deformation. Shear, which is a change in shape without a change in volume, can be elastic in solids but it is inelastic in air. The shear modulus of a gas is zero and accordingly, air cannot support transverse sound waves.

The longitudinal wave motion of sound in air is rather difficult to imagine because it is something outside everyday visual experience. Figure 3.4 may help. In this diagram the sound is taken to be sinusoid (a pure tone), and distance from the source in the direction of the sound wave is shown from left to right. Time is shown from top to bottom. Each row of dots represents the positions of sound particles at a given instant. The top row shows the

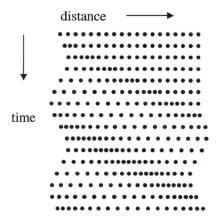

FIG. 3.4 Particle displacement in a longitudinal wave.

particles in the undisturbed field at the moment just before sound has reached the leftmost particle; the other rows show the same particles at successively later times. The way the displacement of any one particle changes with time can be seen by following the movement of the same dot from row to row. In the top row the particles are all at rest. They occupy the equilibrium positions about which they will vibrate when the sound reaches them. It will be seen that as the wave proceeds, each particle in turn is set in motion. Adjacent particles do not move exactly in step with one another because there is a slight delay as the wave advances from one particle to the next. The cumulative effect of this is to create regions in which the particles are packed either more closely or more loosely than they were in the undisturbed state. These changes correspond to compression and expansion of the air. As the particles themselves oscillate sinusoidally in a horizontal direction, the regions of compression and expansion advance from left to right.

The motion of the individual particles is akin to that of a pendulum — they are momentarily at rest at the extreme displacement where the direction of movement reverses, and they have their greatest velocity as they pass through the central (equilibrium) position. Looking along any row in the diagram, we should observe that the particles are closest to each other or most widely separated when their velocity is greatest but that they have their normal spacing at the turning points where the velocity is zero. This is demonstrated more clearly in Figure 3.5. The figure shows two sinusoids, labelled 1 and 2, that are identical except that 2 follows 1 after a small delay. These sinusoids represent the oscillations of two adjacent particles in one of

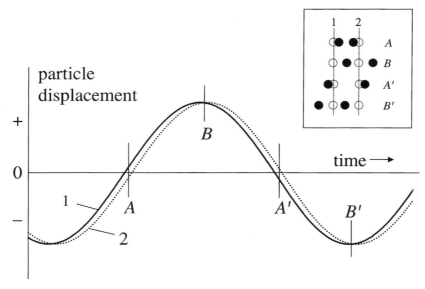

FIG. 3.5   Displacement of neighbouring particles 1 and 2 in a longitudinal wave.

the rows in the previous figure. At A and A', where the particles are close to their equilibrium positions, the curves have their greatest separation vertically. This means that the *difference* in the displacements is greatest. At A the displacement of particle 1 is towards the right of the diagram, in the direction of propagation of the wave, and the displacement of particle 2 is is in the opposite direction. The particles are therefore closest to each other in this part of the cycle, the compression of the air is greatest, and the sound pressure is a maximum. The opposite happens at A'. At B and B', where the particles are approaching their greatest distance from their resting positions, their displacements are almost identical. The particles therefore have the same separation as they had in the undisturbed sound field, the air is neither compressed nor expanded, and the sound pressure is momentarily zero. Note that A and A' are times in the cycle when the particle velocity is greatest and that B and B' are times when the velocity passes through zero as the direction of movement reverses. In general it can be shown that for plane sinusoidal waves the sound pressure and the particle velocity have the same phase; that is, the sinusoids representing pressure and particle velocity are exactly in step with one another. The sound pressure is positive when the particles and the wave are moving in the same direction and negative when they are in opposite directions. We will refer to this again in Chapter 7.

Wavefronts

The air we live in exists, of course, in three-dimensional space, and, to imagine a sound wave properly we should picture it in three dimensions too. As a sound wave progresses from its source, the boundary between the air that is vibrating and the air that has yet to be disturbed is a surface. This surface is called a *wavefront*. The boundary moves outwards from the source with the speed of sound, which is about 344 ms^{-1}. If we confine our attention to a practically small volume, it will not be long before the sound spreads throughout its entirety and, according to the definition just given, the wavefront will soon have passed beyond the region of interest. The idea of a wavefront is, however, a useful one, and it would be helpful to define something less evanescent that exists in regions where a sound field is already established. In pursuit of such a definition it may be observed that the time taken for the wave to travel from the source to the wavefront is necessarily the same at all points on the surface. It follows that the vibrations of all sound particles at the wavefront have the same phase. We can thus extend the definition of a wavefront by using the term to describe *any* surface within a sound field at which vibrations are in phase.

Over moderate distances air is isotropic—it has the same physical properties in all directions. The speed of sound is therefore independent of direction except over distances of several kilometres or more. At distances from a source that are large compared with the dimensions of the source itself, the wavefronts are almost spherical. Moreover, at a sufficient distance the wavefronts will appear flat, just as the earth's surface seems flat if we look at only a small part of it. The sound waves associated with such wavefronts are called *plane progressive waves*. Truly plane waves may not be realizable, but in many circumstances it is possible to produce waves that approximate this condition for practical purposes. Plane waves are important practically and as a theoretical abstraction because they are sound waves in their simplest form.

Frequency, Velocity and Wavelength

We have already described frequency and velocity, and most readers will know that *wavelength* is the distance between corresponding points in successive cycles of the wave. Scientists and writers of technical glossaries frequently strive to provide definitions that can be applied as widely as possible. This sometimes imposes conditions which may not seem necessary at first sight. We have hitherto used *velocity* (or *speed*) to mean the rate at which a sound wave advances. There is nothing wrong with this definition when applied to ordinary sound waves in air. In such waves the speed of sound is quite independent of frequency. There are, however, circumstances when the speed of a wave is *not* independent of frequency. Waves travelling over the surface of water are an example. When speed depends on frequency, the profile of a nonsinusoidal wave is not constant; instead, groups of waves are created that travel with what is called the *group velocity*. This need not be considered further but is mentioned only to show that the assumption of a constant, frequency-independent velocity is not always valid. A careful definition of velocity should therefore allow frequency to be specified. In the previous chapter it was stated that complex vibrations (such as the sawtooth vibration described earlier) can be treated as if they were composed of a number of sinusoidal vibrations added together. Acoustical terms are often defined in the knowledge that they may be applied to the sinusoidal components of a vibration or wave; accordingly, the definitions specify sinusoidal motion.

The velocity of a sinusoidal wave is best described by what is called the *phase velocity*. If we think of a number of points along a line in which the wave is advancing, the phase of the vibration at any instant at one point will be seen at the next point a few moments later, and at the following point later still, and so on. A 'locus' of constant phase can therefore be imagined to be in transit, moving outwards from the origin. Its velocity is the phase velocity. The phase velocity is therefore the velocity at which the wavefronts travel. When describing sound waves in air, the word *velocity* (or *speed*) almost always means the phase velocity.[2] The velocity of sound is therefore the velocity at which the wave advances into the undisturbed air or, generally, the velocity at which wavefronts move.

It can be shown that the elastic and the inertial properties of air determine the speed of sound: it is the square root of the ratio of the modulus of elasticity κ to the density ρ_0.

$$c = \sqrt{\frac{\text{modulus}}{\text{density}}} = \sqrt{\frac{\kappa}{\rho_0}} \qquad 3.2$$

Because air is a good insulator, the flow of heat between adjacent parts of the sound field is usually negligible. The bulk elastic modulus has therefore the form applicable to an adiabatic process, namely, $\kappa = \gamma P$, where

---

[2]'Speed of sound: magnitude of the phase velocity in a free progressive sound wave.' IEC 60050-(801) (1994). In air, the phase and group velocities are identical.

P is atmospheric pressure (see Chapter 1). Therefore,

$$c = \sqrt{\frac{\gamma P}{\rho_0}}$$

The relationship between atmospheric pressure and density is given by the gas laws. The equation of state for an ideal gas was given in Chapter 1 as $PV = RT$, where R is the universal gas constant and V is the volume of 1 mole of the gas. By changing R to a different constant R', we can make the equation apply to a different mass, 1 kg say, so that V is now the volume of air whose mass is a kilogram. By definition, this volume is the reciprocal of the density, so that

$$P/\rho_0 = R'T$$

From these two equations we see that the velocity of sound is independent of atmospheric pressure because a change in pressure produces a corresponding change in density. The ratio of pressure to density is therefore constant provided the temperature remains the same. This ratio does, however, depend on temperature, with the result that sound travels more quickly in warm air than in cool air. In most circumstances the temperature dependence is not very great because T is the *absolute* temperature and c is proportional to its square root. For example, if the velocity of sound at 20°C is 344 ms^{-1} then at 0°C it would be about 3·5% less:

$$344 \times \sqrt{\frac{273}{273 + 20}} = 332 \, \text{ms}^{-1}$$

The change in the speed of sound with altitude is due to the fall in temperature with increasing height above the earth's surface, as shown in Table 1.1.

The *wavelength* of sound in a sinusoidal progressive wave can now be defined as the distance between two wavefronts in which vibrations differ in phase by one complete cycle. A complete cycle corresponds, of course, to a phase change of 2π radians or 360°. Wavelength is usually represented in formulae by the letter lambda (λ).

If a sinusoidal progressive wave is produced by a source whose frequency is f Hz, then f cycles are completed in 1 second and during each cycle the wave advances by a distance equal to one wavelength. The velocity, frequency, and wavelength in a sinusoidal wave are therefore very simply related:

velocity = frequency × wavelength

that is,

$$c = f\lambda$$

Frequency, the number of cycles per second, is the reciprocal of the duration of each cycle, that is, the reciprocal of the period, *T*. Therefore,

$$c = \lambda/T$$

We also write the velocity in terms of angular frequency ω as follows:

$$c = \frac{\omega\lambda}{2\pi} = \frac{\omega}{k}$$

where $k = 2\pi/\lambda$ and $\omega = 2\pi f$. The constant k is called the *phase-change coefficient* or *wavelength constant*. If the wave is 'frozen' at an instant so that the particle displacement is seen as a function of position, this coefficient is the rate at which phase changes with distance from the source. Angular frequency, on the other hand, is the rate of change of phase with time when vibrations are observed at a given place.[3]

Particle Displacement and Velocity in Sinusoidal Waves

It was stated earlier that for any wave travelling in a specified direction the particle displacement depends on $ct - x$ (Equation 3.1). Here, for the sake of argument, we are making the x-axis the direction in which the wave is moving. The wave advances in the positive x-direction, that is, from left to right as x increases. Now, ct and x are distances — you could measure them in metres — whereas a sine function requires an angle for its argument. Multiplying $(ct - x)$ by the phase-change coefficient $2\pi/\lambda$ converts the distance to an angle in such a way that its sine has the required periodicity. The particle displacement can then be expressed as

$$\xi = a \sin\frac{2\pi}{\lambda}(ct - x)$$

---

[3]The constants k and ω are important and frequently used. The reader should take care to understand and remember their definitions. In this text k has been called the *phase-change coefficient*. This name for $2\pi/\lambda$ or ω/c is usually introduced in the context of the propagation and attenuation of sound or electromagnetic waves in ducts and transmission lines. Other names are *wavelength coefficient, wavelength constant, and phase constant*. The name *wave number* also appears, but in other branches of science (notably spectroscopy), *wave number* means $1/\lambda$, not $2\pi/\lambda$.

where a is the displacement amplitude, that is, the greatest particle displacement during the cycle.[4]

Alternative forms of this equation are

$$\xi = a \sin(\omega t - kx) \tag{3.3}$$

and

$$\xi = a \sin 2\pi \left(\frac{t}{T} - \frac{x}{\lambda} \right)$$

The particle velocity at a given location x is the rate at which particle displacement changes with time. Particle velocity u is given by

$$u = \omega a \cos \frac{2\pi}{\lambda} (ct - x)$$

There are two important things to notice here. The first is that if the displacement amplitude is a, then the velocity amplitude u_p is ωa; the second is that the particle velocity leads the displacement by a quarter-cycle, that is, by $\pi/2$ radians or 90°. Recall that $\sin(\theta + \pi/2) = \cos \theta$ and see Figure 3.6.

Sound Pressure, Particle Velocity, and Energy

It can be shown that in plane sinusoidal waves the sound pressure is directly proportional to the particle velocity. The constant of proportionality is $\rho_0 c$, where ρ_0 is the density of air.[5] This important constant is called the *characteristic impedance* of air. We will meet it again in Chapter 7. The relationship between sound pressure and particle velocity can therefore be written formally as

$$p = \rho_0 cu \qquad 3.4$$

Earlier in this chapter we referred to the importance of the air's elastic compressibility and its inertia in the production of a sound wave. Compression and expansion of the air during each cycle of a sound vibration creates a temporary store of potential energy that is available for transmission in the wave until the sound pressure returns to zero. A corresponding kinetic

[4]For simplicity the arbitrary phase, ϕ, included for example in Equation 2.2, has been omitted here and in the corresponding equations. A more general expression would include ϕ or be written as $\xi = a \sin(2\pi/\lambda)(ct - x) + b \cos(2\pi/\lambda)(ct - x)$.

[5]The density ρ_0 is the density of air that is undisturbed by the presence of sound. It is therefore the density at the static atmospheric pressure. The subscript is usually added to distinguish this density from the alternating component of density in a sound field.

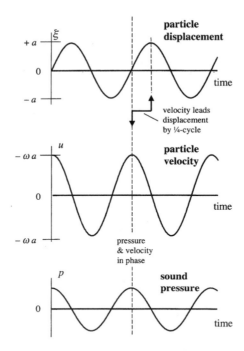

FIG. 3.6 Sound particle displacement and velocity and sound pressure in a plane sinusoidal wave.

energy is associated with the particle velocity. The stored elastic energy E_p per unit volume at any point in the sound field is $\frac{1}{2} \times$ (stress)$^2 \div$ modulus (see Chapter 1), where the stress is the sound pressure. We can therefore write this algebraically as

$$E_p = \tfrac{1}{2}p^2/\kappa$$

Kinetic energy is $\frac{1}{2} \times$ mass $\times$ (speed)^2, so that the kinetic energy E_k per unit volume is the density $\times$ the square of the particle velocity, that is,

$$E_k = \tfrac{1}{2}\rho_0 u^2$$

With some simple algebra and the relationships given in Equations 3.2 and 3.4, it can be shown that the kinetic and potential energies are at all times equal at a given point in the sound field. The total energy per unit volume is given by

$$E = E_p + E_k = \frac{p^2}{\rho_0 c^2} = \rho_0 u^2$$

The energy E in unit volume is called the *sound energy density*. In this expression the sound pressure p and the particle velocity u are the instantaneous values of these variables. The energy density in a plane sinusoidal wave is therefore itself a fluctuating quantity. Seen as either a function of time or position, its variation is the square of a sine function (Figure 3.7). There is an interesting contrast here between the energies in a sound wave and those in the simple harmonic oscillator described in the previous chapter. In the sound wave, as in the oscillator, the average potential energy is equal to the average kinetic energy, but in the oscillator these energies are constantly exchanged so that their sum is unchanging throughout the cycle.

In acoustics we often want to describe varying quantities such as energy density, sound pressure, and so on in terms of their average values. The average is usually required at a given location in the sound field and is made over a specified time. The measurements commonly made with a sound-level meter are time averages at the point where the microphone is located. Averages could in principle also be made at a given instant over a specified distance in a given direction. At the end of Chapter 2 attention was drawn to the importance of root mean square values. The average energy density in plane sinusoidal waves is given by

$$E_{av} = \frac{p_{rms}^2}{\rho_0 c^2} = \rho_0 u_{rms}^2 \qquad \qquad 3.5$$

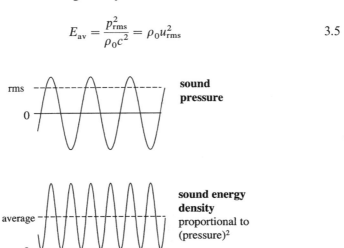

FIG. 3.7 Sound pressure and sound energy per unit volume.

The average sound energy density is an important quantity in so far as the indication on a sound-level meter is directly related to its magnitude expressed in decibels. Readers are reminded that because we are dealing with sinusoidal waves the root mean square values can be obtained from the pressure and velocity amplitudes by dividing by root 2. It is important to notice that in all the preceding expressions the energy is independent of frequency when it is given in terms of sound pressure or particle velocity. A given pressure level registered on a sound-level meter corresponds to the same acoustic energy per cubic metre, regardless of frequency.

The *sound energy flux* is the sound energy flowing per second in a specified direction through a specified area, and *sound intensity* is the sound energy flux per unit area. We would usually be interested in the sound intensity in the direction along which the wave is progressing, and in the following text this direction should be assumed unless it is stated otherwise. The idea of 'energy flow', though found occasionally in scientific writing and frequently in writing that has nothing to do with science, is apt to be imprecise, and an explanation of its meaning in the present context is appropriate. We have just seen that the air in a sound field possesses potential and kinetic energy that it would not have if left undisturbed. As a sound wave advances, energy is therefore acquired by the air in regions hitherto undisturbed. Moreover, whenever sound encounters some object or boundary, some of its energy is absorbed and converted to a nonacoustic form (ultimately heat). So we can imagine energy to be 'carried' from the source to other places and to other objects. The 'flow of energy' (the flux) is the amount of energy so transported every second.

A sound wave advances a distance numerically equal to its velocity c in 1 second, so that an area of 1 m^2 in the wavefront sweeps out a volume numerically equal to c m^3 in the same time. The average energy contained in this volume is therefore equal to $E_{av} \times c$. This energy has been supplied by sound passing through a unit area and is, by definition, the sound intensity, I. Multiplying throughout Equation 3.5 by c gives the acoustic intensity as

$$I_{av} = \frac{p_{rms}^2}{\rho_0 c} = \rho_0 c u_{rms}^2 \qquad 3.6$$

There is a useful comparison to be made between acoustic and electrical quantities. This will be described in more detail in Chapter 7, but it is mentioned here because the formulae for simple electrical circuits are well known and the analogy is helpful in understanding Equation 3.6. Sound pressure and particle velocity are analogous to voltage and current, and the characteristic impedance $\rho_0 c$ is analogous to electrical resistance. The expression for acoustic intensity given in Equation 3.6 therefore corresponds to the expression for the heat generated by an electric current in a resistance (see Chapter 1). In the electrical case, energy provided by the source of the electricity is dissipated in the resistor; in the acoustic case, sound energy provided by the source is dissipated in the sound field as just described.

Numerical Example: Magnitudes of Acoustic Quantities in a Plane Wave

In the foregoing paragraphs we have looked at the anatomy of a sound wave in some detail, so it may be helpful to form an impression of the magnitudes of the physical variables involved, taking as an example sound as it might exist in a real situation. We assume that the formulae applicable to plane sinusoidal waves are a sufficient approximation to the reality. Suppose we are talking about a sound that would register on a meter as 80 dB SPL. Although the meaning of the decibel scale has not yet been explained, most people will know that 80 dB is quite a significant sound level, perhaps near the lower end of the range for much disco or rock band music. The rms sound pressure at this level is 0·2 Pa.

Particle Velocity and Displacement

The characteristic impedance of air at 20°C is 414 Nsm^{-3}, so the particle velocity corresponding to a sound pressure of 0·2 Pa must be $0\cdot2/414 = 4\cdot8 \times 10^{-4}$ ms^{-1}, a speed readily exceeded by the average garden snail.

To find the corresponding particle displacement, we have to nominate a frequency — let's make it 1 kHz — and divide the velocity by the angular frequency. The particle displacement at 1 kHz is therefore $4\cdot8 \times 10^{-4}/2 \times \pi \times 1000 = 7\cdot7 \times 10^{-8}$ m.

It is interesting to compare the rms particle velocity just calculated with that of the air molecules. It can be shown that the molecules of a gas have an rms speed c_m that is directly related to the speed of sound c by

$$c_m^2 = \frac{3c^2}{\gamma}$$

where γ is the ratio of the principal specific heats, which for air is 1·40. The speed of sound at 20°C is 344 ms^{-1}, so the rms speed of air molecules is $344 \times \sqrt{(3/1\cdot4)} = 504$ ms^{-1}. This is about one million times greater than the sound particle velocity just calculated.

Sound Intensity and Energy Density

Substituting in Equation 3.6 gives the sound intensity as $I = (0.2)^2/414 = 9.7 \times 10^{-5}$ Wm^{-2}, that is, about 100 μW per m^2. With the acoustic power coming through a window 1 metre square, all collected and turned to heat, it would take 33 years to boil enough water for a cup of tea.

The energy density for the same sound is $I/c = 9.7 \times 10^{-5}/344 = 3 \times 10^{-7}$ Jm^{-3}. In a large room (1000 m^3), the total sound energy at any one time would therefore be 3×10^{-4} J. This is the energy you would get from a 100-watt lamp after 3 μs.

When we think about sound waves and attempt to describe them as we have done so far in this chapter, we are likely to imagine the air to be in a state of vigorous agitation. It is certainly true that the particles oscillate rapidly, for they do so at the same frequency as the sound, but it may come as a surprise to find just how small the movements are at all ordinary sound levels and how little energy is involved. One might think that the bombardment of the eardrum by immense numbers of air molecules travelling at supersonic speeds would raise a cacophony that would completely obscure the feeble acoustic vibrations, but in fact the thermal noise from air molecules colliding with the eardrum is all but inaudible. Although random molecular collisions are responsible for the static air pressure, which is vastly greater than any conceivable sound pressure, it requires a change in pressure averaged over the whole surface of the eardrum to generate an audible sound. Molecular collisions are haphazard, and fluctuations in the average pressure necessarily occur. But the number of collisions is beyond imagination ($\approx 10^{23}$ per square cm per second) so that the average static pressure is highly stable and its variation at audio frequencies is below the level of audible sound.

We have illustrated here what small magnitudes sound vibrations can have. It would, however, be a mistake to assume that particle displacements and velocities are always so minute. They can be several orders of magnitude greater near to a source, particularly if its dimensions are small compared with the wavelength. At low frequencies the movement of the diaphragm of a loudspeaker, for example, may be large enough to be clearly visible.

Combining Sound Waves

Suppose sounds are coming from two sources so that the waves from one mix with the waves from the other. An important feature of sound propagation is that when two sounds are combined, the particle displacement is the sum of the displacements due to each source acting alone, that is,

$$\xi = \xi_1 + \xi_2$$

The law that allows the addition of components due to each wave is called the *principle of superposition*.[6] It is valid for other physical quantities, such as particle velocity and sound pressure, that are linearly related (simply proportional) to displacement. It should be noted that in general the particle displacements, velocities, and so forth are not in the same direction because they are related to waves that are travelling in different directions; therefore, they must be added as vectors. They are also, of course, functions of time, frequency, and location in the wave. The addition may therefore be a little complicated even if the principle is a simple one. A wonderful consequence of the principle of superposition is that when waves travelling in different directions meet, each continues on its way unaffected by the other. This may not be obvious for sound waves because we are usually unaware of the existence of individual wave trains. Usually we are in the region where the waves mix and we hear only the combination (though we can often identify or 'listen in' to the individual sources). With light waves the principle is more evident: a beam of light from a torch is quite unaffected if it is intersected by a beam from another torch.

Interference

Where waves of the same frequency from different sources are combined, their superposition will generally give rise to a vibration whose amplitude is different from that due to each component acting alone. The process whereby the amplitude of vibrations varies from one locality to another in a sound field because waves have been combined is called *interference*. It often happens that the variation in amplitude is so distributed as to create a regular pattern called an *interference pattern*. This is a well-known phenomenon in optics (Figure 3.8), and readers may find it helpful to refer to elementary books on light to find examples. An important special case concerns the superposition of two waves having the same frequency and amplitude. If each wave is associated with a vibration of amplitude a, then their combination must produce an amplitude somewhere between 0 and $2a$ depending on whether vibrations have the same phase and reinforce one another or have opposite phases and cancel each other. The former is called *constructive interference* and the latter *destructive interference*. The colours seen when water is coated with a thin film of oil provide an example of

[6]Attributed to Christiaan Huygens (1629–1695).

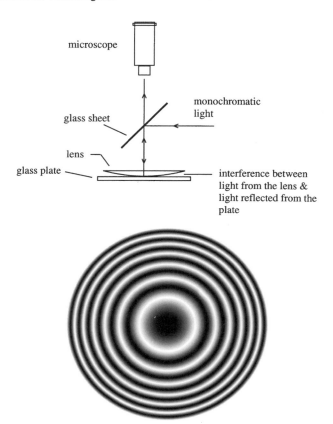

FIG. 3.8   Newton's rings.

destructive interference in which, for certain wavelengths, light reflected from the oil-water boundary arrives at the surface in antiphase with the incident light. The perceived colour is that of white light minus the colour removed by the cancellation.

Standing Waves

When two sound waves of the same frequency and similar amplitudes but travelling in opposite directions combine, they produce a special interference pattern called a *standing wave*. The pattern consists of periodic variation in amplitude arising from alternate constructive and destructive interference along the path of the individual waves. Standing waves are easily created (often unintentionally) whenever sound is reflected back towards its source by, for example, a wall, a person, or an object in the sound field. It is important to know something of the mechanisms underlying the formation of these waves because of their potential to cause serious error in acoustic measurements. The problem is a particular concern in audiometric procedures such as sound field audiometry, where the source is at some distance from the ear.

Figure 3.9 shows two plane progressive waves of equal amplitude and a standing wave formed by their combination. In this example the waves are shown as sound particle displacements at a given instant. The displacements are sine functions varying with position on the x-axis, which has been drawn in the direction of travel of wave 1. Wave 2 travels in the opposite direction. The sum of the vibrations of waves 1 and 2 is shown by the solid-line sinusoid at the bottom of the diagram. This is the standing wave. At successive instants the whole wave train labelled wave 1 moves to the right while the train labelled wave 2 moves to the left. The interference pattern,

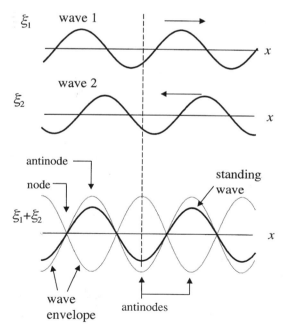

FIG. 3.9 The formation of a standing wave by a combination of waves traveling in opposite directions.

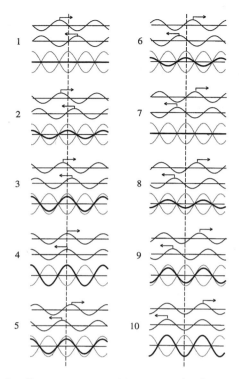

FIG. 3.10   A standing wave seen at ten succesive instants in the cycle.

however, remains stationary — its sinusoidal pattern does not move along the x-axis. The fact that a stationary wave is created by the addition of two moving waves is illustrated in Figure 3.10, which shows wave 1 and wave 2 and the sum of these vibrations at 10 successive instants in the cycle. A place in an interference pattern where the amplitude of vibration is zero (or at least has a minimum value) is called a *node*, and a place where the amplitude is a maximum is called an *antinode*. If, as is generally the case, destructive interference at a node is incomplete, so that a minimum rather than a zero occurs, the node may be described as *partial*. The distance between two successive nodes or two successive antinodes is equal to half the wavelength. Corresponding to the characteristic that is locally zero, we can distinguish different types of nodes (and antinodes), namely, *displacement node, velocity node, pressure node*, and so on.

In diagrams that represent standing waves, it is often helpful to draw two curves that together show the wave envelope. Following the usual convention, dependent variables such as pressure and displacement are displayed vertically while position is shown horizontally. The wave envelope defines, at any given place, the positive and negative limits of vibration, namely, plus and minus the amplitude of the vibration in the standing wave. The envelope therefore consists of a series of 'loops' connecting the nodal points. The antinodes occur at the widest part of each loop.

The creation of the standing wave from the superposition of progressive waves travelling in opposite directions can be demonstrated mathematically as follows. For wave 1 running in the positive x-direction, the particle displacement ξ_1 is, as in Equation 3.3

$$\xi_1 = a \sin(\omega t - kx)$$

while for wave 2 running in the opposite direction it is

$$\xi_2 = a \sin(\omega t + kx + \phi)$$

where ϕ is an arbitrary phase angle. It is easily shown that these sinusoids add to give

$$\xi_1 + \xi_2 = 2a \cos(kx + \phi/2) \sin(\omega t + \phi/2)$$

The magic of this is that x and t are no longer both part of the same sine function but instead are split so that the variation with time is given by one function (the sine term) while the variation with distance is given by a second function (the cosine term). The terms preceding the sine provide the amplitude of the sine function, and this amplitude varies with x. The variation defines the pattern of the standing wave. The displacement and velocity nodes occur wherever the cosine term is zero, that is, wherever the angle $kx + \phi/2$ is an odd multiple of $\pm \pi/2$. The distance between nodes is therefore equal to half the wavelength of the two superimposed waves. The antinodes are found midway between successive nodes, where the cosine term is equal to plus or minus 1. The greatest amplitude in the standing wave is therefore $2a$. It may be noticed that the standing wave exists regardless of the relative phase of the two waves that create it. Changing ϕ moves the pattern along the x-axis but leaves it otherwise unaltered. It may also be remarked that if the parent waves do not have the same amplitude, they nevertheless produce a stationary interference pattern, but of course destructive interference is then incomplete and only partial nodes are to be found.

An interesting property of a standing wave formed from two waves of equal amplitude is that in the region between two adjacent nodes the particles all move in the same direction at any given time. The vibrations have different amplitudes, of course, according to their location on the x-axis, but they have the same phase; that is, they move in step with one

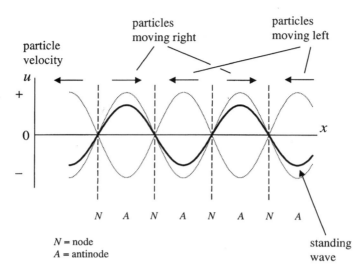

FIG. 3.11   Particle displacements in a standing wave. *N* = node; *A* = antinode.

another. In contrast, particles on either side of a node move in opposite directions (Figure 3.11). If the parent waves have different amplitudes, there is a progressive phase change between adjacent partial nodes. The rate of change of phase with x is nonuniform and is more rapid near the nodes than elsewhere. For the equal-amplitude condition it is instantaneous.

Earlier in this chapter it was stated that the sound pressure in a plane sinusoidal progressive wave is proportional to particle velocity (page 55). We might therefore expect to find that the sound pressure in a standing wave is least at the displacement or velocity nodes where the air is permanently at rest because destructive interference has cancelled the sound vibrations; correspondingly, we might have anticipated that the greatest pressure would be found at the antinodes. It may therefore come as a surprise to find that in the standing waves just described the sound pressure is greatest at the nodal points. In other words, the velocity nodes coincide with the pressure antinodes. This may not seem such a paradox when we remember that the particles on either side of a node are alternately converging towards the node and diverging from it in successive half-cycles. A further explanation is provided in Figure 3.12. The top diagram provides a snapshot of a sound wave and shows its pressure waveform as a function of distance. The direction in which the wave is moving is shown by the arrow on the axis; the direction in which the sound particles are moving is shown by the arrows at the maxima and minima, which are labelled P_+ and

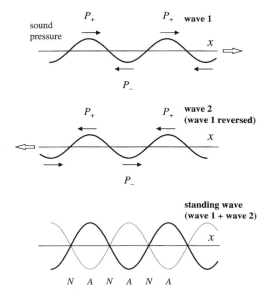

FIG. 3.12 Sound pressure and particle velocity in a standiave. *N* = pressure node, velocity antinode; *A* = pressure antinode, velocity node.

P_- according to whether the sound pressure is positive or negative at the instant of observation. Now mentally make a copy of wave 1, invert it left to right as a mirror image, and place it beneath the original to create wave 2 as shown. The particle velocities are clearly reversed in this process, but the pressures are not. This is because pressure is directionless, and whether it is positive or negative (greater or less than atmospheric pressure) depends on whether the particle velocity is in the direction of propagation of the wave or opposed to it. The reflection reverses both the velocity and the direction of the wave. As a consequence, sound pressure and particle velocity are in phase in wave 1 and in antiphase in wave 2. When the waves combine they produce the pattern of nodes and antinodes shown in the bottom diagram. Pressure antinodes occur at places where the peaks and troughs in the pressure waveform of wave 1 line up momentarily with the corresponding peaks and troughs in wave 2. At these points the particle velocities (shown by the arrows) are in opposite directions. The velocities therefore cancel. To locate the velocity antinodes we must find where waves 1 and 2 will be when the velocities are in the same direction. One-quarter of a cycle after the moment captured in the top diagram, wave 1 will have moved to the right and wave 2 to the left. The pressure peaks in wave 1 and the pressure troughs in wave 2 will then be aligned over the pressure nodes, but at these points it can be seen that the particle velocities are in the same direction. They therefore reinforce to create velocity antinodes.

Standing Waves in Pipes

The relevance of this subject to audiology is, of course, that the external ear is as a first approximation the acoustic equivalent of a pipe. It is closed at one end by the tympanic membrane, which, despite the best efforts of the middle ear transformer, is a considerable impediment to the transmission of sound. A further audiological interest is that pipes (plastic tubes) are used to couple hearing aids to earmoulds, and fine-bore tubes are used to convey sound to probe microphones. In order to give a simple account of the transmission of sound in pipes, it is necessary to place some restriction on the type of pipe to be considered. It will be assumed that the pipe is a cylinder with rigid walls, either open at both ends or closed at one end by a flat, rigid surface. The diameter of the pipe has to be small compared with its length and small compared with the wavelength of sound, but not so small that sound is attenuated through energy loss due to the viscosity of the air. Attenuation is important in tubes used for probe microphones, but its influence on the acoustics of hearing aid tubes is small except when porous elements (filters) are inserted.

The progress of a sound wave in a uniform pipe is much the same as that of a plane wave in free air, but the ends of the pipe, whether open or closed, are discontinuities that cause the sound to be reflected back into the pipe, where the superposition of the incident and reflected waves creates a standing wave according to the principles already discussed. Two kinds of reflection are possible. At a closed end the particle displacement and particle velocity at the solid surface are necessarily zero and the standing wave must therefore have a velocity node at this point. The node is created by the reversal of the particle velocity at the reflection so that the reflected wave is initiated with particle velocities in antiphase with those in the incident wave. The phase of the pressure, however, is not reversed. At the closed end there is therefore a pressure antinode in which the pressure is double that in the incident wave. When sound is reflected from the open end of a pipe, exactly the opposite is true. At the opening the sudden absence of confinement allows the emerging sound wave to expand into the surrounding air to create a spherical wave. The divergence is accompanied by a sudden fall in sound pressure and an increase in the particle velocity. Superposition of an emerging wave with its reflection from the open end creates a pressure node and a velocity antinode at the opening. Reflection from an open end of a pipe reverses the phase of the sound pressure but leaves the phase of the particle velocity unchanged.

If a stable wave system is to be maintained within a pipe, two things are necessary. The first is that the wavelength of the sound must be such that the standing wave pattern matches the length of the pipe, placing a velocity node at a closed end and an antinode at an open end. The second requirement is that sound energy radiated from an open end must be replaced. This is possible if the pipe receives energy directly from a sound source within it or indirectly from the external sound field. The energy supply must have the same frequency as that of the standing wave. This is an example of *resonance*. In addition to the nodes or antinodes at the ends of the pipe, there may also be similar nodes and antinodes within the pipe. For a given pipe, there will exist a series of possible standing wave patterns, each with a different number of nodes depending on the resonance frequency. A given pattern is called a *mode* of vibration. The wavelength becomes shorter and the frequency increases with the number of nodes. The lowest possible frequency is called the *fundamental frequency*. Frequencies for the other modes are multiples of the fundamental.

Figure 3.13 shows the possible standing wave patterns for resonance in closed and open pipes. If the pipe is closed, its length must be an odd

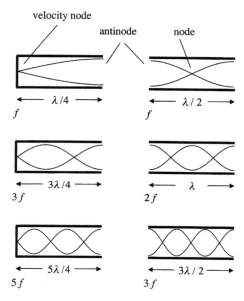

FIG. 3.13 Modes of vibration in closed and open pipes.

number of quarter waves. Allowable frequencies are therefore f, $3f$, $5f$, and so on, where $f = c/\lambda = c/4L$ for a pipe whose length is L. A pipe open at both ends must contain an even number of quarter wavelengths or, to put it another way, any integral number of half wavelengths. Resonances are therefore possible at frequencies f, $2f$, $3f$, $4f$, and so on, where $f = c/2L$.

Frequencies that form a series of the kind just described are called *harmonics*. The lowest frequency (the fundamental) is called the *first harmonic*, the next frequency in the series is the *second harmonic*, and so on. The vibrations of a closed pipe contain only odd harmonics, while those of an open pipe comprise a full harmonic series. The harmonic series is a special case of something more general. Most vibrating systems are capable of vibrating in different modes with increasing frequency above the fundamental. In many cases the various modes can be excited simultaneously to produce what is called a *complex tone* made up of several or many frequencies. The component pure tones are called *partials*. Partials other than the fundamental are called *overtones*. They are not necessarily integer multiples of the lowest frequency. For example, the vibrations of a free metal bar[7] form the series f, $2.76f$, $5.40f$, $8.93f$, and so on. It will be seen from these definitions that in an harmonic series the second harmonic is the first overtone, and so on. This may sometimes be slightly confusing, especially in the context of musical sounds, because musicians often regard the words *overtone* and *harmonic* as synonyms.

Standing Waves in Strings

The vibration of a stretched string is well known to most people. It is the basis of numerous musical instruments, including those in the violin family, the guitar, and the piano. Such instruments are certainly not intended to produce pure tones, and much craftsmanship is devoted to the creation of that pleasing blend of partials that delights the ear. In each case the vibrations of the string and, more important, the vibrations of the instrument as a whole are complex and very difficult to analyse and to explain. Some general observations can nevertheless be made. Usually, after an initial transient at the onset of bowing or plucking or striking with a hammer, the string settles into a more or less steady condition in which standing waves are formed between the bridge and the stopped end (violin and guitar) or between bridgelike fixtures near the extremities (piano). The strings of a piano possess relatively great energy, and their vibrations die

---

[7] A tuning fork is an example of a free bar. The U-bend and the stem are complications. If the fork is struck carefully, the fundamental dominates and the sound is quite pure. Striking the fork violently creates audible overtones.

away quite slowly. The vibrations of a guitar decay more rapidly and those of a violin even more so, but the sound of a violin is normally maintained by bowing, which produces a complex sawtooth vibration in which the string is alternately pulled and released by the bow.

Because by their nature strings are very long compared with their diameter, the only vibrations that matter are those perpendicular to the string itself, and waves travelling along a string are *transverse* waves. Such waves are intuitively easier to understand than the longitudinal waves that we have described hitherto. In its simplest form the string is assumed to have negligible stiffness—if it were removed from the instrument it could be bent easily. In use, the string is of course stretched and under considerable tension (sometimes almost to breaking point). Any transverse displacement that alters the shape of the string locally gives rise to a transverse component of the stretching force directed towards the equilibrium position. The force is proportional to the displacement, provided it is small enough. The situation is therefore similar to that in the harmonic oscillator described in the previous chapter. It is found that the velocity of the transverse wave is almost independent of frequency but increases with the square root of the tension and decreases with the square root of the mass per unit length of the string. Therefore,

$$c = \sqrt{\frac{T}{m}}$$

where T is the tension and m the mass per unit length.

The standing wave pattern must, of course, have velocity nodes at the fixed ends of the string—for example, at the fingerboard and the bridge of a violin. The pattern is therefore analogous to that of a standing wave in a pipe closed at both ends (a condition not included in the previous section). A wave travelling towards a node is reflected in the reverse phase so that the distance between nodes must be an integral number of half wavelengths. This gives rise to the full harmonic series f, $2f$, $3f$, $5f$, and so on, where $f = c/2L$ (Figure 3.14).

It is evident that the fundamental frequency is given by

$$f = \frac{1}{2L}\sqrt{\frac{T}{m}}$$

which may be rearranged to show the ratio of the tension to the mass per unit length of the string as

$$\frac{T}{m} = 4L^2 f^2$$

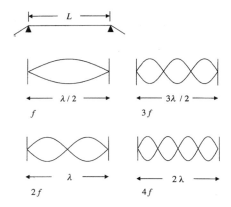

FIG. 3.14 Modes of transverse vibration on a string.

In musical instruments, energy is needed to produce a strong, sustained note. It is intuitively obvious that a taut, massive string has more vibrational energy than a slack, light one. The length and frequency of a string are determined by the size of the instrument and the musical note that is required of the string in the 'open' or unstopped condition. The ratio of tension to mass is therefore fixed for a particular string, but both tension and mass have to be high if vibrations are to have adequate energy. The tension is limited by the strength of the string and of the instrument itself. It is usually necessary to increase the mass of the bass strings by wrapping them with wire or other material. Simply increasing their diameter would produce an undesirable stiffness and would have the further disadvantage of making them difficult to attach.

Beats

So far in this chapter we have considered the addition of vibrations that have exactly the same frequency. Suppose we allow the frequencies to differ. Imagine then that we observe vibrations at a point in the combined field of two sources such that the vibrations due to each source acting singly differ in frequency and amplitude. Because one cycle repeats more rapidly than the other, the two vibrations differ progressively in their relative phase. There will be times when the component vibrations are more or less in step with each other so that they reinforce, and times when they will be out of step and partially cancel. In general the combination can produce a vibration that is quite complicated. The waveform is not a simple sinusoid,

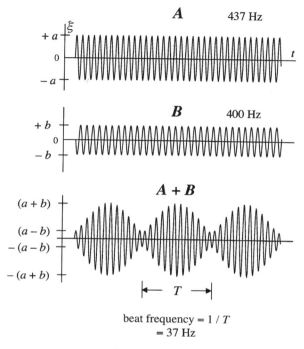

beat frequency = 1 / T
= 37 Hz

FIG. 3.15 Beats.

and it may even be aperiodic in the sense that it has no recurring pattern. But an important special case exists when the frequencies differ by just a small amount. In these circumstances the rate of change of phase is very small, so that the more rapid vibration 'overtakes' the other only gradually. Their superposition then gives an almost sinusoidal vibration whose frequency is the average of the component frequencies and whose amplitude varies slowly between the sum and the difference of the component amplitudes. The word *beats* is used to describe this phenomenon, the *beat frequency* being the rate at which the amplitude maxima recur (Figure 3.15). The beat frequency is equal to the difference between the frequencies of the two original vibrations.

The way that beats are produced can perhaps be understood more easily if presented mathematically.[8] For simplicity we will assume that the vibrations have the same amplitude, and, as we have done previously, we omit arbitrary phase angles. Our two component vibrations at frequencies f_1 and

[8]See Appendix A, Trigonometric Functions, Example 4.

f_2 are then

$$\xi_1 = a \sin \omega_1 t \quad \text{and} \quad \xi_2 = a \sin \omega_2 t$$

where $\omega_1 = 2\pi f_1$ and $\omega_2 = 2\pi f_2$. The vibrations add to give

$$\xi = \xi_1 + \xi_2 = r \sin \tfrac{1}{2}(\omega_1 + \omega_2)t \qquad 3.7$$

where

$$r = 2a\sqrt{\tfrac{1}{2} + \tfrac{1}{2}\cos(\omega_1 - \omega_2)t}$$

It will be seen in Equation 3.7 that the sine term has the average frequency of the original vibrations. When the difference between the two frequencies is small, the cosine term changes slowly to give a slow change in the amplitude r of the sine term. The least value of r is zero and the greatest value is $2a$, that is, twice the amplitude of the components. If the original amplitudes are unequal — a and b, say — then Equation 3.7 is more complicated, but the general principle is the same: the amplitude changes slowly from $a + b$ to $a - b$. Beats are, not surprisingly, more readily audible if these amplitudes are nearly equal so that the contrast between their sum and difference is enhanced.

We have described beats as occurring slowly simply to convey the idea that the compound wave can be treated as a single-frequency sinusoid with a variable amplitude. The word *slow*, however, is a relative term, implying only that the beat frequency is much less than the frequency of the primary tones. If the beat frequency is high enough it may have a tonal quality, giving the auditory illusion that the complex contains a vibration whose frequency is equal to the beat frequency. This has important implications for theories of hearing because, despite appearances to the contrary, there is no acoustic energy at the average and difference frequencies. This point is perhaps a little difficult to understand, particularly as Equation 3.7 clearly has these average and difference terms. What it means is this: if the complex vibration is analysed with a microphone together with an electronic filter that admits only a narrow range of frequencies, sound would be detected with the filter tuned to either of the two primary frequencies, but nothing at all would be detected with the filter tuned to the average frequency or to the difference frequency. The fact that the ear can perceive a tone whose pitch corresponds to the beat frequency implies that the ear can, in some circumstances, hear sounds whose frequencies are not present in the external environment.

We will consider the analysis of complex sounds in Chapter 5, but for now it is worth stating that when a sound is said to 'contain' or 'comprise' certain frequencies it usually means that the waveform can be synthesised by the addition of sinusoids having these frequencies. On this account there can be only two frequencies 'contained' in the two-tone complex that gives rise to beats as just described. It should be noted, however, that the principle of superposition that we have assumed applies only to linear systems in which, for example, sound pressure and sound particle velocity are directly proportional to one another. An important case where linearity cannot be assumed is the ear itself — the displacement of the basilar membrane at a given frequency is not simply proportional to the sound pressure in the external field. The nonlinearity produces vibrations of the basilar membrane at frequencies not present in the sound field. This is an example of what is called *intermodulation distortion*, which we will consider in Chapter 5.

Diffraction

Diffraction or 'bending' of sound by objects in a sound field is an almost universal phenomenon that influences the sound arriving at our ears and the way we perceive it. An acquaintance with this part of acoustics is therefore an essential part of audiology.

It is not easy to give a concise definition of diffraction that covers all possibilities. Here is a rather lengthy explanation taken from a former British Standard:

> When a sound source radiates into a medium containing any irregularity, boundary, or inhomogeneity, the difference in sound pressure observed at any point in the field compared with that which would occur in an infinite homogeneous medium can be described generally as due to diffraction processes.

> In a more restricted sense 'diffraction' is the name given to the disturbance of the sound field due to an object, within a distance of a few wavelengths from the object. Reflection and refraction are special cases of diffraction; reflection occurs when a wavefront, impinging on a boundary between two media, is changed in direction within the first medium; Refraction occurs when the wavefront passes into the second medium; by extension the term 'refraction' includes a progressive change of the direction of the wave normal due to a space gradient of phase velocity caused, for example, by an ordered inhomogeneity in the medium. Scattering [is the name given to] the combined effects of diffraction from an irregular array of objects, and is generally expressed in terms of its average effect at a large distance from the objects.[9]

[9]BS 661 (1969) *A glossary of acoustical terms.* Unfortunately, this excellent document has now been withdrawn.

Most of this definition is clear enough, but the technical language in the last two sentences may need some explanation. The 'wave normal' is the direction in which the sound is travelling, and although a 'space gradient of phase velocity' would not seem out of place in a work of science fiction, what it means here is a change in the velocity of sound between one place and another. An obvious example is the change in the speed of sound with altitude in the atmosphere. The gradient in velocity causes the path of the sound to deviate from a straight line — an example of *refraction.*

At a large distance, the combined effects of diffraction, referred to as *scattering* in this quotation, are more familiar in optics than in acoustics. For example, the halo that can often be seen around the moon or other distant source is caused by the diffraction of light by water droplets or ice particles in the atmosphere. Similar phenomena occur in acoustics but they are seldom obvious. We will consider reflection and refraction later under separate headings, but for now it may be noted that although they are special cases of diffraction, we normally use these terms in a fairly restricted sense. *Reflection* usually refers to the change in the path of a sound wave when it encounters an extensive boundary such as the wall of a room or the side of a building. The corresponding circumstance of the *refraction* of sound as it passes across a boundary is not part of everyday experience if only because we exist in only one medium — air. Refraction by a gradient in atmospheric conditions rather than a boundary is a far more common observation. When we think of reflection or refraction of sound, we often imagine the sound to be transmitted in rays, analogous to rays of light. A *ray* is the direction in which energy flows; it is the direction in which a point on the wavefront is travelling. Reflection and refraction can therefore be shown geometrically by a ray diagram that does not require knowledge of the frequency or wavelength. This representation fails when the sound field is disturbed by objects that are not very large compared with the wavelength or, for large objects, when we are interested in effects close to some discontinuity such as an edge or an opening. Here the wave properties are all-important. We need a word to describe such influences on the sound field, and it is provided by *diffraction* in this restricted but frequently used meaning.

The importance of diffraction in acoustics is apparent when we compare the propagation of light with that of sound. It is almost axiomatic that light travels in straight lines and that objects cast well-defined shadows. Sound, on the other hand, is seemingly unhindered by most objects, so that it is difficult to create a sound shadow even when such a thing would be desirable as a means of blocking unwanted noise. The reason for this difference is apparent when we compare the wavelengths of light and sound. The wavelength of light in the centre of the visible spectrum (green) is about

$5\cdot3 \times 10^{-7}$ metres (530 nm). This is a million times smaller than the wavelength of sound at 640 Hz. Sound travels at about 344 metres per second or 344 millimetres per millisecond. The period of a 1-kHz tone is 1 ms, so the wavelength at this frequency is 344 mm (about 13·5 inches). Even at 20 kHz, the conventional upper limit of audibility, the wavelength is 17 mm. It follows that the wavelength of sound is seldom negligible compared with the size of ordinary objects in the way that the wavelength of light is. We shall therefore find diffraction phenomena to be important throughout acoustics.

The problem of calculating the disturbance caused by an object in the path of a sound wave is a difficult one, and analytical solutions are available only for a small number of special cases. There are, however, some general principles that give a valuable insight into the way diffraction comes about. Foremost among these is *Huygens' principle*, which provides an elegant and intuitively simple method of constructing the new wavefront after diffraction has occurred. According to this principle, all points on a wavefront can be thought of as sources of 'secondary' waves: the envelope of the secondary waves is the new wavefront (Figure 3.16). The vibration at a point outside the wavefront can, in principle, be obtained from the superposition of the vibrations at that point arising from the secondary sources. These vibrations differ in amplitude and phase according to the distance from the wavefront; thus, superposition requires more than simple addition. Vector addition is required, in which the contribution from each secondary source has a magnitude corresponding to the amplitude of vibration and a direction corresponding to its phase. A wavefront is, of course, a continuous surface containing an infinite number of secondary sources, and the superposition therefore involves the addition of an infinite number of terms. This may seem an infinitely difficult problem, but in fact there are some fairly simple mathematical procedures that can provide solutions in particular circumstances. An example is shown in Figure 3.17 for diffraction by an edge. The

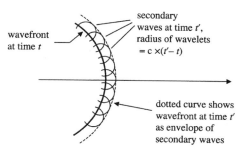

FIG. 3.16 Huygens' construction.

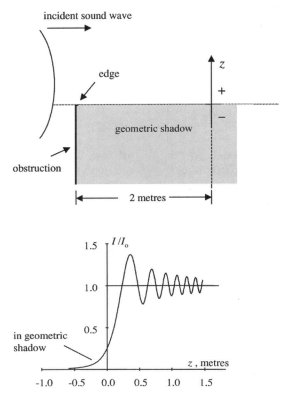

FIG. 3.17 Diffraction of sound by the edge of wall in the path of an advancing wave. The upper part of the figure shows the geometrical disposition of the incident sound and the obstruction; the lower part shows the change in sound intensity with height z at 2 metres from the edge. I_o is the intensity in the incident sound wave. The frequency is 4 kHz.

edge, which is straight, runs perpendicular to the page. It forms the top of an obstruction, a solid wall, in the path of a cylindrical or plane wave. Sound is scattered above the wall and below it into the geometrical shadow as shown. The variation of sound intensity with height above and below the edge is shown by the graph. This variation would be observed in a vertical plane at a horizontal distance of 2 metres from the wall.

The diffraction of sound by solid objects is important in physiological acoustics and in the measurement of sound: diffraction by the head and the body modifies the sound field at the entrance to the ear canal; diffraction by the microphone itself modifies the sound field being measured. The surface of a small solid object reflects the incident sound in all directions, so that the sound is said to be *scattered* by the object. The scattering depends on the size of the object in relation to the wavelength of the sound. Specifically, the ratio of the amplitude of the scattered sound in a given direction to the amplitude of the incident sound is proportional to the volume V of the scattering object and inversely proportional to the square of the wavelength λ and the distance r from the object; that is,

$$\frac{a_s}{a_i} \propto \frac{V}{\lambda^2 r}$$

or, in terms of intensity,

$$\frac{I_s}{I_i} \propto \frac{V^2}{\lambda^4 r^2}$$

It is well known in optics that the intensity of light scattered by minute particles is proportional to $1/\lambda^4$. It is for this reason that the sky, which we see as scattered light, appears blue. The shorter wavelength dominates. Sunsets are red because we see light directly from the sun or its reflection in the clouds. Dust particles selectively scatter the short-wavelength light so that the residue is more or less white light minus blue. It appears red. In acoustics, a complex sound reflected by small objects is richer in the high-frequency components than the incident sound. Correspondingly, scattering objects between the source and the observer preferentially remove the high-frequency components.

Diffraction of sound by the human head and torso is well known to audiologists. The influence of the head on a previously undisturbed field is illustrated in Figure 3.18, but it will be seen that other factors — particularly ear canal resonance — have a more important bearing on the sound pressure at the eardrum. It is worth noting in relation to acoustic measurement that the sound pressure close to a very small object is almost the same as that in the undisturbed field. At the other extreme, where the object is large compared with the wavelength, the sound pressure is doubled at the reflecting surface. This is similar to the pressure doubling at the closed end of a pipe (page 68). It occurs also at the human head if the frequency is sufficiently high. The importance for measurement is that a microphone has to be small compared with the wavelength if it is to register faithfully the pressure in a sound field; otherwise, allowance must be made for the diffraction by the microphone itself.

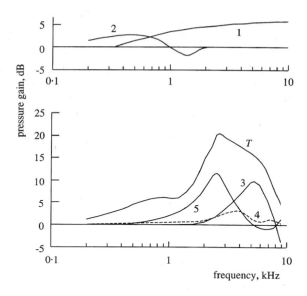

FIG. 3.18 Effect of head, body, and external ear on the sound pressure at the eardrum. Curve *T* shows pressure at the eardrum relative to pressure in a previously undisturbed field with sound incident at 45°. Curves 1 to 5 show components of this pressure gain: 1, spherical head; 2, neck and torso; 3, concha; 4, pinna flange; 5, ear canal and eardrum. Each curve shows the effect of adding the component, assuming the preceding components are in place. (Reproduced with permission E. A. G. Shaw. The external ear. in *Handbook of Sensory Physiology*, Vol. 5/1, eds. W. D. Keidel and W. D. Neff. Springer, Berlin, 1974, p. 468.)

Reflection and Refraction at an Extended Surface

When sound encounters a boundary between two media, some of the sound is transmitted and some is reflected back into the first medium. The ratio of the intensity in the transmitted sound to that in the incident sound depends on the change in the characteristic impedance $\rho_0 c$ at the boundary. Sound travels much more swiftly in solids and liquids than in air, and of course the densities of solids and liquids are very much greater than the density of air. A consequence is that only a small fraction of the incident energy is transmitted as sound passes from air to a solid or liquid, or as it passes in the opposite direction. Most of the energy is reflected.

When sound is reflected from a smooth surface whose extent is considerably greater than the wavelength, the mechanism is similar to the specular (mirrorlike) reflection that is so well known in optics. In these circumstances the propagation of sound can be represented by rays and described geometrically. The same laws apply in acoustics as in optics. For reflection, the angle of incidence is equal to the angle of reflection, and the incident ray, the reflected ray, and the normal to the surface at the point of reflection all lie in the same plane. For refraction, the ratio of the sine of the angle of incidence to the sine of the angle of refraction is equal to the ratio of the velocities in the first and second medium respectively. Expressed symbolically,

$$i = r \quad \text{for reflection}$$

and

$$\frac{\sin i}{\sin r} = \frac{c_1}{c_2} \quad \text{for refraction}$$

The rules that govern the change in direction of light or sound at a boundary can be explained using Huygens' construction. Figure 3.19 shows a plane wave incident on a reflecting boundary XX. Secondary waves arise in turn at points on the boundary between A and C, and their envelope defines the wavefront CD in the reflected wave. The law of reflection is shown as follows. PA and RC are two parallel rays defining the path of the incident sound. AB is a wavefront and therefore perpendicular to the incident rays. A secondary wave starting from A will reach D in the time taken for sound to travel from B to C. Therefore, AD equals BC. At D the tangent to the secondary wave from A is a wavefront in the reflected sound and therefore is perpendicular to the reflected ray AQ. The triangles ABC and CDA are therefore congruent, so that the angles ACB and CAD are equal. It follows that the angle i between the incident ray and the normal AN to the surface at A is equal to the angle r between the reflected ray and the normal.

Figure 3.20 is a similar construction to show refraction. The sound has a speed c_1 in the first medium shown above the boundary XX and a velocity

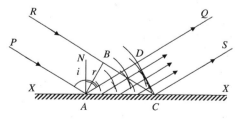

FIG. 3.19 Reflection of sound at a plane surface.

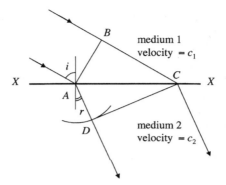

FIG. 3.20   Refraction of sound at a plane boundary.

c_2 in the second medium shown below the boundary. A secondary wave from A reaches D in the time needed for sound to travel from B to C. Again, AB is a wavefront in the incident sound, and because angle ABC is a right angle, angle BAC is equal to the angle of incidence i. Similarly, angle ACD is equal to the angle of refraction r. Therefore,

$$\sin i = \frac{BC}{AC} \qquad \sin r = \frac{AD}{AC} \qquad \therefore \frac{\sin i}{\sin r} = \frac{BC}{AD} = \frac{c_1}{c_2}$$

As mentioned previously, refraction by wind or temperature gradient is a more common observation than the refraction at a boundary just described. It is usual for the wind speed to increase with increasing height. The velocity of sound relative to the ground is the velocity it would have in still air plus the velocity of the wind. Suppose a sound ray is inclined slightly upwards as it leaves the source. Because of the gradient in wind speed, the ray will curve downwards if it travels downwind or upwards if it travels against the wind. If an observer is standing upwind from the source the sound will tend to pass over his head and not be heard, whereas if he is downwind the opposite will happen. Sounds are therefore likely to be heard better at locations downwind of the source (Figure 3.21).

Atmospheric temperature usually diminishes with increasing height. The velocity of sound, which is proportional to the square root of the absolute temperature, is then lower at altitude than at ground level (page 53). Sound rays therefore tend to curve upwards above the observer (Figure 3.22). It can, however, happen that the atmospheric temperature gradient is reversed, at least at low levels. This phenomenon is called *temperature inversion*. This can occur over land on still, clear nights when the earth's

FIG. 3.21 The influence of wind on the propagation of sound.

FIG. 3.22 The influence of temperature gradients on the propagation of sound.

surface loses heat rapidly by radiation. The air, cooled by conduction, sinks so that it does not mix with the warmer air above it. Sound rays in the inverted temperature gradient curve downwards, making distant sources unusually audible.

Questions and Exercises

Practical exercises are marked with an asterisk.

3.1   What is a *sound field*? Approximately what is the frequency range of sounds that are normally audible to humans? Would this range be the same for a cat?

3.2   Distinguish between *static pressure* in the air and *sound pressure*. The root mean square sound pressure in a pure tone is 75 mPa. Express this pressure in terms of standard atmospheric pressure.

3.3   What do we mean by a *sound particle*? Distinguish between sound particles and molecules that are the constituents of matter. How does the movement of a sound particle differ from that of a molecule?

3.4   What is meant by *particle displacement* and *particle velocity*? If, for a 1-kHz tone, the particle displacement amplitude is 3 μm, calculate the rms particle displacement and the rms particle velocity. What is the average particle displacement in any complete cycle?

3.5   Explain in your own words what is meant by a sound wave.

3.6   Distinguish between a *longitudinal* and a *transverse* wave. Which of these terms describes a sound wave in air?

3.7   When a sound wave 'travels', what exactly is the thing that is travelling?

3.8   Refer to Table 1.1. What is the speed of sound at sea level? Give the answer in m/s and then convert it to miles per hour (1 mile = 1·609 km).

3.9*   Here is a simple experiment to find the speed of sound. You need two objects (blocks of wood or Coke cans will do) that can be banged together repeatedly to create a series of impulsive sounds. Stand at a measured distance from a large reflecting surface such as the side of a building. The distance needs to be at least 100 metres. Bang the objects slowly at first and increase the rate gradually until each bang coincides with the echo of the previous one. Have an assistant time the rate of the impulses. The interval between successive impulses is the time needed for sound to travel from you to the reflector and back again. Dividing twice the measured distance by this interval will give the speed of sound. Estimate the uncertainties in the measurements and the accuracy of your result.

3.10   The *Mach number* is the speed of a moving object relative to the speed of sound. A supersonic aircraft is flying at 33,000 ft at mach 1·2. How many minutes will it take to travel 100 miles? (See question 3.8.)

3.11   For a plane sinusoidal sound wave, sketch a graph to show how the sound particle velocity depends on time throughout one cycle. Using the same axis, draw a graph to show the variation of sound pressure. Mark the axis to show the times at which the sound particles are furthest from their equilibrium positions (greatest positive or negative particle displacements).

3.12 What is a *wavefront*? What can be said about the phases of sound vibrations at points on a given wavefront?

3.13 If the frequency of sound is a simple multiple or fraction of 1000 Hz, it is easy to find an approximate value for the corresponding wavelength using the following mental arithmetic. 'The speed of sound is approximately 340 metres per second, which is the same as 340 millimetres per millisecond. The period of an oscillation at 1000 Hz is 1 millisecond. Therefore, sound travels 340 millimetres in one cycle. The wavelength at 1 kHz is 340 millimetres or 34 centimetres.' Use this thinking to find the wavelength of sound at 500 Hz, 2 kHz, and 4 kHz.

3.14 If the speed of sound is $340 \cdot 3 \, \text{ms}^{-1}$, calculate the frequency for which the wavelength is $192 \cdot 2 \, \text{mm}$.

3.15 Write a definition of the *phase-change coefficient* (or *wavelength constant*). The symbol for this constant is k. Find the value of k when the wavelength is $0 \cdot 2$ m.

3.16 Calculate k at 1600 Hz if the speed of sound is $340 \, \text{ms}^{-1}$. What is the phase difference between vibrations at points separated by 17 mm in the direction of propagation of a plane progressive wave?

3.17 This is an exercise in substituting numbers into formulae and calculating the results. If the particle displacement is given by

$$\xi = a \sin 2\pi \left(\frac{t}{T} - \frac{x}{\lambda} \right)$$

calculate its value for a $2 \cdot 3 \, \text{kHz}$ tone at time $t = 1 \cdot 5$ ms, and at a distance $x = 0 \cdot 75$ m from the source when the displacement amplitude is $0 \cdot 31 \, \mu\text{m}$. Take the velocity of sound to be $340 \, \text{ms}^{-1}$. (Remember that the argument of the sine function is in radians.)

3.18 Distinguish between potential and kinetic energy. In a simple harmonic oscillator the sum of the kinetic energy and the potential energy is constant throughout the cycle. Sound energy density is made up of potential and kinetic energy (per unit volume). Is it also constant throughout the cycle?

3.19 Distinguish between sound energy density and sound intensity. In the SI system, what units could be used to express these quantities?

3.20 Sound intensity in a plane wave can be expressed in terms of the characteristic impedance $\rho_0 c$ and the sound pressure or the particle velocity. Compare the formulae for sound intensity with those for the production of heat in an electrical resistor. Calculate the heat energy produced per second in a 15-ohm resistor when (a) the potential difference across it is 3 volts and (b) when it carries a current of 2 amps. Calculate the sound intensity corresponding to (c) an rms sound pressure of 0·015 Pa and (d) an rms particle velocity of 0·012 ms^{-1} ($\rho_0 c = 414$ Nsm^{-3}).

3.21 What is meant by the *principle of superposition*? Why does this principle not apply to nonlinear systems such as the cochlea?

3.22 Explain what is meant in acoustics by *interference* and how it leads to the production of standing waves.

3.23 If, hypothetically, one could view sound vibrations at a 'point' (that is, in a very small region of the sound field), it would not be possible to decide whether the vibrations were part of a progressive wave or a standing wave. What other information would make it possible to decide the difference?

3.24 What are *nodes* and *antinodes*? Why is it sometimes necessary to add qualifiers such as *velocity* node, *pressure* node, and so forth?

3.25* Make an acoustic baffle out of a piece of stiff card or other appropriate material. The baffle can be circular (about 12 inches in diameter) or rectangular if preferred. Cut a 1-inch diameter hole in the centre and attach the baffle to an audiometric earphone after first removing the cushion. The baffle can be fixed to the rim of the earphone using an adhesive such as Blu-tak. Place a sound-level meter with its microphone close to a hard extended surface such as a wall from which sound will be reflected and mount the earphone and baffle in some way (a laboratory stand will do) to face the reflector. Drive the earphone at a moderate frequency, 2 or 4 kHz for example, using an audiometer or laboratory oscillator. Record the change in the indication on the sound-level meter as the distance between the source and the reflector is varied. A series of maximum and minimum values should be observable. Verify that the distance between the earphone and the reflector changes by half a wavelength as the sound-level reading passes from one maximum to the next.

3.26 Describe the formation of standing waves in pipes. Draw a diagram to show the distribution of (a) particle displacement and (b) sound pressure in a pipe that is closed at one end and resonating at its fundamental frequency. Draw similar diagrams to show the particle displacement in the next resonance mode. If the fundamental has a frequency of 1500 Hz, what is the frequency of the next resonance mode?

3.27 If a flute behaves as a pipe that is open at both ends, how long must the instrument be so that the lowest available note has a frequency of 261 Hz? (Speed of sound = 340 m/s.)

3.28 A string is stretched between two fixed points, 850 mm apart. If the fundamental vibration occurs at 130 Hz, what is the velocity of transverse waves carried by the string?

3.29 What are *beats* and how are they produced? A beat is heard when two 1024-Hz tuning forks are sounded simultaneously. It is estimated that there are, on average, 3 beats per second. One of the forks is known to be accurate to better than 0·01%. What is the percentage error in the frequency of the other fork?

3.30 Two electrical signals, one at 25·00 kHz and the other at 25·05 kHz, are added electrically and amplified. An earphone connected to the amplifier fails to produce an audible sound. When the circuit is made nonlinear by clipping positive cycles of the signal at the input to the amplifier, an audible 50-Hz tone is created. Consider the current in the earphone averaged during an interval of several milliseconds and describe how the average changes with time. Hence explain the creation of the audible tone.

3.31 What is meant by *diffraction*? Why are optical shadows a common occurrence while sound shadows are far less noticeable?

3.32 Why is the sky blue?

3.33 A solid 10-cm diameter sphere is placed in a sound field in air. The sound pressure in the unobstructed field was 0·01 Pa. Consider the wavelength of sound in relation to the size of the sphere and state what sound pressure is expected at the surface of the sphere when the frequency is (a) 100 Hz and (b) 15 kHz.

3.34 What is meant by *specular* reflection? When might sound be reflected in this way?

3.35 Sound is heard by reflection from the side of a large building. If the source is 30 metres in front of the building, where does the sound appear to come from?

3.36* Using a ruler and compass, make a drawing using Huygens' method to show the reflection of a parallel beam of sound at a plane boundary. An example is provided in Figure 3.19.

3.37 What is meant by *refraction*? Explain what we mean by the *angle of incidence* and the *angle of refraction* in optics and acoustics. A beam of ultrasound in medium *A* makes an angle of incidence of 35° at the boundary with medium *B*. The speed of sound is 20% higher in *A* than in *B*. What is the angle of refraction?

3.38 Why does sound travel faster in warm air than in cold air? If the speed of sound is 340 m/s at 15°C, what speed will it have at 38°C?

3.39 What do we mean by a *temperature inversion* in the atmosphere? Explain why it causes sound rays from sources near the ground to be bent downwards. Why should this make distant sources more audible than normal?

3.40 Why is a source downwind from an observer less easily heard than a source that is upwind?

4

Sources of Sound

Simple Sources

Any vibrating object, or indeed any localized disturbance of the air, is necessarily a source of sound insofar as the disturbance is passed to other regions by wave transmission. Although sources vary greatly in their construction, their physical size, and the way in which they work, some general observations are possible. It is helpful to begin by considering sources that are physically so small compared with the wavelength of sound that they are, for practical purposes, located at a *point* in the sound field. When describing the diffraction of sound, it was stated that sound passes readily around small objects so that they do not produce well-defined acoustic shadows. As might be expected, a similar relationship exists in regard to sources—a source that is small compared with the wavelength radiates uniformly because there is no region corresponding to a shadow from which sound is excluded. Although the sound field may be nonuniform in regions very close to the source, it will become ever more uniform (independent of direction from the source) as the distance is increased. A *simple source* is one that radiates uniformly; most sources can be regarded as simple when observed from a sufficient distance. It is self-evident that wavefronts arising from a simple source are spherical rather than plane.

One of the most elementary sources that can be imagined is a small sphere whose radius is made to vary sinusoidally to produce a uniform spherical wave. This type of source is a hypothetical construction rather than a practical reality, but the justification for introducing it is that many real sources can be replaced by an equivalent spherical source, at least so far as the sound field at a distance is concerned. The radial vibrations of a sphere produce a sinusoidal movement at its surface, to and from the centre. The velocity of the surface is therefore sinusoidal also. The velocity is directed, of course, at right angles to the surface, and the velocity multiplied by the area of the surface is therefore the volume of air displaced per second. The rate at which air is displaced by the surface of the elementary spherical source varies sinusoidally; the amplitude of this sinusoid is called the *strength* of the source. The strength of a simple spherical source is therefore the maximum rate of volume variation at its surface (in m^3 s^{-1}).

The strength of a source is an important characteristic because it determines the acoustic power that will be radiated. For a simple source whose dimensions are small compared with the wavelength, it can be shown that the average power W in each cycle is given by

$$W = \frac{\rho_0 c k^2 Q^2}{8\pi}$$

where Q is the strength of the source, ρ_0 is the density of air, λ is the wavelength of sound, and $k = 2\pi/\lambda$. It can be seen from this that the power at a given wavelength depends only on the physical properties of air and the strength of the source. It is independent of the dimensions of the source, and indeed any simple source of the same strength will produce the same power, regardless of its construction, provided that its size is small compared with the wavelength.

At a distance r from the source, this power is radiated through a spherical surface whose area is equal to $4\pi r^2$, so the average sound intensity I at distance r from a simple source of strength Q is given by

$$I = \frac{\rho_0 c k^2 Q^2}{32\pi^2 r^2}$$

It is important to notice here that that the sound intensity—the sound energy passing through 1 m^2 each second—diminishes with the inverse square of the distance from the source (Figure 4.1). Measured in decibels, the sound intensity decreases by 6 dB for each doubling of the distance from the source. This is in accordance with the principle of conservation of energy. Air is perfectly elastic and no energy is lost during sound transmission, providing, of course, that the sound field does not contain any sound-absorbing objects. The rate at which sound energy is transmitted through any closed surface surrounding the source must therefore equal the rate at which it is supplied by the source, and sound intensity must necessarily diminish with the increase in area—proportional to the square of the radius—of an imaginary sphere having the source at its centre.

It can be shown that the sound intensity in a spherical wave is proportional to the square of the sound pressure; the relationship between these quantities is the same for both spherical and plane waves, namely,

$$I = \frac{p_{rms}^2}{\rho_0 c}$$

as in Equation 3.6 in the previous chapter. Notice that the analogy with electrical power in a simple resistance still applies. By combining these equations we obtain expressions for the sound pressure in terms of the

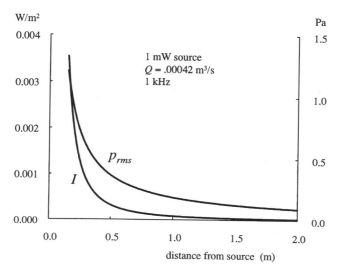

FIG. 4.1 Variation in mean sound intensity and root mean square sound pressure with distance from a simple source.

strength of the source and the radiated power:

$$p_{\text{rms}} = \frac{\rho_0 ckQ}{4\sqrt{2}\,\pi r} \qquad\qquad 4.1$$

and

$$p_{\text{rms}} = \sqrt{\frac{\rho_0 cW}{4\pi}} \times \frac{1}{r}$$

It is important to observe that the sound pressure is inversely proportional to the distance from the source and that for a given radiated power it is independent of frequency. The significance of this is that sound-level meters detect the presence of sound using microphones that respond to sound pressure. The indicated sound level is derived from the average squared value of the sound pressure. From the foregoing discussion it will be seen that the indicated sound level is simply related to the average intensity in the sound field and to the power of the source.

The relationship between intensity and particle velocity in a divergent sound field is, however, more complicated. We find that close to the source the particle velocity is greater than it would be in a plane wave of the same intensity, and as the source is approached, we find that the particle velocity lags the sound pressure instead of being in phase with it. The electrical analogy is with a circuit that has inductance as well as resistance. In general, then, the particle velocity in a spherical wave has a component in phase with the sound pressure and a component that lags the pressure by $\pi/2$. The overall phase difference, determined by the vector sum of these components, therefore lies somewhere between zero at points far removed from the source and $\pi/2$ at points very close to it. In this context what matters is not the absolute distance from the source but the distance relative to the wavelength λ of the sound being produced. A convenient way of expressing distance in relative terms is to multiply by the phase-change coefficient k. Recalling that $k = 2\pi/\lambda$, the distance r from the centre of the source multiplied by k is

$$kr = 2\pi \frac{r}{\lambda}$$

The sound particle velocity is in phase with the sound pressure when $kr \gg 1$ and lags by one quarter of a cycle when $kr \ll 1$. In general the phase difference between pressure and velocity is θ, where $\tan \theta = 1/kr$.

The phase relationship between particle velocity and sound pressure is used to define two kinds of sound field. The *far field* of a source is a region in which pressure and velocity have practically the same phase; the *near field* of a source is a region in which there is a significant difference in phase. The sound intensity corresponding to a particle velocity u in the far field of a source is given by

$$I = \rho_0 c u_{\text{rms}}^2$$

as it is for a plane wave. In the far field, the particle velocity is inversely proportional to the distance from the source, and for a given radiated power it is independent of frequency. In the near field, the increase in particle velocity as the source is approached is more rapid and not independent of frequency (Figure 4.2). For example, at one-tenth of a wavelength from the centre of a simple source, the particle velocity in the spherical wave is approximately 16 times greater than it would be in a plane wave of the same intensity.

Double Sources

Vibrating objects usually have more than one vibrating surface. It often happens that the sound is produced principally at just two surfaces moving in opposite directions, that is, in antiphase. For example, if a loudspeaker is not mounted in a large baffle or in a sealed cabinet, sound is radiated from both the front and back surfaces of the cone. Similarly, a vibrating plate or

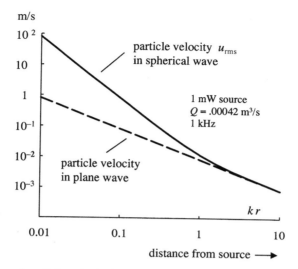

FIG. 4.2 Sound particle velocity in a spherical wave. The dashed line shows the particle velocity in a plane wave of the same intensity.

string has opposed surfaces that are the sources of the sound they produce. It is evident that if such double sources are small compared with the wavelength, the radiation from one surface will almost cancel the radiation from the other and, compared with a single source, a much greater vibration will be needed to produce a given sound. For example, the vibrations of a string or the tines of a tuning fork are often large enough to be clearly visible, but the sound that such objects produce directly is feeble.

A double source comprising two sources of equal strength but opposite phase, separated by a small distance is called an *acoustic doublet* or *acoustic dipole*. The arrangement is illustrated in Figure 4.3. Suppose that each

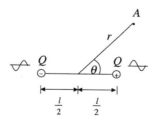

FIG. 4.3 Acoustic double source.

source has strength Q and that the distance between the sources is l. We are interested in the sound pressure at a point A at the distance r and angle θ shown in the figure. The sound pressure can be calculated at any point in the field of the doublet by adding the components due to each source acting alone (principle of superposition). However, it is not possible to obtain a mathematically simple result unless some conditions are imposed. These are that kl is small compared with 1 and that kr is much greater than 1. If only the first condition applies, then the rms sound pressure at A is

$$p_{rms} = \frac{\rho_0 ckQ\sqrt{(kl)^2 + (l/r)^2}}{4\sqrt{2}\,\pi r} \times \cos\theta$$

If both conditions apply,

$$p_{rms} = \frac{\rho_0 ck^2 Ql}{4\sqrt{2}\,\pi r} \times \cos\theta$$

These expressions may look a little complicated, but, for a given wavelength, they each have the form

$$p_{rms} = \text{constant} \times \frac{\cos\theta}{r}$$

As with a simple source, the sound pressure is inversely proportional to the distance from the source, but for the dipole it also varies with the cosine of the angle to the axis, so that the pressure distribution around the dipole has two lobes[1] (Figure 4.4). The sound pressure created by the double source is $kl\cos\theta$ times that for a single source. Because this factor is much less than 1 (in accordance with the stated conditions), the double source is necessarily much weaker than its single-source counterpart. It can be shown that the total power radiated is less by a factor of $(kl)^2/3$.

Infinite Baffles

Radiation from the back of a source such as a loudspeaker can be excluded by means of a baffle. A baffle consists of a plane, rigid surface with an aperture that just accommodates the source. It is not possible to formulate general expressions for the field in front of the baffle, but certain special cases are usually described. Foremost among these is that of a circular source in an infinite baffle. Obviously, an infinite obstruction is a mathematical abstraction, but in practice finite baffles behave in much the same way

---

[1]The lobes are perfectly circular if the sound pressure is plotted in absolute units (Pa instead of dB).

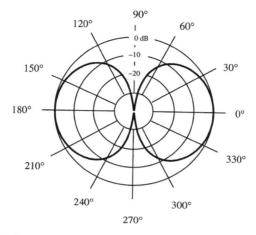

FIG. 4.4 Sound pressure distribution around a double source.

provided that their extent is much greater than the wavelength of the radiated sound. As with the spherical source considered previously, the shape of the source is unimportant so long as it is small compared with the wavelength.

To start, suppose that the source is the small sphere considered previously. This is illustrated in part 1 of Figure 4.5. The sound fields are identical at points A' and A, on either side of the source. Adding the baffle (2) has virtually no effect on the sound field of this simple source because the particle velocity is parallel to its surface at all places close to the surface so that no reflection occurs. It would therefore make no difference to the sound pressure at A if the back of the source were removed and replaced by a fixed (nonvibrating) surface (4). In these circumstances, however, there would be no sound at A'. Note, however, that because of diffraction, sound would be present at A' if the baffle were absent (3). If the sources illustrated in parts 1 and 4 of the diagram have the same radius and vibrate with the same amplitude, it is evident that the hemispherical source has half the strength of its spherical counterpart. But the sound pressure at A is the same in each case. If, therefore, the vibrational amplitude of the hemispherical source is doubled so as to give it the same strength as the spherical source, the sound pressure will also be doubled. In other words, a small source in an infinite baffle produces twice the sound pressure that can be obtained in the unobstructed field of simple source of the same strength.

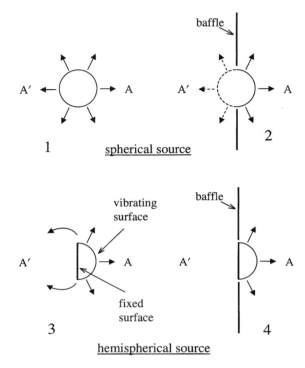

FIG. 4.5   Spherical source and infinite baffle.

Circular Pistons

Suppose the source is a circular plate that just fills the opening in the baffle, and suppose that the plate moves sinusoidally in a direction perpendicular to the baffle. A physical piston just described is not strictly necessary because exactly the same radiation would be produced by a disk of air moving in the same way. The theory that follows would therefore apply equally to the radiation from a pipe provided that its opening was flanged by an extensive baffle. A circular piston opening into an infinite baffle is again an abstraction, but it is important because many real sources behave in approximately the same way and, reciprocally, the directional properties of this type of source are those of a circular microphone placed close to a large plane surface.[2]

---

[2]Microphones used to record conversations in interview rooms are often mounted against a wall so that they receive frontal radiation only.

The mathematical analysis of radiation from a piston is difficult except in some special cases. An important case which clearly has practical application is where the sound pressure is required at large distance from the source. This means that if the distance is r, then kr must be large compared with unity. Again, suppose we have a point A at a distance r from the centre of the piston, and suppose that a line from A to the centre is inclined at an angle θ to the axis, as shown in Figure 4.6. The diagram is symmetrical about the axis, and, in polar coordinates, A is the point (r, θ). It can be shown that the sound pressure at A is given by

$$p_{rms} = \frac{\rho_0 ckQ}{2\sqrt{2}\,\pi r} \times F(ka \sin \theta)$$

where a is the radius of the piston. To understand this equation, compare it

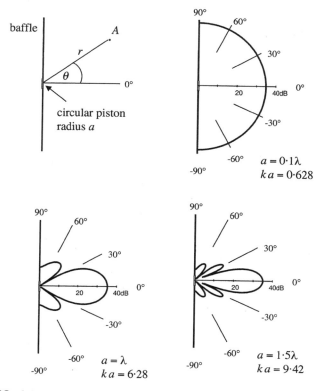

FIG. 4.6 Intensity distributions for piston in an infinite baffle.

with Equation 4.1. The first few terms are the same, except that the 4 in front of the root sign has been replaced by a 2, implying that, other things being equal, the sound pressure is double that of a simple source of the same strength. The last term, $F(\ )$, is a *function* whose value depends on ka multiplied by the sine of the angle from the point A to the source. The value of this function can be obtained from mathematical tables.[3] For points on the axis where θ is zero, the function is equal to 1, so that the pressure is exactly double what would be obtained from a simple source. At other points the pressure always has a lower value depending on the angle to the axis and the wavelength of sound. The sound intensity is proportional to the square of the sound pressure, and its distribution around the piston source has one or more lobes depending on the wavelength (Figure 4.6). Notice that at low frequencies the radiation is almost uniform, but at high frequencies a far greater proportion of sound is radiated centrally.

The directional behaviour of a source can be expressed in terms of what is called the *directivity factor*. This is defined as the ratio of the sound intensity I_0 on the axis at a distance r from the source to a reference intensity I_{ref}. The latter is defined as the total radiated power W divided by the area of a sphere of radius r. So, the directivity factor D is defined as

$$D = \frac{I_0}{I_{ref}}$$

where

$$I_{ref} = \frac{W}{4\pi r^2}$$

The quantity $d = 10 \log D$ is called the *directivity index* or *directional gain*.

It is evident from the definition just given that D is equal to 1 for a simple source. It can be shown that the directivity factor of the piston source approaches 2 when ka is much less than 1, but it is approximately equal to $(ka)^2$ for large values of ka.

Real sources are unlikely to have the same directivity as the hypothetical piston. One reason for this is that the moving part — the diaphragm of a loudspeaker, for example — is likely to be flexible, so that the motion is not uniform over the whole surface. Real baffles are, of course, not infinite, and indoors scattering from other reflecting objects makes the source less directional. As a generalization, it may be said that the directivity is greater at high frequencies than at low frequencies. At points away from the axis, the intensity of high-frequency sounds is likely to be considerably reduced. This can be a problem in public address systems outdoors, where scattering

[3]The directivity function $F(x) = 2J_1(x)/x$, where $x = ka \sin \theta$ and $J_1(x)$ is the Bessel function.

by neighbouring objects has little effect. This is one reason why such systems often employ groups of speakers pointing in different directions.

Practical Sources

This section considers sound sources in common use, paying particular attention to those that have audiological applications.

Tuning Forks

The tuning fork is a surprisingly complicated source in that it is a far from trivial matter to present a complete analysis of its manner of vibration and sound production. It has, however, some very useful properties. The sound it produces is a pure sinusoid; its frequency is highly stable, and its simple, rugged construction and small size make it ideal as a portable instrument. Its principal use, of course, is as a frequency standard for tuning musical instruments, but it has also been adapted for diagnostic use in clinical audiology. From a mechanical point of view the fork can be regarded either as a single 'free-free' bar or as two 'fixed-free' bars attached to a heavy base (Figure 4.7). In its fundamental mode, a straight free-free bar vibrates about two nodes located symmetrically about its centre. Bending the bar into a U causes the nodes to move closer together, and the additional mass and stiffness of the stem has a similar effect. The stem shares the vibration of the central antinode. When the fork is regarded as two fixed-free bars, it is

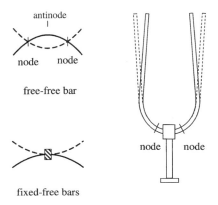

FIG. 4.7 The tuning fork.

apparent from the symmetry that forces on the stem cancel in the lateral direction but that components in the direction of the axis of the stem remain.

The tines of a tuning fork can vibrate in modes other than the fundamental. As mentioned previously (Chapter 3), these modes give rise to overtones whose frequencies are not harmonics of the fundamental, and a harsh tone is produced if the fork is struck with sufficient force to add significant vibration at the modal frequencies. The overtones tend, however, to be more highly damped than the fundamental, and when used carefully a tuning fork produces a very pure tone. Its frequency is highly stable, perhaps to 1 in 10^5 if its temperature is kept constant. Even if temperature does change, its effect on frequency is to reduce it by only about 1 in 10^4 for each degree Celsius temperature rise.

The width of the tines is of course very small compared with the wavelength of the radiated sound, and as a source of airborne sound the tuning fork is rather feeble. A further complication is that each tine has an inner and outer surface, so that it is in effect a double source, and the fork as a whole has two such double sources. The fork is therefore a seemingly unsatisfactory source for air conduction audiometry. Nevertheless, sufficiently reproducible results can be obtained if it is used in a standard way, that is, close to the ear with the tines in line with the axis of the external ear canal.

In otological applications the vibration of the stem of the tuning fork is particularly important because it is used as a source of bone-conducted sound. The origin of this vibration is not as obvious as it might appear. If the fork is regarded simply as a free-free bar then the motion of the stem is derived from the central antinode as just described, but if the stem is relatively massive then the fixed-free description is more appropriate, in which case the antinode does not exist. A general explanation can, however, be provided as follows. In our discussion of Newton's laws in Chapter 1 it was pointed out that 'it is impossible to move the centre of mass of any closed system by any action occurring entirely within it'. If the fork is unrestrained, its centre of mass will remain entirely motionless in spite of the vibration of its various components. Imagine that the fork is upright as shown in Figure 4.7. When the tines bend, the centre of their combined mass will move very slightly upwards or downwards relative to the mass of the stem and lower parts of the fork. (If this is not obvious, it will become so if the bending is imagined to be much greater than would normally be.) In order to keep the centre of mass of the entire fork motionless, the stem must therefore move vertically in the opposite direction. It can be shown that if the tines move sinusoidally at the fundamental frequency, the movement of the stem has a component at the fundamental frequency and components at higher frequencies. The fundamental is usually dominant, but there may be a significant component at twice this frequency. Some low-frequency forks (128 Hz and less)—for example, those used by neurologists to elicit the sense of vibration—are likely to have stem vibrations that are predominantly at the double frequency, but the standard otologists' tuning forks with frequencies from 250 Hz upwards produce negligible stem vibration at frequencies other than the fundamental. The double-frequency component can be reduced by giving the tines a slight 'set' so that they are not exactly parallel to each other.

The stem of a tuning fork vibrates with an amplitude that is very small compared with that of the tines. One can regard this state of affairs as analogous to a lever in which the tines have a large mechanical advantage over the stem. As a consequence their motion is not greatly damped by holding the stem in the fingers or even by pressing it against a hard surface. The technical description would be that the fork is mechanically a high-impedance source when its output is taken from the base (as in a bone conduction test). To produce a strong sound from a tuning fork it is necessary to place its stem on some large surface that can serve as a sounding board. The vibrations are of course damped by this action, but not excessively so, for they may remain audible for several seconds.

Loudspeakers

Loudspeakers radiate sound into a more or less unconfined space, as distinct from earphones, which are sources coupled directly to the ear. In nearly all loudspeakers the source is a vibrating diaphragm that is driven either electrodynamically by passing a current through a coil attached to the diaphragm, or electrostatically. Most speakers are of the electrodynamic type. The diaphragm is usually a shallow, stiff cone made of a lightweight material such as paper (Figure 4.8). It is supported around its circumference by an annulus of corrugated material that allows axial movement but is relatively resistant to movement in a radial direction. Additional support is often provided centrally, again using corrugated material. The coil (known as the *voice coil*) is cemented to the centre of the diaphragm It lies in a narrow cylindrical gap between the pole pieces of a powerful permanent magnet. The magnetic field is radial, so that when current passes through the coil, an axial force is generated. The diaphragm together with the voice coil and the suspension system constitute a mechanical oscillator as described in Chapters 2 and 7. Vibrations are damped, partly by mechanical losses in the suspension, but more importantly by interaction with the surrounding air and the creation of sound waves. The movement of the coil

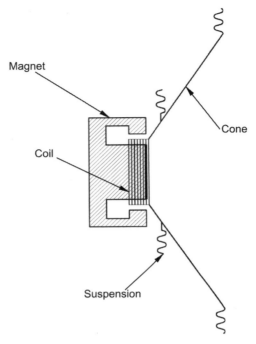

FIG. 4.8 Direct-radiator loudspeaker. (Adapted from Figure 10 in L. E. Kinsler and A. R. Frey. *Fundamentals of Acoustics*, 2nd ed. John Wiley & Sons, New York, 1962, p. 248.)

in the magnetic field creates a voltage proportional to the velocity of the coil. This is similar to the back emf in an electric motor, and its effect is generally to reduce the output of the speaker, particularly if the amplifier that drives it has a low impedance so that it provides the same voltage regardless of load.

The design of a loudspeaker is a difficult task requiring compromises to be made according to the mechanical, acoustical, and electrical specifications. It is virtually impossible to design a speaker that will perform efficiently and that will have a uniform, undistorted acoustic response over the entire auditory range. Most 'hi-fi' systems have two or more speakers, each restricted to a range of frequencies in which their performance has been optimized. Efficiencies are low — often just a few percent — and most of the electrical energy supplied ends up as heat in the voice coil.

An obvious difficulty, particularly troublesome at low frequencies, is that a simple diaphragm will radiate from both the front and the back surfaces.

This problem is often overcome by mounting the speaker in a cabinet so that sound from the back surface of the cone is absorbed. If the enclosed speaker is placed near a wall or other extended surface, it then becomes, to a first approximation, a piston source in an infinite baffle, at least for sounds of short wavelength.

The low-frequency performance of a speaker can be improved by fitting a horn. This is often done for speakers intended for public address systems. The action of the horn is equivalent to that of a transformer in providing a better match between the electomechanical properties of the diaphragm and the acoustic load. In effect the sound pressure on the diaphragm in the throat of the horn is raised relative to the pressure at the mouth of the horn, and importantly, the pressure is more nearly in phase with the velocity of the diaphragm than would be the case if the horn were removed. This leads to a considerable increase in efficiency. The high acoustic output required in public address systems is of course out of place in audiological applications, and loudspeakers for audiological use are invariably of the direct-radiation type. Hornlike tubing is, however, sometimes used in hearing aids to connect the earphone and the ear canal.[4]

Audiometric Earphones

The construction of an audiometric earphone is usually similar to that of a small loudspeaker but the design requirements are quite different. A loudspeaker is intended to radiate into an open space, whereas an earphone is coupled directly to the ear so that its output is to the volume of air trapped beneath the earcap (earphone cushion) and the pinna together with the ear canal and the middle ear, as described in Chapter 7. The requirement is for a uniform frequency response in a real ear, ear simulator, or acoustic coupler and for very low distortion over a very wide range of output power. The radiated acoustic power is generally much less than that required of a loudspeaker, and comparatively small displacements of the diaphragm are needed. The design objectives are achieved in the construction shown in Figure 4.9. The diaphragm is made of metal foil to produce a rigid conical piston that moves as a unit without breaking up into alternative modes of vibration at high frequencies. The suspension is quite stiff, having only one simple corrugation. The conical backplate and the enclosing plastic case prevent sound being radiated from the back surface of the diaphragm, and the felt pads in vents cut in the

---

[4]The connection is sometimes called a Libby horn (after E. R. Libby). It improves the high-frequency response of the hearing aid.

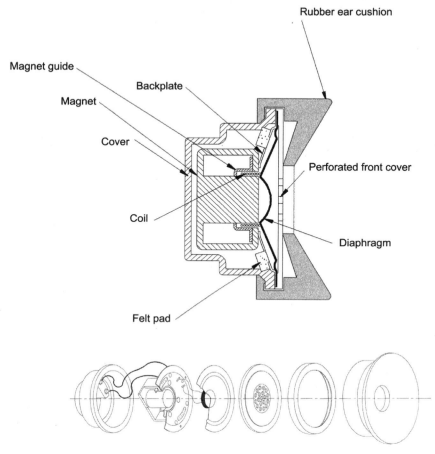

FIG. 4.9 An example of an audiometric earphone.

backplate provide acoustic damping. A typical frequency response is shown in Figure 4.10.

Hearing Aid Earphones

Hearing aid earphones, also known as *receivers* because they receive an electrical signal and convert it to sound, are either of the external type, mounted directly on the earmould and intended mainly for use with body-worn aids, or of the miniature type used in behind-the-ear and

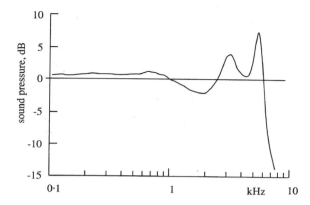

FIG. 4.10 Audiometric earphone (TDH-39). Typical frequency response in a 6-cc coupler for a sinusoidal input at a constant voltage level. The sound pressure level is shown, arbitrarily, relative to the output of the earphone at 1 kHz.

in-the-ear models. Design considerations include the conflicting require-ments of low power consumption and high output sound pressure. The frequency response is intentionally far from uniform. It makes use of natural resonances in the earphone to give response characteristics considered desirable for a hearing aid, compensating both for hearing loss and the presence of an earmould. The displacement of the diaphragm should be a reasonably linear function of the drive current, but the near-perfect linearity found in an audiometric earphone is unnecessary and a small amount of distortion is acceptable. Small size and low cost are important.

Figure 4.11a shows the construction of a miniature receiver designed for use in a behind-the-ear hearing aid. The driving force is produced by an armature which is made from a flat strip of ferromagnetic material bent to the shape of a letter U, rather like a tuning fork. One limb of the armature lies in an air gap between the poles of two permanent magnets that produce a field perpendicular to the surface of the strip. The other limb is attached to the magnet and pole piece assembly. The armature passes through a coil so that its free end is magnetized with a polarity that depends on the direction of the alternating drive current in the coil. The armature is therefore attracted to one or other of the permanent magnets. Its movement is coupled to the diaphragm by the drive rod as shown in the diagram. The diaphragm operates within a very small air volume, so it can produce a high sound pressure. Sound is transmitted through a small outlet pipe attached to the top cup and then through a connector that joins the plastic pipe to the earmould.

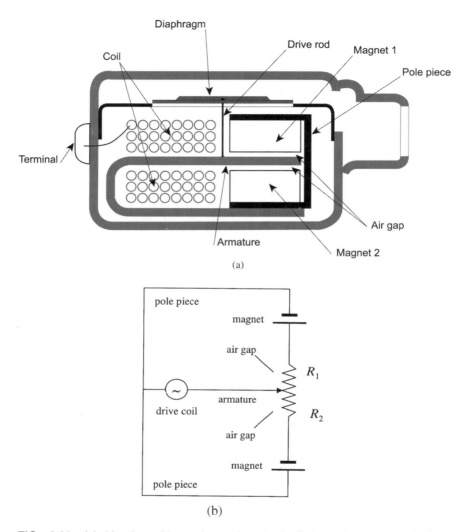

FIG. 4.11 (a) Hearing aid earphone (receiver). Balanced armature design. (Based on a drawing supplied by Knowles Electronic Company.) (b) Equivalent electric circuit representing the magnetic circuit of the earphone.

The design of the earphone is somewhat more subtle than it might appear at first sight. Its magnetic components form what is called a *magnetic circuit*. This is analogous to an electric circuit in which voltage (electromotive force) represents a magnetic potential (measured in ampere-turns) and in which electric current represents magnetic flux (in webers). The magnetic equivalent of resistance is called *reluctance* (in henries). An electric circuit equivalent to the magnetic circuit of the earphone is shown in Figure 4.11b. The permanent magnets, represented by batteries, provide a steady (dc) magnetic flux. The pole piece has a very low reluctance and can therefore be represented as a direct connection between the outer ends of the magnets and the fixed end of the armature. The free end of the armature is equivalent to the moving contact in a potential divider represented by the resistances R_1 and R_2 that correspond to the magnetic reluctances of the two air gaps. Ideally, the tip of the armature should lie centrally between the magnets to make the reluctances of the air gaps equal ($R_1 = R_2$). There is then no dc flux in the armature. In this condition the armature is said to be *balanced*. The drive coil generates an alternating magnetic potential and is therefore represented as an alternating-voltage source. The transducer can operate either as an earphone or as a microphone. As an earphone, it works as just described; as a microphone, displacement of the tip of the armature reduces one air gap and increases the other, causing a magnetic flux to flow in the armature. This generates a voltage difference across the ends of the coil. (If the transducer is used as a microphone, the alternating source is replaced by a galvanometer in the equivalent electric circuit.)

The balanced armature design offers good linearity and high efficiency in a compact arrangement. A particular advantage is that the coil itself does not move (as it would in a conventional loudspeaker or dynamic microphone), so its mass can be disregarded. The components of the earphone can be modified at manufacture to reduce the output at low frequencies or to reduce the high-frequency peaks, according to the user's requirements. A typical response curve is shown in Figure 4.12.

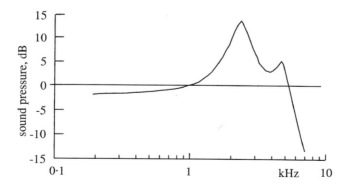

FIG. 4.12 Hearing aid earphone (Knowles EH series). Typical frequency response in a 2-cc coupler for constant sinusoidal input current, 0.7 mA rms. The sound pressure level is shown, arbitrarily, relative to the output of the earphone at 1 kHz (typically 100 dB re 20 μPa).

Audiometric Bone Vibrator

A bone vibrator held against the mastoid can provide direct bone-conducted stimulation of the cochlea for audiometric purposes. Similar vibrators are used in bone conduction hearing aids for patients who lack an adequate air conduction pathway. The construction of a typical audiometric vibrator is shown in Figure 4.13. Vibration is transmitted through a plastic case in contact with the patient's head. An armature, screwed to the case, is

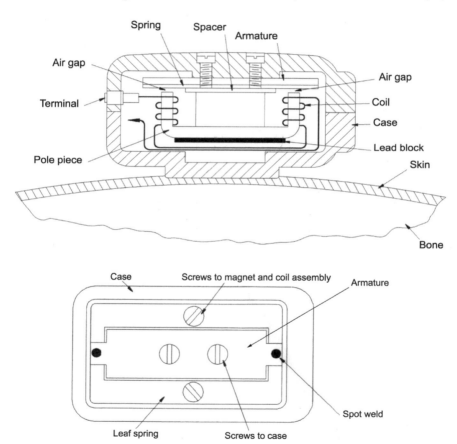

FIG. 4.13 Audiometric bone vibrator (Radioear B71). The main diagram is a cross section through the vibrator; the subsidiary diagram shows details of the spring that connects the case and armature to the magnet. (Based on a drawing supplied by the Radioear Corporation, New Eagle, Pennsylvania.)

attracted to the poles of a permanent magnet. The pole pieces pass through coils carrying the drive current, whose effect is to add an alternating component to the magnetic field. The magnet and coil assembly is attached to a lead block to increase its mass. The assembly is suspended by a leaf spring which maintains the separation between the poles and the armature. The case and armature can be represented by a mass m, and the magnet, lead block, and coils by a mass M. These masses are connected mechanically by the spring. If the vibrator were suspended freely, m and M would oscillate about their centre of mass, which would remain stationary. In use, however, the movement of the smaller mass is restricted by its contact with the head so that the centre of mass is obliged to move, creating a reaction that constitutes the desired auditory stimulus. The response of a typical vibrator is illustrated in Figure 4.14. The force generated for a given drive voltage and the corresponding stimulus both vary considerably with frequency.

An undesirable feature of the type of vibrator just described is that it is a source of airborne sound. The vibrator is, however, a fairly small object compared with the wavelength of sound over most of its useful range, and radiation from one side of the vibrating case tends to cancel radiation from the opposite side. The use of foam earplugs to attenuate the air-conducted sound is sometimes recommended for bone conduction audiometry at frequencies from 3 kHz upwards.

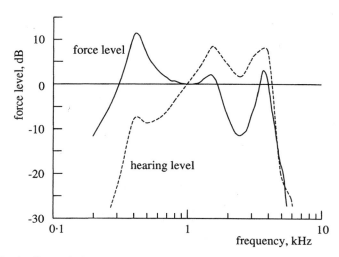

FIG. 4.14 Audiometric bone vibrator (Radioear B71). Typical frequency response on IEC mechanical coupler for a constant rms input voltage. The corresponding hearing level is also shown. Values are relative to the output at 1 kHz.

Questions and Exercises

Practical exercises are marked with an asterisk.

4.1 What is meant by a *simple source*? What is the shape of the wavefronts in the neighbourhood of this kind of source?

4.2 Explain what is meant by the *strength* of a simple source. What is the relationship between the strength of a source and the acoustic power that it produces? Sketch a graph to show this relationship at a constant frequency.

4.3 How does the strength of a simple acoustic source change with frequency if the power it produces is constant? Calculate the strength of a simple source needed to produce 1 mW at (a) 125 Hz and (b) 4 kHz ($c = 343$ ms^{-1}; $\rho_0 c = 415$ Nsm^{-3}).

4.4 How do sound intensity and sound pressure vary with distance from a simple source? Calculate the root mean square sound pressure at a distance of 2·2 m from a simple source radiating 60 μW ($\rho_0 c = 415$ Nsm^{-3}).

4.5 Distinguish between the *near field* and the *far field* of an acoustic source.

4.6 What is an *acoustic doublet*? What can be said about the sound pressure on a line that is perpendicular to a line joining the centres of the sources that make up a doublet (i.e. in the direction $\theta = 90°$)?

4.7 For a given dipole source, what is the ratio of the rms sound pressure at $\theta = 75°$ to the rms pressure at $\theta = 0$?

4.8* In this experiment we attempt to measure the radiation around an acoustic dipole. An audiometric bone vibrator is approximately a dipole source of airborne sound, at least for frequencies below 6 kHz. Suspend a Radioear B71 audiometric bone vibrator by its supply cable so that the driving face is vertical. To do this, cut a wine cork in half longitudinally, pass the cable between the two parts, and then bind them together with tape. It may help if the cable is first stiffened by binding it to a straight piece of thick wire, made from a coat hanger perhaps. Hold the cork vertically in a laboratory clamp, with the cable and bone vibrator hanging below. Place a sound-level meter with its microphone at the same height as the vibrator at a distance of 20 cm. Drive the vibrator at 1 kHz or at a higher frequency using an audiometer or other oscillator and measure the air-radiated sound.

Turn the cork to change the orientation of the vibrator relative to the microphone and plot the polar distribution of sound pressure. Try this for different frequencies and with the microphone at different distances from the source. If standing waves are a problem, use warble tones or narrow bands of noise instead of tones. The orientation of the vibrator can be estimated visually, but for better precision you can fix a pointer or a scale to the cork to measure its orientation. If the cable has been stiffened as suggested, the vibrator will turn through the same angle as the cork.

4.9* This is a laboratory experiment requiring an artificial mastoid and filters. Place the stem of a tuning fork on an artificial mastoid and compare the output at the stated frequency of the fork with that at twice this frequency. You may observe a significant vibratory output at the double frequency when using at 125- or 250-Hz fork.

4.10* Hold a piece of card with its edge parallel and very close to one of the tines of a tuning fork. Observe that the fork sounds much louder when the card is at about 45° to the direction of vibration. It then restricts the circulation of air around the tine. Placing the card between the tines or in line with them has little effect. This demonstration, attributed to Stokes in 1868, is described by Rayleigh in *The Theory of Sound*, Vol. 2, p. 306.

5
Nonsinusoidal Waveforms

The notion that real systems can undergo perfect sinusoidal oscillation is essential in the description of acoustic phenomena. Our descriptions rely heavily on terms such as angular frequency, phase, amplitude, wavelength, and so on that have no meaning except in the context of such oscillation. The fact that analysis is based on sounds that are not part of the real acoustic environment should not surprise us. Simplification is at the heart of the scientific method and remains acceptable so long as it enhances understanding and leads to descriptions that accord with observations of actual events.[1] The problem for acoustics is to extend the ideas developed for harmonic oscillation so that they may be of value in the description of more general kinds of vibration.

If we recorded examples of the sounds that we normally encounter and displayed them on an oscilloscope, we would usually find the waveforms to be irregular. Their appearance would not be entirely random and we would probably find sections where the vibrations were more or less periodic, but it would be difficult to discern any overall pattern. The problem of analysing such waveforms has been solved mathematically by applying transforms in which the original wave is replaced by sine and cosine waves. The mathematics is perhaps not as simple as a familiarity with these transforms may lead us to believe, but computer processing has made transformation a practical reality, and it has become a standard analytical tool. There are many applications. In audiology the analysis of otoacoustic emissions and the digital processing of speech signals are well-known examples. Because of the fundamental importance of frequency analysis in understanding non-sinusoidal waveforms, we will start with a rather general and nonmathematical account of the Fourier transform.

The Fourier Transform

The transform has its origins in Fourier's work in the early 19th century on heat conduction. The essence of the transform is that it takes any function

---

[1]According to anecdote, a theoretical paper on the hazard to aircraft from bird strikes began with the words 'consider a spherical chicken...'.

of one variable, let us say x, and converts it to a different, but related, function[2] of another variable s. In other words,

$$f(x) \text{ has the transform } F(s).$$

When s is specified, the function $F(s)$ supplies sine and cosine terms whose amplitudes and arguments contain s. The transformation is a purely algebraic process, and x and s may or may not represent tangible physical quantities, but in many applications, including acoustics, x often represents time and s often represents frequency.

The importance of the Fourier transform is therefore that it allows us to describe the response of physical systems either as a function of time or as a function of frequency. The functions are $f(t)$ and $F(\omega)$, where t is time and ω is angular frequency. When we obtain $F(\omega)$ from $f(t)$, we say that $F(\omega)$ is the Fourier transform of $f(t)$. It is also possible to carry out this process in reverse so that $f(t)$ can be obtained from $F(\omega)$. This involves calculation of the *inverse transform* of $F(\omega)$. The two functions should therefore be considered in pairs. From a purely mathematical point of view it is true that not all functions have transforms, but fortunately those that are transformable include all that describe real physical events. If a time function can be seen on an oscilloscope or plotted on a chart recorder, then it does have a Fourier transform.

To explain the process in more familiar terms, suppose that we drive an audiometric earphone from some electrical source that generates a rectangular voltage pulse at its terminals. This is exactly what happens when we use acoustic clicks as a stimulus for auditory evoked potentials. Now, the pressure waveform inside the ear canal will not be a faithful replica of the voltage waveform at the earphone. In fact, the two waveforms may have little resemblance, because the sound pressure that is created by the voltage pulse depends on the dynamics of the earphone and the acoustics of the ear canal. The time course of the sound pressure in the ear could be recorded using a probe microphone, or the measurements could be made on an ear simulator rather than a real ear. Either way, we arrive at a description of the output of the earphone as a function of time. This is called the *time domain* response. But in audiology we usually want to know what frequencies are involved so that we can relate the physiological response to the audiogram or to some other frequency-related feature of hearing. The prospect of converting the time response to a frequency description is therefore inviting. Applying the transform to the time function leads to a description in the *frequency domain*. We can visualize $F(\omega)$ as a continuous spectrum in which sinusoids of all frequencies are represented. If all the

---

[2]The use of the term *function* is explained in Appendix A.

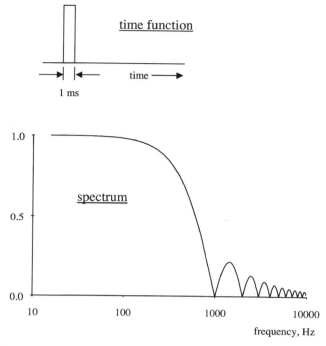

FIG. 5.1 Fourier spectrum for a 1-ms rectangular pulse in the frequency range 15 to 10,000 Hz. The ordinate is the amplitude of the spectrum.

sinusoids were added we would recover the original time function. The magnitude of $F(\omega)$ tells us the relative contribution to the time function coming from sinusoids whose frequencies are in the neighbourhood of some chosen value of ω.

Figure 5.1 shows the frequency spectrum corresponding to the Fourier transform of a rectangular pulse. To give the pulse some physical meaning, we may imagine again that it describes a voltage waveform applied to an earphone. Suppose that the voltage is turned on at time zero and turned off 1 ms later, having remained constant in the interval. It can be seen that low frequencies make an important contribution, while at higher frequencies the spectrum consists of a series of diminishing 'heaps' at intervals of 1 kHz (the reciprocal of the duration of the pulse). Notice that the first minimum in the spectrum occurs at 1 kHz. If the pulse width is reduced, the first minimum moves to a higher frequency. A short pulse therefore has a more uniform spectrum than a long one (Figure 5.2); an infinitely short (but infinitely strong) pulse would theoretically have a completely uniform spectrum. This

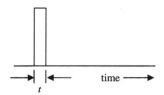

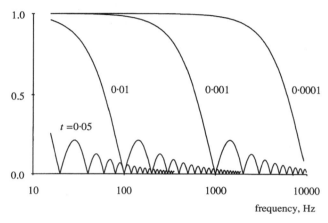

FIG. 5.2 Fourier spectrum for a rectangular pulse. This figure shows how the spectrum changes with pulse width. The parameter is the pulse width *t* in seconds.

is a general feature of all transient signals — the shorter their duration, the broader their spectra.

The Fourier transforms of two other time functions are shown in Figures 5.3 and 5.4. The first of these is a simple exponential decay, the voltage on a capacitor perhaps, whose charge is instantaneously established at time zero and then allowed to leak away through a resistor. The second is a damped sinusoidal oscillation. Once initiated, the oscillations diminish exponentially at a rate depending on the amount of damping. In this example the spectrum is, as one would expect, greatest at frequencies close to the frequency of oscillation. The widening of the spectrum around the peak increases with the amount of damping. On the other hand, if the damping is progressively reduced towards zero so that vibrations die away ever more slowly, the distribution around the peak becomes narrower until all that remains is a line at the frequency of the oscillator. We will consider line spectra later on.

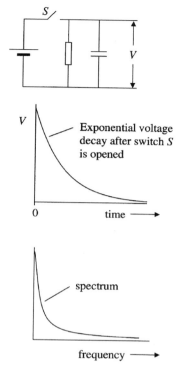

FIG. 5.3 An exponential signal and its Fourier transform.

The Fourier transform provides an alternative way of describing events, allowing them to be pictured in either the time or frequency domain. The frequency view is particularly useful when we want to know how a system will respond to a given input. By *system* we mean anything for which there can exist a definable input and output. The input might be a voltage or a sound pressure or some other physical entity, and the response likewise. The earphone example given earlier is typical. Its input is a voltage, and its output is a sound pressure in the ear canal. The problem is to relate the output to the input. This may well be difficult, but we are often nearer a solution if the input can be described in terms of its spectrum and the system described in terms of its frequency response. The mathematical steps involved need not concern us here, but we might intuitively expect the response at any one frequency to depend on the system's frequency response multiplied by the 'amount' of the spectrum of the input at that frequency.

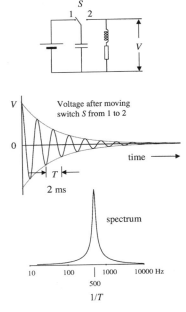

FIG. 5.4   A damped oscillation and its Fourier transform.

Describing the Spectrum

The word *spectrum* is used as a general term to imply that something is distributed according to frequency. In optics we talk of the *visible spectrum*, meaning the distribution of optical frequencies and their associated colours. The same idea applies in other branches of physics and, of course, in acoustics. When describing a sound spectrum, the distributed quantity is usually sound pressure or something closely allied to it. The distribution may or may not be continuous, but we will assume for the moment that we are dealing with a continuous spectrum in which *all* frequencies are possible within a specified, and possibly infinite, range. Because the spectrum involves harmonic (sinusoidal) oscillation, we usually describe it in terms of amplitudes or root mean square values, or as power (*power spectrum*). The latter can be found by squaring the rms values.

A specified range of frequencies constitutes what is called a *frequency band*. It is often useful to express the range as the ratio of the highest frequency, f_2, to the lowest frequency, f_1. The *octave* is a convenient unit for this purpose. The bandwidth is one octave if $f_2/f_1 = 2$. It is a third-octave if

$f_2/f_1 = 2^{1/3}$. When bandwidths are expressed as ratios,[3] or as fractions of an octave, the centre frequency f_c of the band is the geometric mean of the limiting frequencies; that is, $f_c = \sqrt{f_1 f_2}$. For example, the centre of the one-octave band between 4 kHz and 8 kHz is $\sqrt{32} = 5.66$ kHz.

The total acoustic energy due to all component vibrations within the band is called the *band power*. If expressed in decibels,[4] it is the *band power level*. The corresponding sound pressure is the *band pressure* or *band pressure level*. A bandwidth of 1 Hz is a significant special case because the band power level or pressure level in a 1-Hz band is sometimes called the *spectrum level*. Unfortunately, not all authorities agree on this definition. However, there is a better description of the spectrum that denotes something similar to the spectrum level except that the bandwidth is made increasingly narrow until it becomes infinitesimally small. Of course, the power in such a band is also infinitesimally small, but if we divide the power in the band by the bandwidth, we obtain, in the limit, a finite quantity called the *power spectral density* or, in decibels, *the power spectral density level*. Similar definitions apply to specified quantities, such as sound pressure, that are related to power. We therefore find *spectral density* defined as the mean square of the sound pressure (or other defined quantity) divided by the bandwidth, as the bandwidth approaches zero. Again, *level* may be appended to denote the decibel equivalent. The difficulty is that the name *spectrum level* is a permissible synonym of *spectral density level*. If the word *density* is included, then an infinitesimal bandwidth is implied, but if it is omitted, there may be some ambiguity. Fortunately, a 1-Hz band is so narrow that the two quantities usually have the same numerical value whatever definition is used. The spectral density describes the spectrum at a particular frequency, namely, the frequency at the centre of the infinitesimal band. Its usefulness is seen if we consider a small band of frequencies over which the spectrum is sensibly uniform. Multiplying the spectral density in this part of the spectrum by the bandwidth gives the total power in the band.

Bandwidth and spectrum level are important characteristics of a signal. A sinusoid has zero bandwidth and is easily handled by amplifiers and transducers. A transient is more difficult, particularly if it has a short duration. Its generation and transmission require equipment whose operating bandwidth and frequency response are commensurate with the spectrum of the signal. It should be remembered that the bandwidth of any real equipment is finite, depending particularly on the response characteristics of

---

[3]Frequency is often shown on a logarithmic scale so that equal distances along the frequency axis correspond to equal frequency ratios. See Appendix A.

[4]Some of the descriptions that follow will inevitably make use of decibels—part of acoustic terminology that we have not hitherto explained. Information about decibels is provided in the next chapter and in Appendix B.

transducers rather than the performance of electronic amplifiers. In audiology we have the further complication that the acoustics of the outer ear or the mechanics of the head are almost certain to modify the signals received from earphones and bone vibrators. What leaves the audiometer as a clean electrical pulse is likely to end up as a stimulus with a very different waveform. This is a particular difficulty for work involving evoked potentials, where short-duration stimuli are nearly always required. It is to a lesser degree a problem in the measurement of click-evoked otoacoustic emissions, where, ideally, one would like the stimulus to have a uniform spectrum in the range 500 to 6000 Hz. In practice it is often necessary to settle for something less perfect.

Signals and Noise

The words *signal* and *noise* usually have obvious meanings in context, and we have already used *signal* in the foregoing text without definition. It is difficult to give entirely satisfactory definitions of these terms, but in physics generally — and that includes acoustics — the word *noise* usually means something unwanted that interferes with the detection or measurement of some other quantity, namely the signal. But these are very general statements. Usually in acoustics a signal is something with an obvious pattern, and often, but not always, the noise is haphazard. The linguistic difficulty is that we need a word to denote the object of interest whether it be a sound pressure at a microphone or a voltage at some point in an amplifier, and it would be inconvenient to change the word according to the category of waveform. In these circumstances we accept that *signal* means the sound pressure or voltage under discussion even if its waveform has no pattern. We speak of 'signal processing' rather than 'noise processing'. The distinction between a desired signal and an undesired noise is clear, however, when we talk of *signal-to-noise ratio*. This is defined as the power in a sample of the signal divided by the power in a corresponding sample of the noise. The ratio is important because the ability to detect or to accurately measure the signal is usually directly related to it.

So far we have considered only signals such as a pulse or a decaying sinusoid whose waveforms are entirely predictable. Such signals are said to be *deterministic* either because their waveforms have been defined mathematically or because they correspond to observable physical events occurring in carefully controlled circumstances. If '*x*' (voltage, sound pressure, particle displacement, etc.) is known at any instant, its value at any future time is entirely predictable and its history is written. Truly deterministic signals are rare except as a mathematical abstraction, but real signals are often a reasonable approximation. If we recorded the waveform of a note played on a piano, we would be justifiably surprised to find a very different waveform on repeating the experiment. But there are signals whose waveform has no discernible pattern. The sequence of values of 'x' that constitutes the waveform is *random* and no prediction can be made as to what value of 'x' will follow another.

The concept of randomness is central to statistics. Its essence is not that x can have *any* value (it cannot) but that the series of numbers formed by taking successive values of x has no pattern. To explore this idea, think of the random numbers made available by electronic calculators and computers. Strictly speaking these are pseudo-random numbers; if we knew the algorithms well enough, it might be possible to calculate one number given the preceding one, and indeed the sequence of numbers from a given starting point does recur. For most practical purposes, however, it is fair to say that no underlying pattern is recognizable. The numbers are, nevertheless, constrained. They usually fall in the range 0 to 1 (or just less) and, being digitally generated, the least possible nonzero difference between any two numbers depends on the precision of the calculation or display. The important thing to realize is that although the numbers appear in an haphazard sequence when called one after the other, the ensemble (or *population*) to which they belong does have stable features. Its range, mean, and standard deviation are known and can be estimated from samples. The defining property of electronically generated random numbers of this kind is that all values in the range 0 to 1 are equally likely. The probability distribution is uniform. Equal likelihood is not, however, a prerequisite for randomness. The height of the next man we encounter in the street is one of a large set of random values, but he is unlikely to be a dwarf or a giant. What is important so far as random waveforms are concerned is that the signal at any given moment should be independent of the signal at any other moment. As we shall see, such independence is unobtainable in real physical systems but it can be approached as an approximation.

As we have said, noise is often associated with random events. We therefore find *random noise* defined as 'Oscillation due to the aggregate of a large number of elementary disturbances with random occurrence in time' (IEC 60050-801). In many instances this kind of noise has its origins at a molecular or even subatomic level. For example, the random collisions of air molecules with the diaphragm of a microphone are a source of noise that places an absolute limit on the least sound that can be measured.[5] Similarly,

---

[5]This statement is not quite true, because certain signals, particularly recurrent ones, can be recovered even if they are below the noise level of the measuring equipment. For example, the auditory brainstem response is small compared with the electrical background recorded on the surface of the head, but it can be detected by signal averaging.

a fluctuating electrical potential difference can be observed across the ends of a resistor because of the random collisions between charged particles (electrons) and the atomic matrix of the conducting material. Larger potentials are generated at the junction of semiconducting materials in diodes and other electronic components. Diode noise, greatly amplified of course, provides the masking noise for audiometry. It would, however, be wrong to think that all noise is random or at least approximately so. It is often conspicuously nonrandom, as in the case of 'mains hum' from the electrical supply. But random noise is an important category that is treated extensively in engineering texts. It is also the basis of the signal-detection theories of hearing.

The waveform representing noise in a specified interval does have a Fourier transform. The spectrum is a characteristic of the noise. If the spectrum is uniform, that is, if the spectral density is the same at all frequencies, we can call the noise *white noise* by analogy with white light, which is a more or less uniform mixture of light containing all wavelengths in the visible part of the electromagnetic spectrum. White noise is an example of *wide-band noise*, so called because it contains a wide range of frequencies. At the other extreme, where the spectrum is significant over only a small range of frequencies, we say that the noise is *narrow-band noise*. This type of noise is important in audiometry because it has the least loudness for a given amount of masking. Other commonly used types of noise are *high-pass noise* and *low-pass noise* having, respectively, only high or low frequencies in their spectra. These forms of noise are often needed for test purposes, particularly in diagnostic audiology, and they can be derived from wide-band noise by filtering. Indeed the word *pass* as just used implies that the specified characteristics have been obtained by passing the noise-generating signal through a filter.

Filters

In the most general sense, a *filter* is a device whose purpose is to increase or decrease an oscillation according to its frequency. If the signal is complex (nonsinusoidal), then the modification depends on the distribution of frequencies in its spectrum. Filters have many applications, particularly in mechanics, optics, electronics, and acoustics. Their role in electronics is especially important because electronic devices are used to handle signals from a great variety of sources, including microphones and electrical sources used in sound production, measurement, and recording. In this brief reference to filters we will consider only the electronic kind.

Filters have an input and an output. If the input signal is a sinusoid, then the output is also a sinusoid whose amplitude and phase depend on the frequency. The ratio of the output signal to the input signal at a given frequency is called the *transfer factor* of the filter.[6] Filters are designed either to reject or to pass signals whose frequencies fall in a specified band or whose frequencies are above or below specified limits. As mentioned earlier, the attributes *band pass*, *high pass*, and *low pass* describe the characteristics of filters. A well-designed filter produces little gain or attenuation and no change of phase in signals falling in its pass region. Beyond these limits the amount of attenuation is either modest or severe, depending on design, and the phase change is likely to approach $\pm \pi/2$ for remote frequencies. The change of phase is often unimportant in acoustics and unlikely to be perceived by the ear, which is rather insensitive to phase, but it can be troublesome in other circumstances. Filters are needed to reduce the noise that accompanies evoked potentials, and a change in phase can lead to a change in the observed waveform, adding to the difficulty of interpretation. The phase problem can be overcome using digital filtering in which the Fourier spectrum of the signal is derived and then modified in accordance with the required filter characteristics. The output is obtained from the modified spectrum by calculating the inverse transform. With modern electronics this sequence of events can be accomplished so rapidly that the time required for processing has often no practical importance. This is what is meant by *real-time* processing. The ability to process signals at such a speed is an essential part of digital hearing aid technology.

There are several mathematical techniques for determining the effect of a filter on the signals that pass through it. The easiest of these, conceptually, is as follows. The Fourier transform of the input signal is obtained so that the signal is replaced by sinusoids. The sinusoid at each frequency is then multiplied by the transfer factor to give sinusoids whose amplitude and phase differ from those in the original signal according to the design of the filter. The inverse transform is then applied to derive the output signal. Figure 5.5 shows the result of this type of calculation applied to the filtering of a rectangular pulse by a simple low-pass filter comprising a resistance and capacitance arranged as shown in the diagram. The filter attenuates the high-frequency components of the input signal so that the rapid changes in voltage at the beginning and end of the pulse are not transmitted.

It is interesting to consider what happens when seemingly random noise is modified by a band-pass filter. The response characteristics of the filter influence the rate at which the output signal is likely to change from moment to moment. Rapid changes are associated with high-frequency components of the signal, and these components are, by design, selectively

---

[6]The change in phase can be taken into account by expressing the transfer factor as a complex number in a manner similar to complex impedance. (Impedance is discussed in Chapter 7.)

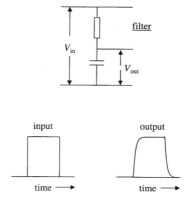

FIG. 5.5 Low-pass filter. The input is a rectangular pulse. The shape of the pulse after filtering has been found using the Fourier transform and its inverse as described in the text.

attenuated at frequencies above the pass band. The same is true in the opposite sense for low-frequency components. If the bandwidth is large these influences are unlikely to be evident on visual inspection of the waveform (using an oscilloscope, for example), but if the bandwidth is narrow the output has a distinctly sinus quality (Figure 5.6). Because all physical instruments are band-limited to some degree, it follows that a truly random waveform can never be observed.

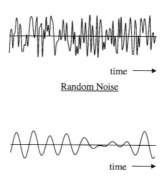

FIG. 5.6 Band-pass filtering applied to random noise. The lower diagram shows the same noise after narrow-band filtering.

The Speech Spectrum

Speech is produced by the flow of air from the lungs. In voiced speech this flow is interrupted by the vocal folds to produce *laryngeal tones*. The vocal folds open and close at rates of from 75 to 500 per second, depending on intonation and voice quality. The average is about 120 interruptions per second for men and 240 for women. The frequency content of speech depends on the laryngeal tone and on the filtering effect of the throat, mouth, and nasal cavities, which are continually modified in the act of speaking. Resonances in the vocal tract reinforce sounds within particular frequency bands. Several characteristic bands known as *formants* are associated with vowels and diphthongs. The bandwidth in which resonance occurs decreases with increasing duration of the utterance, so that the longest vowels may be associated with bands of only a few hertz. Consonants involve the production of turbulent airflow at some point in the vocal tract, either by forcing the exhaled breath through a constriction or by completely blocking the airflow and suddenly releasing it. They have been described as transient or transitional sounds as distinct from the vowels, which have a more sustained character. In terms of their frequency content, they are not clearly distinguishable from vowels except that they are generally weaker and tend to occupy a wider range of frequencies.

The frequency characteristics of particular speech sounds can be examined with the help of a speech spectrograph. This instrument samples speech a few words at a time and generates a visible record showing how the sound energy in a series of specified frequency bands varies from one moment to the next. The record, which runs from left to right with increasing time, shows the energy in each of the bands arranged vertically in ascending order of frequency. The relative energy in each band at a given time is indicated by modulating the display so that the strongest signals appear as black areas while weaker signals are given progressively lighter shades. Figure 5.7 is an example.

The speech spectrograph provides information about individual speech sounds, but for a broader picture it is usual to turn to the so-called long-term spectrum. This is used to assess the performance of communica-tions systems such as telephones and radio transmissions and as a basis of the formulae used to select hearing aids. The long-term spectrum is based on samples of speech that contain a large number of words. It shows how, on average, the acoustic energy in the speech varies with frequency. The frequency scale can have various forms, but for hearing aid work it is often convenient to employ a simple logarithmic scale. It is then logical to express the spectrum as the energy in bands, such as third-octave bands, that have

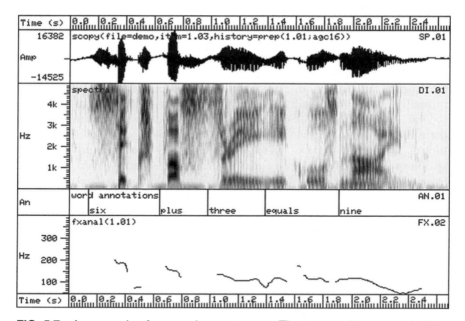

FIG. 5.7   An example of a speech spectrogram. The top panel is a record of the sound pressure created by a person speaking the words 'six plus three equals nine'. The main panel shows the spectrum as described in the text. The bottom panel shows the level of the fundamental speech tone. (This figure was generated by the Speech Filing System of the Department of Phonetics and Linguistics at University College, London. It is copied, with permission, from an example that appears on their Web page, http://www.phon.ucl.ac.uk/resource/sfs/.)

a constant relative bandwidth throughout the frequency range. An example of the long-term speech spectrum is shown in Figure 5.8.

Periodic Signals

A periodic signal is one that repeats itself over and over without change from one cycle to the next. It may at first be rather surprising to find that a periodic signal does not have a Fourier transform. The reason for this is that anything that is truly periodic would have to be in existence, unchanging, for all time. The transform of such a signal would give rise to an infinitely wide spectrum that could not be defined (in technical language, its Fourier integral does not converge). However, no physically observable signal is periodic in the restricted sense implied here. All real signals have a

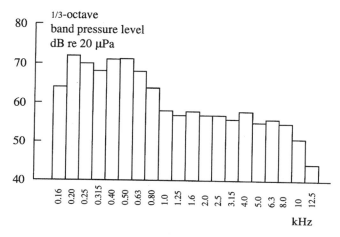

FIG. 5.8 The speech spectrum. (Based on data from Figure 2 in D. Byrne. The speech spectrum. *Br. J. Audiol.* 11(2): 40–46, 1977.)

beginning and an end, or at least they change in some way as time passes. We think of events as being periodic if they recur practically without change for the duration of our interest in them, and it is often useful to obtain a spectrum representation of such real signals.

It was remarked earlier that the spectrum of a damped harmonic oscillation becomes narrower as the amount of damping is reduced. A sinusoid that exists indefinitely is of course a fine example of a periodic signal, and in the limit its spectrum degenerates to a line of zero bandwidth. If we look at a more complex signal we see a similar pattern. Suppose that we have a rectangular wave—the sort of thing that is easily produced in the laboratory with an ordinary signal generator—and suppose that we arrange things so that the signal decays progressively with time. Now consider a sample of the signal taken over a time that is long compared with the period of each cycle. The spectrum of the sample is a continuous band of frequencies, but it has a structure related to the detail of the waveform. It would look different, for example, if the rectangular waves were replaced by triangles to make a sawtooth pattern. If the rate at which the signal decays is made smaller, we find that the peaks and valleys in the spectrum are accentuated; ultimately, as the rate of decay approaches zero, the spectrum is reduced to a series of lines, as illustrated in Figure 5.9.

In general, the Fourier spectrum of a periodic signal is a line spectrum called a *Fourier series*. Any periodic signal that corresponds to a real physical process can be represented by such a series. The line spectrum

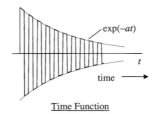

Time Function

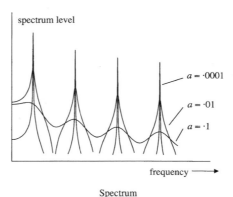

Spectrum

FIG. 5.9 Diagram showing how the continuous spectrum of a signal becomes a line spectrum when the signal is truly periodic. In this example the signal is a rectangular wave whose magnitude decays exponentially with time. The parameter *a* is the decay constant. As this approaches zero, the peak-to-peak value of the wave becomes constant to give a waveform in which cycles repeat exactly.

comprises sine and cosine terms whose phase is arbitrarily made equal to zero at whatever moment we take to be the origin of the time axis. The combination of these terms by simple addition produces the periodic waveform. If the waveform has a period equal to T, the sine and cosine vibrations in the series have frequencies that are multiples of the fundamental frequency $1/T$. In other words, the Fourier series is an harmonic series. The number of terms in the series is infinite, but the amplitudes diminish as the series progresses, and the sum of the vibrations is finite.[7] The series also contains a constant (think of it as a cosine wave of frequency zero) whose value is the average of the waveform over one cycle.

---

[7]An infinite series can have a finite sum. Achilles does catch Zeno's tortoise even though the process requires an infinite number of (diminishing) steps. For example, the series $1 + 1/2 + 1/4 + 1/8 + \cdots = 2$.

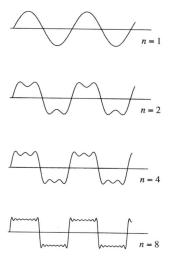

FIG. 5.10 Fourier series: construction of a rectangular wave by adding sinusoids. In each example, *n* is the number of terms that have been added.

The Fourier series corresponding to a periodic rectangular wave whose magnitude lies between plus and minus *a* is

$$A \sin \omega t + \frac{A}{3} \sin 3\omega t + \frac{A}{5} \sin 5\omega t + \cdots$$

where $\omega = 2\pi/T$ and $A = 4a/\pi$. This happens to be a series containing only sine terms because the cosine terms have zero amplitude. The constant is also equal to zero because the wave is symmetrical (its positive and negative parts have the same size and shape). The addition of the component sinusoids and the reconstruction of the rectangular wave are illustrated in Figure 5.10.

Sampling

So far we have considered only signals that are continuous in the sense of being uninterrupted. Even signals, such as the rectangular wave, which have sudden changes in level from one instant to the next are continuous inasmuch as they are defined at any instant we care to nominate. There are many occasions, however, when signals are recorded not as a continuum but as a series of samples, taken usually at regular intervals. The processing and display used in many scientific instruments and in computers often make use of sampled data, but sampling does of course predate the electronic era. If we are asked to draw a graph to represent some function, $y = 2 + 3x^2$ for example, we would start by constructing a table of values of y corresponding to arbitrarily chosen values of x. We would then use the data in the table to plot a series of points on a sheet of graph paper. Because in this example we already know something of the underlying function, we would be justified in joining the points with a smooth line. If, on the other hand, we had been presented only with a table of values and had been given no information about the source of the data, the construction of a smooth line between the points would be an act of faith based on some assumption about the relation between the y and x. The point to bear in mind is that, in general, sampling provides only incomplete information and we need to be careful how we interpret it. A particular difficulty arises if the sampling rate is too low in comparison with the frequencies present in the sampled waveform. When this happens we may get the false impression that a low frequency is present when in fact the signal is at a much higher frequency. This phenomenon is called *aliasing*. The way it comes about is illustrated in Figure 5.11. To prevent aliasing, the sampling rate must be at least twice the highest frequency present in the signal. Before sampling, it is a usual practice to filter out components that do not meet this requirement.

Although there are exceptions, sampling usually involves the conversion of the sampled data to digital form (analogue-to-digital or A-to-D conversion). This again leads to a loss of information because the digital signal can have only one of a set of possible values rather than any value that happens to be present in the waveform at the moment of sampling. The digital representation is in the form of a binary number. To see what this means, consider first an ordinary decimal number such as the number 137. This

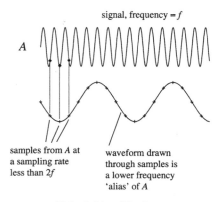

signal, frequency $= f$

A

samples from A at a sampling rate less than $2f$

waveform drawn through samples is a lower frequency 'alias' of A

FIG. 5.11 Aliasing.

number can be written as $(1 \times 10^2) + (3 \times 10^1) + (7 \times 10^0)$. We see therefore that the greatest number that can be represented by, say, three digits is 999, that is, $10^3 - 1$. In general, the largest value that can be represented by a decimal number having n digits is $10^n - 1$. Binary numbers are similar to the familiar decimal numbers except that their base is 2 rather than 10. For example, the binary number 1101 is equivalent to the decimal number $(1 \times 2^3) + (1 \times 2^2) + (0 \times 2^1) + (1 \times 2^0) = 13$. If there are n binary digits (*bits*), then the highest possible number is $2^n - 1$. An analogue-to-digital converter using 8 bits would provide values from 0 to $(2^8 - 1)$, that is, from 0 to 255. If the original signal increased smoothly with time, the digitally sampled signal would increase in steps corresponding to a change of $1/255$ times the greatest signal level to be accommodated. Therein lies a difficulty, because although a step of $1/255$ may seem quite small (less than 0·5%), the signal will not generally occupy the whole range of the converter; furthermore, some provision has to be made for dealing with positive as well as negative values. Usually one of the 8 bits is used to this end, leaving only 127 levels in the positive or negative parts of the signal. Digital conversion may therefore give rise to a significant coarseness in the sampled data.

Sampled data converted to digital form are ideal for machine processing. This allows numerical methods to be used to carry out mathematical operations that would otherwise be impossible. Examples include frequency analysis and filtering.

The Discrete Fourier Transform

The discrete Fourier transform works with sampled rather than continuous data. It leads to a spectrum containing a finite number of sine and cosine waves which add to give the original waveform as it exists at the sampled points. The corresponding spectrum is therefore a line spectrum. The number of terms in the spectrum (including a constant, zero-frequency term) is equal to the number of samples. By analogy with the discrete points that make up a line plotted on a sheet of graph paper, we can regard the discrete Fourier spectrum as the 'graph paper equivalent' of the standard (continuous) Fourier transform. If the number of samples is sufficient, we find that the two transforms give very similar results. An example is shown in Figure 5.12. Here the time function is a single pulse consisting of a quarter cycle of a cosine wave. The function is zero at all times before and after the pulse. The curve in the lower part of the figure shows the continuous Fourier spectrum, and the circles show the discrete spectrum. The latter is based on 512 samples of the time function. For clarity, the diagram shows only every third term up to the 126th in the series.

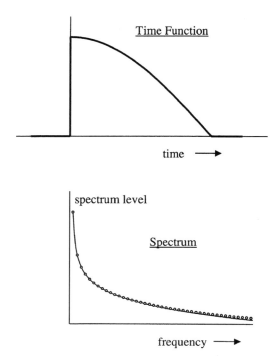

FIG. 5.12 Comparing the ordinary (continuous) Fourier transform (line) and a discrete Fourier transform (circles) derived from 512 samples of the time function.

As with the Fourier series, the terms that make up the discrete transform are harmonically related; that is, they are multiples of a fundamental frequency. If the data contain N values sampled so that the first value occurs at time zero and the Nth value occurs at time T, then the spectrum contains a constant together with $N-1$ terms at frequencies $1/NT$, $2/NT$, $3/NT$, and so on. Now although all these terms would be required for a mathematical reconstruction of the original samples, only terms having frequencies up to $\frac{1}{2}N/NT$, that is, $\frac{1}{2}T$ are physically meaningful. The remaining terms have frequencies corresponding to less than two samples per cycle and so are inadmissible according to the criterion stated previously. They are needed in a mathematical sense, but they do not provide information about the fine structure of the waveform. The remarkable thing about the discrete transform is that no matter how few samples are taken (except that the minimum number is two), addition of the sine and cosine terms gives a time function

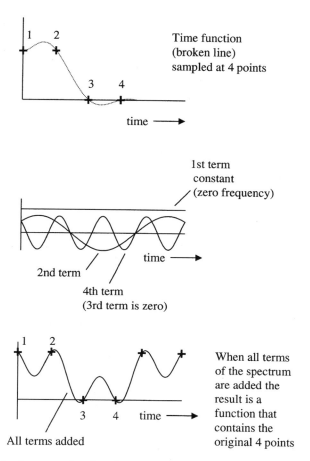

FIG. 5.13 An example of a discrete Fourier transform having four terms.

that passes exactly through the sample points. An example using just four samples is shown in Figure 5.13. As demonstrated in this example, addition of the sine and cosine terms leads to a periodic function containing the original samples. It also contains replicas of these samples if the addition is continued. The discrete Fourier transform should therefore be thought of as a cyclic process that repeats itself at intervals equal to the time over which samples were taken.

The discrete Fourier transform is immensely important in signal analysis. One of the principal reasons for this is that algorithms have been invented that enable the transform and its inverse to be calculated very rapidly using microprocessor technology. For this reason the transform is often called the *fast Fourier transform* (FFT). Audiological applications include frequency analysis of otoacoustic emissions and, occasionally, auditory evoked potentials, and digital filtering. The transform is an essential part of the processing used in digital hearing aids.

The Co-Power Spectrum

When explaining the terminology used to describe a spectrum, it was stated that the way in which acoustic power is distributed in a signal is given by the power spectrum of the signal. Because power is proportional to the square of the amplitude of vibration, an acoustic power spectrum consists of terms whose magnitudes are proportional to the square of the sound pressure at each frequency. The co-power spectrum is a variation on this theme. Its importance for audiology is that it has become widely used as a way of showing the energy distribution in otoacoustic emissions.

Let us assume that we have recorded the sound pressure in the ear canal over a defined period (perhaps 20 ms) following an auditory stimulus. If this recording were just a single curve showing the change in sound pressure during the 20-ms period, we could obtain an ordinary power spectrum by squaring the terms in its Fourier transform. A co-power spectrum, however, requires two time functions as its starting point. We therefore begin by making two independent measurements that lead to two versions, A and B, of the pressure waveform. Now, however careful we are, the two curves will not be identical because they will be contaminated by different samples of the noise that inevitably accompanies any measurement. This means that Fourier analysis of the two waveforms will lead, at each frequency, to two vibrations, one having a pressure amplitude p_A and the other having amplitude p_B according to their origin. These results could be combined by calculating $p_A p_B$ instead of $(p_A)^2$ or $(p_B)^2$, but a more useful outcome can be obtained if we also take into account the relative phase of the two vibrations. Writing them as $p_A \sin \omega t$ and $p_B \sin(\omega t + \phi)$ gives a term related to acoustic power of the form $p_A p_B \cos \phi$. This is the magnitude of the co-power spectrum at the angular frequency ω. Its importance is that its value is greatest when the two waveforms are highly correlated so that component vibrations in A have the same phase as those in B. The phase difference ϕ is then zero and its cosine is equal to unity. The presence of noise impairs the correlation and reduces the magnitude of the co-power spectrum.

Recovering Signals in Noise:
Signal Averaging

There are many techniques that aid the detection of signals in the presence of competing noise. The choice of method often depends on what is known about the signal and about the noise. Biological signals such as evoked potentials are often intimately related to the stimulus that produced them. Their waveforms are repeated each time the stimulus is presented, and they appear after exactly the same delay (latency) on each occasion. The competing noise that makes these signals difficult to observe is generally unrelated to the stimulus; it is this difference that makes it possible to enhance the signals at the expense of the noise in order to improve the signal-to-noise ratio. A favourite method is signal averaging. It is widely used in the observation of evoked potentials such as the auditory brainstem response and in the recording of otoacoustic emissions that follow an aural stimulus. To understand how averaging works, it is first necessary to know something about the noise. So that the discussion is not too general, we shall limit it to the observation of otoacoustic emissions, but the same principles apply to evoked potentials.

The noise reaching the microphone in the ear canal has several sources, including environmental noise in the test area and respiratory, muscular, and possibly vascular noise from the patient. Added to this is the electronic noise generated within the microphone and its amplifiers and other circuit components. The noise we observe is, then, an aggregate of contributions from the various sources. If we sample this noise at intervals of a few tens of milliseconds (the time between each stimulus), we expect to obtain a set of values that are practically independent of one another. We may not know much about the statistics of such samples, but because the microphone and its amplifiers are sensitive only to alternating signals, we can expect the average of the samples to have a value that approaches zero as the sample size is increased. In statistical language, we would say that the samples of the noise are taken from a population whose mean is zero. Now suppose we sample the noise on, say, 100 occasions, each at a fixed time after the stimulus. Let m_1 be the average of the 100 values so obtained. We can repeat this process to obtain similar average values, m_2, m_3, and so on. In general these averages will not be zero, but they will have positive and negative values that deviate from zero by varying amounts according to chance. A statistical measure of the likely deviation is called the *standard deviation*. It is proportional to $1/\sqrt{N}$, where N is the size of the sample (100 in this case). The average acoustic power in the noise is proportional to the square of the standard deviation (the *variance*); that is, it is proportional to $1/N$.

Suppose now that within the noise there is some unchanging component, namely, the signal that we are attempting to identify. Because the signal is constant, its average value will not depend on the size of the samples. Averaging (signal + noise) therefore gives a result that consists of a constant measure of the signal and an average of the noise which diminishes as the sample size is made larger. The ratio of the power in the signal to that remaining in the noise is proportional to N. If the signal is associated with a sound pressure p_S in the ear canal, and the noise, after averaging, is associated with a sound pressure p_N, then p_S/p_N is proportional to $\sqrt{N}$. The improvement in the observed waveform is therefore proportional to the square root of the number of samples in the average. Figure 5.14 provides an illustration. Instead of taking a physiological signal as an example, an arbitrary but easily recognized signal (a decaying cosine wave) has been used. The diagram shows the progressive recovery of this signal as the number of samples in the average is increased.

Linearity and Distortion

Distortion affects components such as transducers, amplifiers, and filters. It is a failure to produce at the output a faithful copy of the signal at the input. Although distortion may exist in both linear and nonlinear systems, it is usually associated with nonlinearity. Thus, we should begin by examining what is meant by linearity.

A system is said to be linear if a sinusoidal input leads to a sinusoidal output. If a linear system is supplied with a sinusoidal input, the waveforms of the input and output signals have the same shape (both are sinusoids) and it follows that when the amplitude of the input is halved, or divided by some other factor, the amplitude of the output is reduced in the same way. A graph of the output against the input is therefore a straight line, in other words, it is linear. A system that is linear for an input of a specified frequency and amplitude may not be linear for an input at a different frequency or a greater amplitude. Many systems are practically linear for small inputs but nonlinear for large inputs. An obvious example is that of an amplifier whose output is sinusoidal until the input is increased to a point at which peak clipping sets in. Linearity is essential to the principle of superposition and therefore to the principles of frequency analysis described earlier. Although we have not made it explicit, the assumption has been that when sinusoidal vibrations are added, the particle displacement (or other related quantity) is the sum of the displacements that would be produced by each of the component vibrations acting separately. This requires linear behaviour.

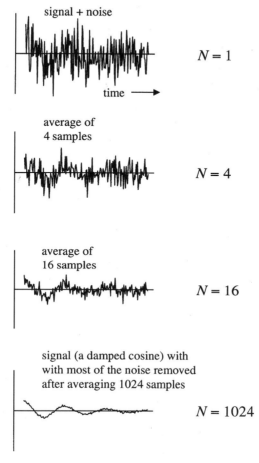

FIG. 5.14 Signal averaging.

A perfectly linear system can produce distortion if its frequency response is not uniform or if phase changes are introduced that depend on frequency. The consequence is that when the input is not a sinusoid, the output waveform will not be quite the same as the input waveform. These forms of distortion are known as *frequency distortion* and *phase distortion*. The former is not necessarily undesirable. An essential characteristic of hearing aids, for example, is that they have a nonuniform frequency response that provides selective amplification of sounds in certain frequency ranges. Nonlinear distortion, on the other hand, is usually something to be avoided.

It exists whenever the output of a system is not directly proportional to the input. Even if a sinusoidal signal is presented, a graph of the output against the input will not be a straight line.

One of the consequences of nonlinearity is a form of distortion called *harmonic distortion*. This occurs when the input is a periodic signal, and it leads to the existence of frequency components in the output that are not present in the input. Harmonic distortion is best measured by examining the response to a sinusoidal input. The input then contains a single frequency, f. The distorted output is a periodic, nonsinusoidal signal that can be represented by a Fourier series whose components have the harmonic frequencies, $f, 2f, 3f, + \cdots$. With the help of filters, or by some other method of analysis, we can measure the amplitudes (or rms values) of these sinusoidal terms and, in the acoustic case, obtain the corresponding sound pressures $p_1, p_2, p_3, + \cdots$. It is then customary to express the distortion as a percentage using the following formula:

$$\text{total harmonic distortion} = \sqrt{\frac{p_2^2 + p_3^2 + p_4^2 + \cdots}{p_1^2 + p_2^2 + p_3^2 + p_4^2 + \cdots}} \times 100\%$$

The distortion due to the nth harmonic is then

$$\text{harmonic distortion} = \sqrt{\frac{p_n^2}{p_1^2 + p_2^2 + p_3^2 + p_4^2 + \cdots}} \times 100\% \approx \frac{p_n}{p_1} \times 100\%$$

It can be seen that these formulae express the root mean square sound pressure due to the unwanted harmonics relative to the root mean square sound pressure in the output signal as a whole.

As an example of harmonic distortion, consider peak clipping, which is likely to occur when an amplifier is overloaded. It is often brought about deliberately in hearing aids to limit the output. Its effect in each cycle is to prevent the output from exceeding some specified value (Figure 5.15). In its most extreme form, infinite peak clipping, the signal is greatly amplified before clipping so that the output waveform is almost rectangular. We can easily calculate the harmonic distortion produced by infinite clipping because we already know the Fourier series for a rectangular wave. The result is

$$\text{total harmonic distortion} = \sqrt{\frac{(\frac{1}{3})^2 + (\frac{1}{5})^2 + (\frac{1}{7})^2 + \cdots}{1 + (\frac{1}{3})^2 + (\frac{1}{5})^2 + (\frac{1}{7})^2 + \cdots}} \times 100$$

$$= \sqrt{\frac{0 \cdot 234}{1 \cdot 234}} \times 100$$

$$= 44\%$$

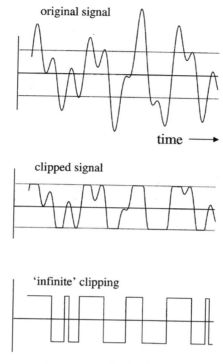

FIG. 5.15 Peak clipping.

This is the greatest distortion that can be produced by peak clipping. Even in hearing aid applications, we would normally want far less distortion, though it is interesting to note that infinitely clipped speech is readily intelligible to people with normal hearing. A total harmonic distortion of less than 10% is probably acceptable. For audiometers, where a good-quality signal is required, distortion should be less than 2·5% for air-conducted tones and less than 5·5% for bone-conducted tones.

Another manifestation of nonlinearity is the creation of distortion products. Suppose that the input is the sum of two sinusoidal vibrations whose frequencies are f_1 and f_2. It is found that the output contains vibrations having these primary frequencies together with vibrations at frequencies $(mf_1 - nf_2)$, where m and n are integers. An important example of this phenomenon is found in the cochlea. The response of the basilar membrane is highly nonlinear, possibly because of motor activity in the outer hair cells. If the ear is presented simultaneously with two pure tones, the vibrations within the cochlea will include tones whose frequencies are of the kind just

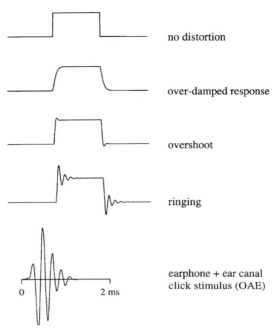

no distortion

over-damped response

overshoot

ringing

earphone + ear canal
click stimulus (OAE)

0   2 ms

FIG. 5.16 Examples of transient distortion in response to a rectangular pulse. The last example shows the pressure waveform in the ear canal following the application of a rectangular voltage pulse to a miniature earphone used to elicit otoacoustic emissions.

described. When such tones, arising from nonlinearity in the cochlea, are detectable in the external ear, they are called *distortion product otoacoustic emissions*. The amplitudes of distortion products depend on the amplitudes and frequencies of the primary tones and on the nature of the nonlinearity. In studies of cochlear physiology some products are more readily observable than others. A favourite is one whose frequency is $(2f_1 - f_2)$.

The kinds of distortion described so far exist when the input to a system is a steady signal. When the input is a transient such as a pulse, or when in effect it is a series of pulses as in a rectangular wave, a type of distortion called *transient distortion* is often observed. It can occur when the frequency response is not sufficiently uniform, and it shows itself as a lag in the response, as an overshoot, or as a 'ringing', depending on the nature of the signal and the frequency characteristics of the system. It is often seen in the response of transducers such as earphones, and in audiological work it is often influenced by the coupling between the ear and the earphone. Examples are shown in Figure 5.16.

Modulated Signals

If we wish to send an audio signal from one place to another using, for example, radio waves, it is usually necessary to do so by modulating a sinusoidal *carrier wave*. This can be done in several ways, the most common of which are *amplitude modulation* (AM) and *frequency modulation* (FM). The carrier frequency is always much greater than the frequency of the audio signal. In amplitude modulation, the amplitude of the carrier is made to vary with the amplitude of the audio signal but the frequency of the carrier is kept constant. In frequency modulation, it is the other way about: the frequency of the carrier is varied according to the amplitude of the audio signal but the amplitude of the carrier is unchanged.

Amplitude Modulation

The amplitude modulation of a carrier by a sinusoidal audio signal is shown in Figure 5.17. Suppose that the unmodulated carrier signal has a voltage amplitude V_c and an angular frequency ω_c. At any time t the carrier voltage is therefore given by

$$V = V_c \sin \omega_c t$$

audio frequency
signal

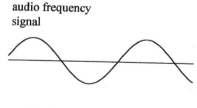

amplitude
modulated carrier

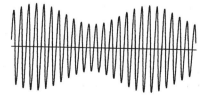

FIG. 5.17 Amplitude modulation.

Let the audio signal have an amplitude V_a and angular frequency ω_m and let the signal have a cosine variation such that its voltage E is given by

$$E = V_a \cos \omega_m t$$

When the carrier is modulated, its amplitude V_c is multiplied by $(1 + m \cos \omega_m t)$, where m is a number called the *modulation depth* or *modulation index*. This number is proportional to the amplitude V_a of the audio frequency signal. It is generally less than 1 for all signal amplitudes that are likely to be encountered. It may be expressed as a percentage. The modulated carrier voltage is therefore

$$V = V_c(1 + m \cos \omega_m t) \sin \omega_c t$$

Therefore,

$$V = V_c \sin \omega_c t + mV_c \cos \omega_m t \sin \omega_c t$$

The product of the sine and cosine terms can be replaced by the sum of two sine terms.[8] Also, because m is proportional to V_a, it can be replaced by kV_a, where k is a constant, to show the dependence on the amplitude of the audio signal. With these changes we have

$$V = V_c \sin \omega_c t + \tfrac{1}{2}kV_aV_c \sin(\omega_c - \omega_m)t + \tfrac{1}{2}kV_aV_c \sin(\omega_c + \omega_m)t$$

This result tells us that an amplitude-modulated signal contains a component at the carrier frequency together with components at the carrier frequency minus the audio frequency and at the carrier frequency plus the audio frequency. In the foregoing description we have considered modulation at a single angular frequency, ω_m. In general, the audio information will be contained in a band of frequencies from, let us say, f_1 to f_2 hertz. This means that in addition to the carrier frequency f_c, the modulated signal will have components in frequency bands $(f_c - f_1)$ to $(f_c - f_2)$ and $(f_c + f_1)$ to $(f_c + f_2)$. These two ranges are called, respectively, the lower and upper *sidebands*. Together, they define the range of frequencies that the transmission system must be able to carry in order to convey the audio information. It is, however, interesting to note that all the audio information is contained in either the upper or the lower sideband, so that it is not strictly necessary to include both. Single-sideband transmission in which the carrier and one sideband are suppressed is often used in radiotelephony to conserve bandwidth and to make more efficient use of the available power in the transmission.

---

[8]See Appendix A, Trigonometric Functions, Example 3.

Frequency Modulation

Frequency modulation is more difficult to accomplish than amplitude modulation and requires a higher carrier frequency and wider bandwidth. Its advantage over amplitude modulation is that audio signal reception is insensitive to the unintended changes in carrier amplitude that inevitably occur in transmission.

In frequency modulation, the angular frequency ω_c of the carrier frequency is made to vary with changes in the amplitude of the audio signal. The angular frequency ω of the modulated carrier signal is

$$\omega = \omega_c(1 + nV_a \cos \omega_m t)$$

where n is a constant (in units of radians per volt) and where the other symbols have the same meanings as before. It can be shown that the voltage of the modulated carrier is then given by V, where

$$V = V_c \sin(\omega_c t + \alpha_f \sin \omega_m t)$$

In this expression α_f is called the *modulation index*. It is equal to $nV_a\omega_c/\omega_m$. Unless the modulating term is very small, the waveform is distinctly nonsinusoidal. This makes it difficult to produce a diagram that illustrates frequency modulation in a realistic way, particularly when the modulating signal is a sinusoid. The process is more easily depicted if a rectangular wave is used instead[9] (Figure 5.18). As with amplitude modulation, sidebands are produced, but they contain more than just the sum and difference frequencies. The frequencies present in the modulated signal are f_c (the carrier), $f_c \pm f_m$, $f_c \pm 2f_m$, $f_c \pm 3f_m$, and so on. An alternative to frequency modulation is *phase modulation*. It produces much the same result.

Pulse-Width Modulation

Pulse-width modulation is used with a carrier whose waveform is a train of high-frequency rectangular pulses that are either positive or negative according to the sense of the audio signal. The width of each pulse is made proportional to the magnitude of the audio signal, as shown in Figure 5.19. In this diagram the audio signal is a sinusoid, but only a small part of it is shown so that the variation in pulse width can be seen clearly. Demodulation is achieved by averaging the pulse train over a time that is long compared with the interval between pulses but short compared with the

[9]The modulation of a carrier by a rectangular wave produces a signal having the same form as that used to transfer information between Modems (a word formed by joining the two terms *modulation* and *demodulation*).

audio signal

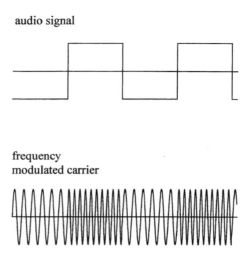

frequency
modulated carrier

FIG. 5.18 Frequency modulation.

period of the audio signal. Pulse-width modulation is used in the output stages of class D hearing aid amplifiers. Its advantage over direct amplification of the audio signal is that little or no current flows in the amplifier when this signal is absent because the pulse width is then zero. Amplification is therefore efficient and battery life is preserved. The output transducer (miniature earphone) can itself act as the demodulator because it is quite unresponsive to the high-frequency carrier signal but does of course respond to the audio frequency average.

audio frequency
signal

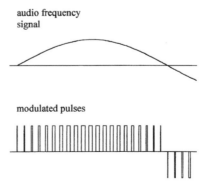

modulated pulses

FIG. 5.19 Pulse-width modulation.

Questions and Exercises

Practical exercises are marked with an asterisk.

5.1 Distinguish between the *time domain* and the *frequency domain* in the description of acoustic or electrical signals. Describe in a non-mathematical way the action of the Fourier transform and its inverse.

5.2 What is meant by a *spectrum*? What is the difference between a *line spectrum* and a *continuous spectrum*? Which type of spectrum is associated with a periodic signal?

5.3 What is meant by *sound power spectral density*? Calculate the lower and upper frequency limits of a third-octave band centred on 1 kHz and find the total sound power in the band if the spectrum is uniform and the power spectral density is $3 \mu W \ Hz^{-1}$.

5.4* What is a *random number*? The variance of the sampled values of a signal is proportional to the power in the signal during the interval in which the samples were taken. The following statistical experiment mimics the measurement of power in a random signal whose magnitude at any instant is uniformly distributed between 0 and 1. (We have not specified a physical unit — it could be a volt or a pascal, for example.) Using a calculator or spreadsheet, generate random numbers in the range 0 to 1. Find the mean and variance of a sample comprising N of these numbers. Continue to add data to your sample and draw a graph showing the progress of the mean and variance as N increases. The mean should approach 1/2 and the variance should approach 1/12, as the sample size becomes large. *Note:* the *variance* is the average squared deviation from the mean. If you have N values of a number x, the mean M and variance V are given by

$$M = (x_1 + x_2 + x_3 + \cdots + x_N)/N$$

$$V = [(x_1 - M)^2 + (x_2 - M)^2 + (x_3 - M)^2 + \cdots + (x_N - M)^2]/N$$

Some calculators and spreadsheets have statistical routines that can calculate the mean and variance for you at each stage.

5.5 A sawtooth wave consists of a voltage that rises slowly from zero to a maximum value and then falls rapidly back to zero before repeating the cycle. Draw a diagram to illustrate this waveform. Using the same time scale, draw two further diagrams to show, respectively, the effect of low-pass and high-pass filtering.

5.6* Laboratory equipment is needed. Use an audiometer or other supply as a source of wide-band noise. Make a direct electrical connection to the output of the source and display the noise on an oscilloscope. Observe the effect of inserting narrow-band filters (third-octave will do).

5.7 A voltage waveform consists of a continuous train of rectangular waves in which the voltage alternates between plus and minus 1 volt. The average voltage in any whole number of cycles is zero. If the wave is represented as a Fourier series of sine terms, what are the amplitudes of the first and third terms in the series?

5.8* This experiment requires a spreadsheet, such as Microsoft Excel, with mathematical functions and chart plotting capability. The aim is to show the Fourier synthesis of a triangular wave from a series of harmonically related sine waves. Excel formulas (in the square brackets) are used here because they are self-explanatory, but you can adapt them to suit other languages if necessary. Where you see 'etc.', copy the formula to the remaining cells in the column. Make the following entries in the spreadsheet:

A1, $\pi/64$, [=PI()/64]. This is just to store π and the constant (1/64).
A5 to A133, the numbers 0 to 128, ('n').
B5 to B133, the angles $n\pi/64$ radians, [=A5*A1, etc.].
C5 to C133, $\sin(n\pi)$, [=SIN(B5), etc.].
D5 to D133, $(1/2) \times \sin(2n\pi)$, [=(SIN(2*B5))/2, etc.].
E5 to E133, $(1/3) \times \sin(3n\pi)$, [=SIN(3*B5)/3, etc.].
F5 to F133, $(1/4) \times \sin(4n\pi)$, [=SIN(4*B5)/4, etc.].
G5 to G133, $(1/5) \times \sin(5n\pi)$, [=SIN(5*B5)/5, etc.].
In column H add the contents of corresponding cells in columns C and D, [=C5 + D5, etc.] (first two terms of the series).
In column I add the contents of corresponding cells in columns E and H, [=E5 + H5, etc.], (first three terms of the series).
In column J add the contents of corresponding cells in columns F and I, [=F5 + I5, etc.] (first four terms of the series).
In column K add the contents of corresponding cells in columns G and J, [=G5 + J5, etc.], (first five terms in the series).

Chart the numbers in columns H to K on the *y*-axis using the numbers in column B on the *x*-axis. You can, of course, extend the synthesis to include higher terms. If so, it will be quicker to include more than one term in each column. The general term is $(1/r) \times \sin(rx)$, where *x* is the angle in column B and $r = 1, 2, 3, \ldots$.

5.9 What is meant by *aliasing*? A complex signal is to be sampled at 8 kHz before conversion to digital form. A low-pass filter is to be used to prevent aliasing. What is the highest frequency that should be passed?

5.10 What is meant by *modulation*? Explain the difference between *frequency modulation* and *amplitude modulation*. The amplitude of a 100-kHz carrier signal is modulated sinusoidally at frequencies up to 8 kHz. What are the frequency ranges of the sidebands?

6
Measuring Sound

General Principles

Sound has many attributes that are amenable to physical measurement, and even perceptive qualities such as pitch and loudness can often be expressed in physical units by reference to established psychometric scales. The attributes most important for audiology fall into two broad classes: those that are energy related and those that are frequency related. Frequency analysis was considered in the previous chapter; so far as measurement itself is concerned, there is little of fundamental importance that we need consider. Once a representative signal has been obtained, frequency measurement and analysis are for the most part the subject of electronic equipment and signal processing technologies rather than acoustics and we shall not give them further attention except to note that frequency meters, wave analysers, and the like can receive their input directly from sound-generating and sound-measuring equipment. Energy-related measurements, on the other hand, are an important part of acoustic science. Because such measurements usually lead to a result expressed in decibels, it is appropriate to begin by describing how decibel scales are constructed and used.

Decibels

It is conventional and convenient to express the sound power radiated by a source, or the energy- or power-related quantities measured in a sound field, with reference to a logarithmic scale.[1] The fundamental scale unit is the *bel*, but it is seldom used. One tenth of it, the *decibel* (symbol dB), is the popular choice whose use is virtually ubiquitous throughout acoustic measurement.

Three reasons are often given for preferring a logarithmic scale:

1. The response of the human ear is said to be logarithmic. What this usually means is that loudness is proportional to the logarithm of sound intensity.

[1]Help with logarithms and a brief description of logarithmic scales are given in Appendix A.

2. When a signal passes through a system, the gain or attenuation in successive stages can be treated as addition or subtraction instead of multiplication and division.

3. Acoustic measurements encompass an unusually large range of magnitudes, and logarithmic scaling is a very convenient method for dealing with the difficulty that this creates.

There is some truth in the first of these, but the power law for loudness is valid for only part of the intensity range of sounds normally encountered, and loudness depends as much on frequency and bandwidth as it does on acoustic intensity. The second reason is valid but not compelling; it is the third that is the most important. The need to accommodate wide-ranging values is not just a problem for acoustics. Logarithmic scales exist elsewhere — to express the brightness of stars (stellar magnitudes) or the strength of earthquakes (the Richter scale), for example.

The basis of the decibel scale is the logarithmic expression of the power W_n of an acoustic source relative to a reference power W_0. We say that the power of the source is n decibels above that of the reference if

$$n = 10 \log_{10} \frac{W_n}{W_0} \qquad\qquad 6.1$$

It is important to understand that the decibel is not an absolute unit of acoustic power but an expression of the ratio of one acoustic power to another. Once the reference power is specified, a scale of acoustic power relative to that reference is established and the word *level* may be used to show that a particular value is referred to the scale. The number n then denotes a *sound power level* relative to W_0. If, for example, a source has an output of 300 μW, its sound power level relative to a microwatt is

$$10 \log_{10} \frac{300 \times 10^{-6}}{1 \times 10^{-6}} = 10 \log_{10} 300 = 24\cdot 8 \text{ dB}$$

The sound power radiated through a unit area is the sound intensity, I. The intensity at any point in a sound field is proportional to the power radiated by the source, and therefore I can replace W in Equation 6.1, with the result that

$$n = 10 \log \frac{I_n}{I_0}$$

For sound measurements in air it is customary to choose 10^{-12} Wm^{-2} as the reference intensity. Unless some other value is specified, this reference may be assumed, thus making n a *sound intensity level* relative to the reference intensity.

The squared values of field quantities such as sound pressure and particle velocity are often proportional to the power of the sound source, so that W in Equation 6.1 can be replaced by p^2 to give

$$n = 10 \log \frac{p_n^2}{p_0^2} \qquad\qquad 6.2\text{a}$$

Remembering that $\log x^m = m \log x$, we notice that this can also be written as

$$n = 20 \log \frac{p_n}{p_0} \qquad\qquad 6.2\text{b}$$

Equations 6.2a and 6.2b are therefore alternative ways of expressing what is known as *sound pressure level*. The reference pressure is $20\,\mu\text{Pa}$. Sound pressure level enjoys a special place in acoustic measurement and particularly in audiology, where decibel scales are found in abundance. Its name is usually abbreviated to *SPL*. When these initials follow the abbreviation for the decibel, a sound pressure level referred directly to $20\,\mu\text{Pa}$ without frequency weighting or other qualification is implied. As an example, the sound pressure level corresponding to a sound pressure of 0.3 mPa is

$$20 \log \frac{0.3 \times 10^{-3}}{20 \times 10^{-6}} = 20 \log 15 = 23.5 \text{ dB}$$

To give an example of the conversion from decibels to absolute units, the sound pressure corresponding to 50 dB SPL is found as follows:

$$50 = 20 \log \frac{p}{p_0} = 20 \log \frac{p}{20 \times 10^{-6}}$$

Therefore,

$$\log \frac{p}{20 \times 10^{-6}} = 2.5$$

so that

$$\frac{p}{20 \times 10^{-6}} = 10^{2.5}$$

Therefore,

$$p = 20 \times 10^{-6} \times 10^{2.5} = 6.32 \times 10^{-3} \text{ Pa}$$

The choice of $20\,\mu\text{Pa}$ as the standard reference sound pressure may seem curious, but the explanation is as follows. The sound intensity in a plane wave is $p^2/\rho_0 c$, where ρ_0 is the density of air and c the speed of sound. The product $\rho_0 c$ is the characteristic impedance, which has a value close to 400 Nsm^{-3}. Accordingly, a sound pressure of 20 μPa corresponds to an intensity of $(20 \times 10^{-6})^2/400 = 10^{-12}$ Wm^{-2}, which is the reference level for sound intensity given earlier. The sound pressure level at the threshold of hearing for a 1-kHz tone is approximately[2] 0 dB SPL.

In sound production and measurement there is often occasion to measure some nonacoustic quantity (usually electrical) that is related to sound pressure. The current flowing in the coil of a loudspeaker or the voltage produced by a microphone are examples. We can extend decibel notation to include such quantities. For example, voltage levels can be defined in a way exactly analogous to sound pressure level, namely,

$$n = 20 \log \frac{V_n}{V_0}$$

If, for instance, we record a voltage of 5 mV, we could express the result as

$$20 \log \frac{5 \times 10^{-3}}{1} = -46 \text{ dB re 1 volt}$$

Decibel notation provides a convenient way of making comparisons. To give an example, suppose that we measure the sound pressure at two places in the field of a simple source. At 100 mm from the source, the sound pressure level is, say, 50 dB. What value would be expected at a distance of 3 metres? Let the sound pressure levels at the two distances be n_1 dB and n_2 dB, corresponding to absolute pressures p_1 and p_2. Then

$$n_1 = 20 \log \frac{p_1}{p_0} \qquad n_2 = 20 \log \frac{p_2}{p_0}$$

The difference in sound pressure level is therefore

$$n_2 - n_1 = 20 \left[\log \frac{p_2}{p_0} - \log \frac{p_1}{p_0} \right] = 20 \log \left[\frac{p_2}{p_0} \times \frac{p_0}{p_1} \right] = 20 \log \frac{p_2}{p_1}$$

If the radiation consists of free, spherical waves, we expect the sound pressure to be inversely proportional to the distance from the source, so that $p_2/p_1 = 0 \cdot 1/3$. Substituting the numerical values then gives

$$n_2 - n_1 = 20 \log \frac{0 \cdot 1}{3} = -29 \cdot 5 \text{ dB}$$

The sound pressure at 3 metres is therefore 29·5 dB less than the sound

---

[2]It is actually slightly greater. The minimum audible field for a 1-kHz tone (plane wave, frontal incidence, binaural) is 4·2 dB.

pressure at 0·1 metres; that is, $50 - 29·5 = 20·5$ dB SPL. Notice that the reference pressure cancels in the calculation of the difference $n_2 - n_1$.

Decibel notation is so helpful that it is used liberally in acoustics and electrical engineering, but it has to be admitted that basic principles are frequently overlooked in the pursuit of convenience. When is a decibel not a decibel? Here is an example. The gain of an amplifier or, indeed, the pressure gain between the eardrum and the oval window may be given in decibels. Although it is quite correct to say that voltage gain is V_2/V_1 or that pressure gain is p_2/p_1, these ratios are not equal to the power delivered by the amplifier divided by the power supplied to its input, or to the power delivered to the cochlea divided by that received at the eardrum. Suppose that instead of an amplifier we had a simple electrical transformer. It comprises just two coils of wire and a ferromagnetic core, and although the output voltage differs from the input voltage, no energy is added. In fact, some energy is lost as heat in the coils because of their resistance. Now, a decibel expresses the ratio of one acoustic or electrical power to another, and to talk about the gain of an amplifier in decibels is questionable unless we are careful to state that we really do mean watts out divided by watts in. The same caution applies to the middle ear — it is a transformer, not an amplifier. It improves sound transmission to the cochlea but does not contribute energy itself. Although the improper use of the decibel may be criticised, some would think it pedantry to insist that only true power ratios may be so expressed. Departure from the path of true righteousness is allowed if it is done thoughtfully.

Another area where care is needed is in the measurement of quantities that are not averages. Looking ahead to the description of sound-level meters, it may be noted that such instruments measure the mean squared value of the sound pressure. This is proportional to acoustic power or intensity averaged over the 'integrating time' of the meter. In order to obtain a correct result, the reference sound pressure must therefore be a root mean square pressure. There is no difficulty in this except that at the turn of a switch the meter might be made to measure a sound level corresponding to the peak sound pressure instead of the rms, and a greater numerical value would be displayed. The power in the sound field has not changed. The change in the display (3 dB for a sinusoid) should be interpreted as a decibel expression of the ratio of the peak power to the average.

Other Decibel Scales Used in Audiology

Audiology uses a number of decibel scales that do not have $20\,\mu\text{Pa}$ as the reference sound pressure. Among the most important are those related to the threshold of hearing. We should be careful to distinguish between scales that represent the level of a stimulus and those that are characteristic of the listener. The former include hearing level and sensation level.

Hearing level (dB HL) in its most general meaning is the level of a stimulus referred to its level at what is taken to be the threshold of hearing for normal listeners. When the stimulus is a pure tone presented in a specified manner in a free field or by an earphone, the normal threshold of hearing is specified by international agreement. For other kinds of stimulus, where no national or international consensus exists, the user has to determine the reference threshold by experiment with a statistically adequate number of otologically normal individuals.

Considerable effort has been made to standardize the threshold for pure tones, especially those presented through earphones. This has led to definitions of hearing level that are somewhat narrower than the one just given, with specific reference to earphones and bone vibrators. To give the hearing level scale some practical meaning, it is necessary to define its zero with respect to sound pressure levels or vibratory force levels measured in acoustic or mechanical couplers. This involves the specification of what are called *reference equivalent sound pressure* (or *force*) *levels*. This subject will be discussed in more detail in Chapter 8.

Sensation level (dB SL) is the sound pressure level of a stimulus referred to its level at the threshold of hearing of the individual subject. The hearing level of a stimulus is greater than the sensation level by a number of decibels equal to the subject's hearing threshold level (see later in this section).

Other scales describe the level of the sound itself rather than as a property of the individual listener. One of these is *loudness level*, which is measured in *phons*. A full definition is given later in this chapter. Another is called *effective masking level* or just *masking level*.

Masking level (dB) is, for a specified masking sound, the number of decibels by which the hearing threshold level of a normal listener would be raised by the presence of the masker. In order to calibrate audiometers, the sound pressure levels corresponding to the 0-dB masking level have been standardized for certain classes of band-pass noise as they affect the hearing thresholds for pure tones (see Chapter 8).

Examples of scales related to characteristics of the listener are hearing threshold level and uncomfortable loudness level.

Hearing threshold level (dB HTL) expresses the hearing threshold of an individual relative to the hearing threshold of normal listeners. It is numerically equal to the level in dBHL of a stimulus that is just audible to the individual. An audiogram is a graph of the subject's hearing threshold levels plotted against frequency.

Uncomfortable loudness level (ULL) expresses an individual's tolerance of loud sounds. It is numerically equal to the hearing level of a stimulus at the point where it is judged to be uncomfortably loud. The reference level is the normal threshold of hearing. The uncomfortable loudness level therefore relates the listener's perception of loudness to the hearing level scale. This makes it easy to show uncomfortable loudness levels on the audiogram.

Microphones

Microphones provide the primary input to all systems used in recording, amplifying, and measuring sound. In one way or another they convert sound pressure into an electrical signal. Designs vary greatly according to application, but certain features are common. Nearly all microphones have a diaphragm that moves in response to the alternating force created by the action of the sound pressure over its surface. In some cases the movement is passed to a transducer such as a piezoelectric crystal or to a coil moving through a magnetic field; in others the diaphragm itself is part of the transducer. The diaphragm and its attachments constitute a mechanical system having significant mass and elasticity and therefore capable of free oscillation. Its motion may have to be damped by adding resistive elements to control the response in the neighbourhood of the resonance frequency.

The ideal microphone would have a uniform frequency response with no phase distortion or nonlinear distortion throughout its operating range. In reality, compromises have to be made. A distinctly nonuniform response can be tolerated in, for example, telephones, dictating machines, and pagers in exchange for desirable electrical characteristics such as high sensitivity and low electrical impedance. Indeed, in these applications a restriction of the bandwidth to frequencies important for speech might be advantageous in reducing noise and in simplifying the requirements for the amplification and transmissions of signals. Hearing aid microphones are often designed to provide a nonuniform response appropriate to the hearing loss. It is in precision acoustic measurement that demands are most stringent. Here the ability to produce a true electrical facsimile of the input waveform is often the principal requirement.

We have hitherto used the terms *response* and *sensitivity* quite freely without definition, making the assumption that their meanings are self-evident. These terms are, however, used many times over in describing the characteristics of microphones and associated measuring systems, and formal definitions are therefore appropriate. The following are taken from IEC 60050:

response of a device or system, the motion or other output that results from a stimulus (excitation) under specified conditions. The kinds of input and output being utilized must be specified.

sensitivity (of a transducer), the quotient of a specified quantity describing the output signal of the transducer by another specified quantity describing the corresponding input signal.

When we apply these terms to the performance of a microphone, we can use *response* to mean the output voltage from the microphone as a result of sound pressure on the diaphragm. The term can apply at a chosen frequency or over a range of frequencies. For the latter, we talk of a *frequency response*, although for brevity the word *frequency* can often be omitted. The word *sensitivity* means specifically the ratio of the output voltage to the input sound pressure. The sensitivity M of a microphone is therefore the rms voltage V at its output divided by the rms sound pressure p being measured; that is,

$$M = V/p$$

The Condenser Microphone

The condenser microphone is almost unrivalled as a transducer for high-quality measurements of sound pressure and the accurate reproduction of pressure waveforms. The simplicity of its form and the wonderful perfection of its construction put it among the most beautiful of man-made things. The fundamental reason for its importance in acoustic measurement is its high stability and uniform frequency response. Its diaphragm, which is very light and unencumbered by attachments, is very responsive to changes in pressure. Below the resonance frequency (which may be 10 kHz or more), the movement of the diaphragm is controlled mainly by the elasticity of the diaphragm itself and that of the air in the microphone capsule, to give a displacement proportional to sound pressure. The electrical output of the microphone is proportional to this displacement and therefore to sound pressure; accordingly, the response is inherently uniform over a wide range of frequencies.

The construction of a condenser microphone is shown schematically in Figure 6.1. It consists of a taut metal diaphragm in close proximity to a fixed backplate. These components form the electrodes of a parallel-plate condenser whose capacitance changes with the deflection of the diaphragm. Standard measuring microphones have nominal diameters of 1, $\frac{1}{2}$, or $\frac{1}{4}$ inches. The diaphragm itself is often made from metal foil (e.g. aluminium, stainless steel, or nickel) with a thickness of typically 2 to 8 μm, stretched

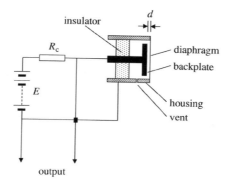

FIG. 6.1 The condenser microphone.

to a tension of the order of 10^4 Nm^{-1}. The gap between the diaphragm and the backplate is very small, perhaps 15 to 30 μm. It will be evident from these figures that the construction of the microphone calls for precision engineering of a high order. The diaphragm is connected electrically to the capsule, but the backplate, which forms the second electrode, is insulated from it. An insulation of typically 10^{15} ohms is achieved by mounting the backplate in quartz, or polytetrafluoroethene (PTFE), or sometimes ruby. It is necessary to vent the microphone capsule to prevent the creation of a static pressure difference across the diaphragm. The vent, consisting of a capillary with a wire insert, offers a high acoustic resistance so that the incident sound has its effect on only the outer surface of the diaphragm. The acoustic isolation is not absolute, however, so for very low frequencies (below 10 Hz) the sound pressure inside the capsule becomes significant. Venting and other deliberately engineered leakage determines the lower limit of the frequency range in which the microphone can work, regardless of the quality of the electrical connections. An example of a condenser microphone is shown in Figure 6.2.

Because the condenser microphone is so important as a primary standard for acoustic sound measurement, it is worth considering its operation in some detail. The capacitance of a parallel-plate condenser is given by

$$C = \varepsilon_0 A/d$$

where ε_0 is a constant (the permittivity of air), A is the surface area of each plate, and d is the distance between the plates. The condenser of the microphone is charged by the application of a polarizing voltage, typically of the order of 200 volts. Call this voltage E. If the capacitance of the

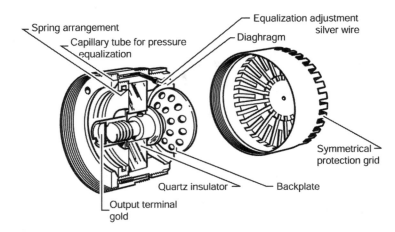

Spring arrangement
Capillary tube for pressure equalization
Equalization adjustment silver wire
Diaghragm
Symmetrical protection grid
Quartz insulator
Backplate
Output terminal gold

FIG. 6.2   An example of a condenser microphone. (Reproduced with permission of Brüel & Kjaer Sound & Vibration Measurements A/S, to whom the copyright belongs.)

microphone is C_m, the charge Q stored on it is then given by

$$Q = C_m E = \varepsilon_0 AE/d$$

We consider first what happens under open-circuit conditions, that is, without leakage and with all electrical connections to the capacitor removed. In these circumstances the charge must be constant. Suppose that a positive sound pressure acts on the diaphragm, causing an inward deflection x which reduces the distance between the plates to $d - x$. This will be accompanied by a voltage change V so that the total voltage on the plates becomes $E + V$. Because the charge remains unchanged, it follows that

$$\frac{\varepsilon_0 AE}{d} = \frac{\varepsilon_0 A(E + V)}{d - x}$$

Therefore,

$$V = -Ex/d$$

where the minus sign shows that the voltage on the condenser decreases when x is made larger.

To use the microphone, its output must be connected to an amplifier (a preamplifier) and to a charging circuit that replaces the charge lost through leakage and as current supplied to the amplifier. A simplified circuit is shown in Figure 6.3. In this circuit the source of the charging voltage is

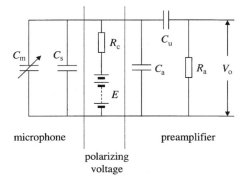

microphone   preamplifier

polarizing
voltage

FIG. 6.3 Simplified circuit for condenser microphone, polarizing voltage, and input to preamplifier.

shown schematically as a battery, though in practice an electronic power supply would be used. The charge is delivered through a very high resistance, R_c. Stray capacitance in the microphone capsule and in the connectors and cables is represented by C_s. The input capacitance and resistance of the preamplifier are shown as C_a and R_a. The capacitance of the microphone itself is C_m. The only other component is the coupling capacitor C_u, which has no effect except to remove the charging voltage from the input to the preamplifier.

In the absence of sound, all voltages in the circuit are constant and no current flows in any component once the capacitors are charged. When sound is present, the varying voltage on the condenser causes small alternating currents to flow in all parts of the circuit. The sum of these alternating currents is the current flowing through the microphone capacitance, C_m. The microphone can therefore be treated as a simple alternating-current generator whose open-circuit voltage is V and whose internal impedance[3] is that of the microphone capacitance. Accordingly, the circuit in Figure 6.3 is equivalent to the one drawn in Figure 6.4. It can be shown that the voltage V_0 at the input to the preamplifier is given by

$$V_0 = E \times \frac{-x}{d} \frac{C_m}{C} \times \frac{j\omega RC}{1 + j\omega RC}$$

where $C = C_m + C_s + C_a$ and $R = R_c R_a / (R_c + R_a)$. If the charging resistance and the input resistance of the preamplifier are very high, then

---

[3]*Impedance* is a generalized form of resistance that applies to components carrying alternating currents. The impedance of the condenser is $-j/\omega C_m$. See Chapter 7.

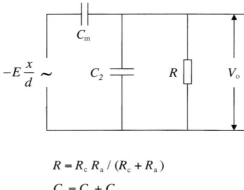

$$R = R_c\, R_a\, /\, (R_c + R_a)$$
$$C_2 = C_s + C_a$$

FIG. 6.4   Equivalent circuit of a condenser microphone. This circuit is equivalent to the one shown in Figure 6.3. The microphone has been replaced by an equivalent voltage source $-Ex/d$ and a capacitance C_m. The other components are as shown.

$\omega RC \gg 1$ at all frequencies in the normal operating range of the microphone. In this case the fraction to the right of the second multiplication sign in the previous expression is practically equal to unity, so that we are left with

$$V_0 = E \times \frac{-x}{d}\frac{C_m}{C} \qquad\qquad 6.3$$

We see from this equation that the output voltage of a condenser microphone is proportional to the displacement of its diaphragm. The relationship between displacement and sound pressure depends on the internal mechanical and acoustic properties of the microphone, but at frequencies well below resonance we can expect displacement to be independent of frequency and to have the same phase as the pressure. At high frequencies, however, the pressure waveform is less faithfully reproduced because the mass of the diaphragm and the air in the capsule become significant. In a well-designed measuring microphone, the vibration of the diaphragm in the neighbourhood of the resonance frequency is heavily damped by the flow of air through small holes in the backplate and by other mechanisms. The controlling force is then resistive so that displacement and output voltage lag the sound pressure, causing phase distortion. Above resonance the vibration is increasingly mass controlled, causing the output to fall and the phase lag to increase (Figure 6.5).

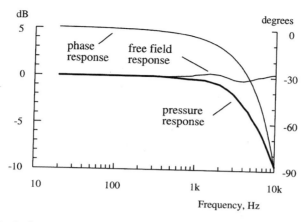

FIG. 6.5 Typical pressure, free-field, and phase characteristics of a 1-inch microphone (Brüel & Kjaer type 4145). The microphone in this example is intended for free-field use. The fall in the pressure response above 1 kHz compensates for diffraction in the sound field to give a uniform free-field response for frontally incident plane waves (see Figure 6.16). The phase response is based on a calculation that assumes critical damping. It shows that the displacement of the diaphragm lags the sound pressure. The lag is 90° at the resonance frequency (10 kHz); it increases with frequency beyond the range shown in the graph, approaching 180° in the limit. (Redrawn from diagrams supplied by Brüel & Kjaer.)

There are three further observations can be made on the theory of the condenser microphone. The first is that a capacitive load, associated with connectors and cables and with the preamplifier itself, influences the output voltage through the total capacitance C as shown by Equation 6.3. An increase in this capacitance reduces the output but does not alter the frequency response insofar as there are no frequency-dependent terms in the equation.

The second observation is to call attention to the assumption that the microphone capacitance C_m has been treated as a constant in the equivalent circuit (Figure 6.4). This is of course an approximation, because if the capacitance were exactly constant the microphone would have no output. Although the separation of the diaphragm and backplate is very small, the deflection of the diaphragm is even smaller, but at extremely high sound levels it is not quite true to say that the microphone is simply a voltage source in series with a fixed capacitance. The inherent nonlinearity associated with a capacitance that depends on a changing dimension is a source of distortion. For most measuring microphones, however, distortion is likely to be negligible in all audiological applications.

Our third observation is that the output voltage from the microphone is inverted if the polarity of the charge is reversed (if E is replaced by $-E$ in Equation 6.3). In audiological work we frequently wish to examine the waveform of the transient signals used as the stimulus for evoked potentials and otoacoustic emissions. When the trace on an oscilloscope is deflected upwards from the baseline, is the sound pressure positive or negative? There is no simple answer or convention. For equipment from one major manufacturer at least, a positive-going output is usually (but not always) associated with a negative sound pressure.

Electret Microphones

The polymer 'cling film' used for food wrapping adheres to surfaces by electrostatic attraction. Charged films known as *electret* can be used in condenser microphones to supply the polarizing charge, thereby making it unnecessary to have an external polarizing circuit. The charged film either replaces the conventional metal diaphragm or is attached to the surface of the backplate. Miniature electret microphones are widely used in hearing aids. In the example of a hearing aid microphone shown in Figure 6.6, the electret is applied to the backplate. The preamplifier (a junction field-effect transistor) is contained within the microphone capsule, keeping stray capacitance to a minimum and providing the user with a microphone/ preamplifier combination in a single convenient package.

Piezoelectric Microphones

Certain crystalline and ceramic materials generate a voltage when subjected to mechanical strain. The phenomenon is known as the *piezoelectric effect*. It provides a useful and often inexpensive mechanical-to-electrical conversion in many applications, including microphones. The transduction works in both directions: the dimensions of a piezoelectric element change when a voltage is applied across its surfaces (Figure 6.7).

The piezoelectric effect is particularly strong in crystals of Rochelle salt (potassium sodium tartrate tetrahydrate), and this material is used as the transducing element in low-cost *crystal* microphones. It provides good sensitivity but is easily damaged by heat and moisture. Piezoelectric ceramics (barium titanate and particularly lead zirconate titanate [PZT]) are less sensitive but far more stable. They can be manufactured in almost any size and shape to suit many different applications, including vibration and force transducers and the so-called *ceramic* microphones that are often used in sound-level meters.

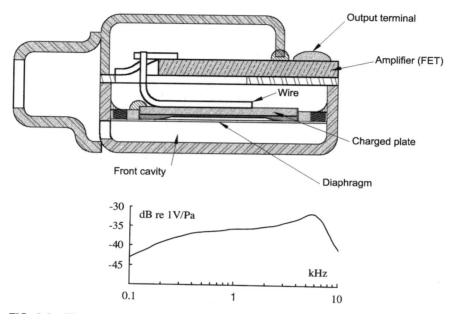

FIG. 6.6 Electret hearing aid microphone—schematic cross section and typical frequency response. (Based on a drawing supplied by Knowles Electronic Company.)

Crystals and ceramics, being solids, are very dense and unyielding compared with air, so that a 'mechanical transformer' is often needed to produce an adequate mechanical strain in response to sound pressure. A favourite technique is to make the piezoelectric element from two parts resembling a bimetallic strip. An element made in this way is called a *bimorph*.[4] If a straight bimorph element is bent to form a curve, the inner section is compressed and the outer section is stretched and a voltage difference is generated between the exposed surfaces. Considerable mechanical advantage can be obtained by mounting the bimorph as a cantilever and applying the action of the microphone diaphragm to the free end (Figure 6.8).

The Moving-Coil Microphone

The microphones described so far have a number of disadvantages over other types. In the first place they are not particularly robust—a high-

---

[4]*Bimorph* is a registered trade name.

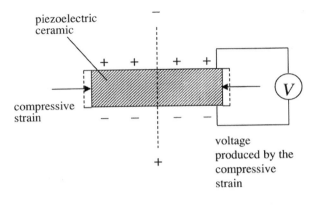

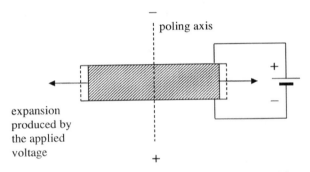

FIG. 6.7 Piezoelectric ceramics. In this example the material is a rectangular block whose cross section is shown in the diagram. When the block is compressed by applying a force to opposite sides, a voltage appears between the upper and lower faces. Conversely, applying a voltage between these faces causes the block to expand laterally. The material is internally polarized during manufacture. The direction of polarization is called the *poling axis*.

quality condenser microphone certainly has to be treated with respect. Another problem is that they produce relatively small amounts of electrical energy and present their amplifiers with a high impedance. The problem is greatest for condenser microphones, which always require a preamplifier, but even ceramic types may be intolerant of lengthy cables between themselves and the amplifier. The moving-coil microphone, on the other hand, is a robust and fairly massive instrument that can take a certain amount of rough handling. Its output impedance is very low (a few tens of ohms) so that it is unaffected by the capacitance of connectors and cables. A useful increase in output voltage can be obtained using a step-up transformer.

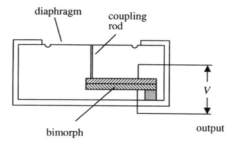

FIG. 6.8 Ceramic microphone.

In principle, the moving-coil microphone works like a moving-coil earphone or loudspeaker operated in reverse. The microphone is also called an *electrodynamic* (or just *dynamic*) microphone. As with a loudspeaker, the diaphragm supports a coil that moves in through the radial field of a permanent magnet, and therefore the induced voltage is proportional to the velocity of the coil. It follows that the frequency response will be uniform if the velocity of the diaphragm is proportional to the sound pressure acting on it. As will be explained in the next chapter, this means that the mechanical impedance at the diaphragm has to be resistive. The movement of the diaphragm is controlled by the elastic compliance of its mountings, by its inertia, by interaction with the air inside the microphone capsule, and by porous damping elements. It is possible to design these components in such a way that the movement of the diaphragm is opposed by a force that is proportional to its velocity throughout most of the intended frequency range (Figures 6.9 and 6.10). Dynamic microphones are often contained in a spherical housing to control diffraction effects.

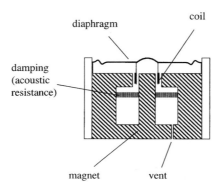

FIG. 6.9 Schematic cross section of a moving-coil (electrodynamic) microphone.

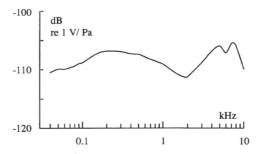

FIG. 6.10   An example of the frequency response of a moving-coil microphone.

Directional Microphones

Ribbon Microphone

The ribbon microphone, like the dynamic microphone, generates an electrical signal through the movement of a conductor in a magnetic field, but the mechanical and acoustic elements are designed to give a response that depends, not on the sound pressure itself, but on the rate at which sound pressure changes with distance, that is, on the pressure gradient in the sound field. To understand what is meant by pressure gradient, imagine the sound field as it exists at a given instant. The pressure in the field varies from one point in the field to another. The rate of change of pressure with distance in a specified direction is the pressure gradient in that direction. We have previously measured distance from the source of sound along the direction of propagation of the sound wave and have called this the x-direction. If a wave is travelling along the positive x-axis away from the source, we can write the sound pressure as

$$p = a \sin(\omega t - kx)$$

where a is the pressure amplitude and ω and k have their usual meanings (see Chapter 3). Readers familiar with calculus will see from this that the rate of change of pressure with the distance x is given by

$$-ak \cos(\omega t - kx)$$

The pressure gradient is therefore an alternating quantity having the same harmonic variation as other acoustic quantities such as sound pressure or particle velocity, and because $k = \omega/c$, it must have a root mean square value that is proportional to frequency. Pressure gradients in directions parallel to a wavefront are likely to be small compared with gradients in the direction of the propagation and we should expect the response of a pressure-gradient microphone to be highly directional unless the sound field is diffuse.

The ribbon microphone consists of a corrugated metal ribbon stretched lightly between the poles of a magnet (Figure 6.11). Both surfaces of the ribbon are exposed to sound, but the presence of a baffle between the front and back of the microphone forces the sound to travel a greater distance to reach the back of the ribbon than to reach the front, assuming it is facing the source. The delay introduces a phase difference between the pressures on the two faces of the ribbon, and there is therefore a net force available to set it in motion. This force is proportional to the frequency of the sound so long as the dimensions of the microphone are small compared with the wavelength. The voltage generated between the ends of the ribbon depends on the velocity at which it moves through the magnetic field, so to obtain a uniform frequency response it is necessary to make this velocity proportional to the reciprocal of the frequency. The frequency terms will then

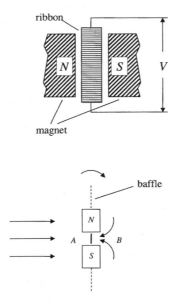

FIG. 6.11 Velocity-gradient (ribbon) microphone. The baffle shown in the lower part of the figure is represented schematically. In reality it is formed by the magnet assembly and microphone housing. In the orientation shown, sound incident directly on face *A* of the ribbon has to pass around the baffle to reach face *B*. This creates a difference in sound pressure across the ribbon, causing it to move in the magnetic field to produce the output voltage *V*.

cancel. If the suspension is very compliant and the damping negligibly small, the movement of the ribbon is controlled primarily by its mass. In other words, the resonance frequency of the ribbon is designed to be below the working range of the microphone (below what is audible). In these circumstances its velocity in response to the pressure difference across it is, as required, proportional to $1/\omega$.

It can be shown that the output of a ribbon microphone corresponds more closely to the velocity of sound particles than to sound pressure. For this reason it is sometimes called a *velocity* microphone. In a plane wave the distinction between a velocity-sensitive microphone and one that is pressure sensitive is immaterial because particle velocity and sound pressure are proportional to one another and in the same phase. This is not true, however, in a divergent field — close to the source of spherical waves, for example — and in these circumstances a velocity microphone will produce a greater output than a normal microphone. The two instruments would give completely different indications in a standing wave. For instance, at a pressure node an ordinary pressure-sensitive microphone would give a low indication, whereas a velocity microphone would give a high indication because pressure nodes coincide with velocity antinodes.

The ribbon microphone is not particularly important for audiological purposes, but it is useful to know about it because of its interesting directional properties, particularly when used in conjunction with the dynamic microphone. Similar principles apply to the directional microphones used in hearing aids, although their construction is entirely different. It can be shown that when the ribbon microphone is exposed to plane waves, its response varies as the cosine of the direction from which the waves are incident, that is, as the cosine of the azimuth θ. This is illustrated in Figure 6.12. If v is the output voltage of the microphone when the sound pressure is p, then we can say that its sensitivity to directly incident sound is v/p. Call this ratio M_0, adding the subscript to indicate that it is the sensitivity when the microphone is facing the source of sound (azimuth equals zero). The sensitivity for sound arriving from the direction θ is then $M_0 \cos \theta$. Now suppose we have an ordinary dynamic microphone whose response is nondirectional. This uniform response can be guaranteed, at least in the horizontal plane, by positioning the microphone so that its diaphragm is horizontal. Let sensitivity of the nondirectional microphone be M_u. If the two microphones are connected in series so that their outputs add, we obtain the combined sensitivity M_c, which is

$$M_c = M_u + M_0 \cos \theta$$

The trick now is to adjust the sensitivities so that they are the same for

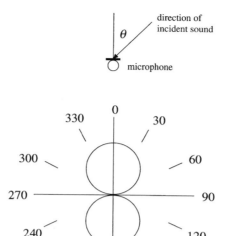

FIG. 6.12 Polar response pattern for a pressure-gradient microphone.

directly incident sound, that is, make $M_o = M_u$. (The method of adjustment is unimportant—it could be done electrically.) For this condition, the overall sensitivity is

$$M_c = M_0(1 + \cos \theta)$$

In combination, the two microphones therefore have a directional response proportional to $(1 + \cos \theta)$. The polar plot of this function is called a *cardioid* (Figure 6.13).

Directional Hearing Aid Microphones

The directional microphones used in hearing aids are, like the ribbon microphone, sensitive to the sound pressure gradient. The aim is to enhance the response to sound arriving from the front or side of the listener and suppress the response to sound arriving from behind. A cardioid or similar response is therefore appropriate. Directional systems may employ one or more microphones and receive inputs from two or more locations in the sound field. We will consider a simple first-order system in which the input is taken from just two points. A typical directional microphone used in

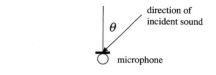

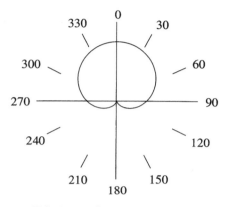

FIG. 6.13 Cardioid response.

hearing aids has two chambers, one on each side of the diaphragm. They are connected by small tubes to openings in the case of the aid. The openings, which we will call the front and rear ports, are approximately aligned with the source of sound when it is directly in front of the listener (azimuth $\theta = 0$).

Suppose that the distance between the ports is x. When sound comes from directly in front of the microphone, the time difference between its arrival at the front and rear ports is x/c. Call this time t_x. It can be shown that the root mean square value of the difference in sound pressure between the two ports is proportional to ωt_x at frequencies for which x is small compared with the wavelength; if the source is at an azimuth θ, the pressure difference is proportional to $\omega t_x \cos \theta$. We therefore find that if a microphone has separate ports supplying the front and rear of the diaphragm, its directional response is bipolar, like that of the ribbon microphone described previously.

An ingenious feature of the hearing aid microphone is that it incorporates an internal acoustic circuit that retards the phase of sounds entering the rear chamber. The phase delay can be achieved by placing an acoustic resistance in the acoustic path. The resistance, in conjunction with the compliance of the air in the chamber and the compliance of the diaphragm, brings about

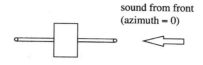

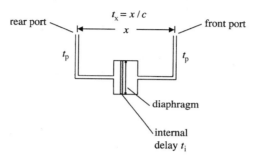

FIG. 6.14   Directional hearing aid microphone.

a phase change that is proportional to the frequency of the sound.[5] The effect is the same as would be obtained by increasing the time taken for sound to travel to the rear surface of the diaphragm, and it is characterized by an 'internal delay time', t_i. Because of the internal delay, the microphone will respond to sound pressure even if the distance between the ports is reduced to zero, in which case the response would be nondirectional. A cardioid response can therefore be obtained by separating the ports by a distance that makes the external delay equal to the internal delay. This is the equivalent to matching the sensitivities of the ribbon and dynamic microphones as described earlier. *Supercardioid* and *hypercardioid* responses are generated if the internal delay is respectively 0·58 and 0·33 times the external delay. These responses resemble the cardioid except that the sensitivity to sound from behind the microphone is increased.

It may be observed that a significant time delay occurs in the tubes that connect the microphone to the external ports in the hearing aid. If the tubes have the same length, then the delays cancel; if their lengths differ, then the

[5]The electrical equivalent is a series RC circuit where the input is applied across both elements and the output is taken across the capacitor. Acoustic resistance and compliance and equivalent circuits are explained in the next chapter. The complete equivalent circuit of the microphone will also have several components including inductances. See, for example, E. V. Carlson and M. Killion, *J. Audio Eng. Soc.* 22: 92–96, 1974.

internal delay is effectively increased or decreased according to which tube is the longer.

In order to reach a better understanding of the way in which the microphone favours sound from the front, consider the case in which the internal and external delays are the same and the ports are aligned with the path of the sound, as shown in Figure 6.14. The connecting tubes, which are of equal length, each impose a delay t_p. First, suppose that sound arrives from the front. After reaching the front port, the sound takes time t_p to reach the front surface of the diaphragm and time $t_x + t_p + t_i$ to reach the rear surface of the diaphragm. The time difference is therefore $t_x + t_i$, which is equal to $2t_i$ because the internal and external delays have been made the same. Now suppose that the source is behind the microphone. After reaching the rear port, the sound requires a time $t_p + t_i$ to reach the rear of the diaphragm and a time $t_x + t_p$ to reach the front of it. The time difference is $t_i - t_x$, which is zero. The sound pressures on each side of the diaphragm are therefore equal and the forces on it cancel. In principle, then, the microphone should be unresponsive to plane sound waves coming from behind.

Although hearing aids with directional microphones may have well-defined polar characteristics when tested in a free field, their performance when worn is less predictable, partly because of diffraction by the head and partly because sound from any one source is likely to be reflected from numerous obstacles that are inevitably present in the neighbourhood of the user. Nevertheless, directional microphones can give a worthwhile improvement in the signal-to-noise ratio in the amplified sound when the noise is fairly diffuse and the speaker is to the front of the listener.

Probe-Tube Microphones

Probe microphones allow sound pressures to be measured in confined spaces that would be inaccessible to conventional microphones and in harsh environments such as the interior of an automobile exhaust silencer. An important feature of the probe microphone is that its presence has a negligible influence on the sound field being measured. Applications include in situ measurements on headphones and hearing protectors, and measurements on machinery, musical instruments, loudspeakers, and telephones. An important audiological application is the measurement of sound pressure within the ear canal.

The probe microphone usually comprises a 'conventional' microphone coupled acoustically to the measuring point by a fine-bore tube. The tube is likely to produce considerable attenuation, so that the sound pressure at the diaphragm is significantly less than at the open end of the tube. For example, a 0·6-mm diameter tube, 100 mm in length, will reduce the pressure by approximately 20 dB at 1 kHz. Attenuation generally increases with increasing frequency, but resonances within the tube may give rise to unwanted peaks and valleys in the frequency response. It is possible to construct a microphone that is carefully matched to the acoustic properties of the tube so that the frequency response is smooth and reasonably constant up to a frequency of a few kilohertz,[6] but for audiological work an adequate probe microphone can be made by attaching a short length of soft plastic tubing to a miniature hearing aid microphone. The response may be sufficiently damped by virtue of the small diameter of the tube, but if necessary, additional resistance can be added. One way is to insert a very small piece of absorbent cotton. An example of a probe microphone is shown in Figure 6.15.

Probe-tube microphones are used routinely to measure sound pressure in the ear canal when hearing aids are fitted. In this work it is not usually important for the microphone to have a perfectly uniform frequency response because the test signals are usually pure tones rather than complex signals. Correction for the frequency response is therefore straightforward.

Microphone Calibration

In this section we consider methods for calibrating microphones. Although some of what is described will apply generally, our chief purpose is the calibration of precision microphones used for measuring sound rather than amplifying or recording it. It should be borne in mind that microphones are evidently not used in isolation but as the first part of a sound-measuring system or sound-level meter. Sound-level meters will be described later, but it may be noted that calibration often involves the meter as well as the microphone, the entire system being calibrated as a whole. Electronic instruments are usually stable over long periods and their calibration is generally straightforward. All that is needed on a day-to-day basis is perhaps an adjustment of the overall gain to obtain a correct indication in response to a standard reference voltage connected at the input in place of the microphone. Microphones, on the other hand, are considered less reliable (though in fact they are often surprisingly stable), and it is usual practice to recalibrate them periodically and to check their performance (in conjunction with the measuring equipment) each time they are used. A further remark to be made here is that microphone calibration is a

[6]For example, Brüel and Kjaer type 4182.

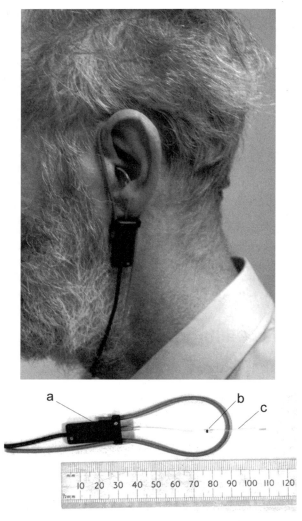

FIG. 6.15 An example of a probe-tube microphone for measuring the sound pressure in the ear at a point approximately 6 mm from the eardrum. The housing (*a*) contains two microphones, one to measure the external sound field and one that is attached to the probe tube (*b*). A moveable collar (*c*) on the tube is used to mark the length to be inserted in the ear. The tube, made of soft, flexible material, has an external diameter of approximately 1 mm and a bore of 0.5 mm. The microphone assembly shown here is part of the Audioscan model RM500 manufactured by Etymonic Design for hearing aid tests and real-ear measurement. (Reproduced by permission of Etymonic Design, Inc.)

technically challenging and interesting subject in its own right, and in this section at least a passing reference will be made to the methods that may be used even though they are not necessarily a part of everyday audiological practice.

We start by reviewing some technical terms that relate to the performance characteristics of a microphone and the type of measurement for which it is designed. Microphones may be used either in a sound field occupying generally some large volume with bounding surfaces that are usually unspecified, or they may be used within a small closed volume whose dimensions are precisely specified. A sound field is described as a *free field* if its boundaries have a negligible influence in the region where measurements are being made. In free-field applications, the sound will often be propagated as plane waves. At the other extreme, sound is repeatedly reflected from the bounding surfaces, with the result that the average sound intensity is the same in all directions. A sound field where this condition applies is called a *diffuse field*. In field measurements it is usual practice to specify field characteristics (such as sound pressure level) as they exist in the absence of the microphone and all amplifiers and equipment associated with it and, naturally, the observer. There are many applications that require sound to be measured in a closed volume. This is particularly true in audiological work, where the performance of audiometric earphones and hearing aids is usually assessed by measuring their output in specially constructed acoustic couplers. The measurements in a closed volume are usually referred to as *pressure field* measurements (or just *pressure* measurements), and microphones designed for this purpose are called *pressure microphones*, notwithstanding that most microphones are in fact sound pressure transducers anyway.

As previously stated, the *sensitivity* of a microphone is the ratio of the output voltage to the input sound pressure. If the measurements are made in a closed volume, the sensitivity is called the *pressure sensitivity* of the microphone, and the sound pressure at the diaphragm is the same as the sound pressure elsewhere in the volume. If, however, measurements are made in a free field, the sensitivity is described as the *free-field sensitivity*. The *diffuse-field sensitivity* of the microphone is defined in a similar way. The pressure p used to establish the sensitivity in a free or diffuse sound field is the pressure in the undisturbed field and may differ significantly from the pressure that is present at the diaphragm after the microphone is placed in the field. This modification of the sound field is an example of diffraction. It was stated in Chapter 3 that the disturbance of a sound field by an object within it depends on the dimensions of the object in relation to the wavelength of the sound: objects that are very small compared with the wavelength have little influence on the sound field, objects comparable to the wavelength have considerable influence, and very large objects reflect sound like a mirror and cast acoustic shadows.

Direct calibration of the free-field or diffuse-field response of a microphone is a difficult laboratory procedure requiring a special test environment. It is generally easier to determine the pressure response and, for sound field measurements, to make corrections that allow for diffraction. Once a microphone has been calibrated by the manufacturer or standardizing laboratory, its pressure response can be used as the basis for calibrating the entire measuring system. Users will usually calibrate their equipment to establish its pressure response at one frequency only (typically 250 or 1000 Hz) and make corrections to determine the response at other frequencies using data supplied by the manufacturer. If the microphone is to be used in a sound field, further or alternative corrections are needed. Because the sound field corrections make allowance for the diffraction produced by the microphone, they depend on the direction of the incident sound. The corrections are often supplied in an abbreviated form showing, for example, the frequency response in a free field for one angle of incidence only. This direction is the one that should normally be used. It is likely to be either 0° (diaphragm facing the source) or 90° (diaphragm parallel to the direction of propagation of sound), or unspecified if the field is diffuse. Note that there is no standard guidance about this; it should not be assumed, as sometimes happens, that a particular orientation is inherently the 'correct one'. Examples of the free-field corrections for a condenser microphone are shown in Figure 6.16. These corrections are to be added to the pressure response. Some microphones intended for free-field measurement have pressure responses that are designed to cancel the free-field corrections and give a nearly uniform frequency response over the working range.

It is necessary to understand that even the best microphones will have a significantly nonuniform frequency response if the frequency range is sufficiently extended. It is therefore necessary to restrict observations to those frequencies and types of measurement for which corrections are known and can be applied, or for which corrections can be ignored. This is one reason for the use of the weighting networks and other filters to be described later. Audiologists are fortunate in this regard because the upper frequency limit needed for most audiological applications is about 8 kHz or less. Good-quality microphones, if small enough ($\frac{1}{2}$-inch, for example) are likely to have a practically uniform response up to this frequency.

Sound Calibrators

Sound calibrators enable a known sound pressure to be applied to a microphone at one, or sometimes more, frequencies. As stated earlier, this

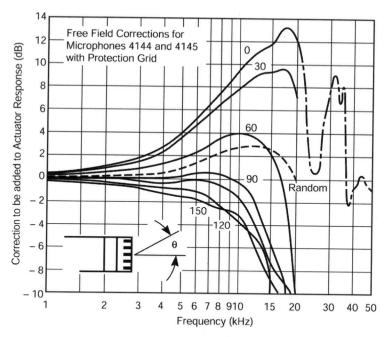

FIG. 6.16 Examples of corrections that should be added to the pressure response of a microphone (Brüel & Kjaer type 4145) to obtain its response in a free field. See also Figure 6.5. (Reproduced with permission of Brüel & Kjaer Sound & Vibration Measurements A/S, to whom the copyright belongs.)

allows the pressure response of the measuring system to be determined at the calibrating frequency, but it does not provide the overall frequency response. Two kinds of calibrator are commonly used. The first of these consists of a small loudspeaker that is acoustically coupled to the microphone by a small cavity. In its simplest form the calibrator relies on the calibration of the loudspeaker to provide a standard sound pressure. A better performance can be achieved if the sound pressure in the cavity is monitored by a reference microphone and thereby controlled so that it maintains the same value regardless of the acoustic load (equivalent volume) presented by the microphone under test. The merits of this type of calibrator are its small size and portability, its robust construction, and its good stability and accuracy. Simple calibrators (with no reference microphone) are usually supplied with hearing aid test equipment so that it may be calibrated before use.

The second type of calibrator is the pistonphone. This consists of a cylindrical cavity into which the microphone is inserted. Two small pistons project radially from a central structure within the cavity. They are driven by a rotating cam attached to the shaft of an electric motor whose speed is electronically controlled. The shape of the cam is such that the pistons move sinusoidally, producing an oscillatory change in the volume of the cavity and a corresponding sinusoidal variation in pressure. The pressure variation (the sound pressure in the cavity) depends on the volume displacement of the pistons relative to the volume of the cavity and on atmospheric pressure (recall that the elastic modulus of a gas is γP for an adiabatic change). Because these quantities are known, the sound pressure can be calculated. The pistonphone therefore provides an absolute method of calibrating a microphone at a single frequency. An accuracy of better than ± 0.2 dB can be achieved. In audiological work, pistonphones are often used to establish the calibration of equipment which is then used to calibrate audiometers.

Electrostatic Actuator

The electrostatic actuator provides a laboratory method for calibrating condenser microphones. It allows the shape of the frequency-response curve to be measured but does not give an accurate value for the absolute sensitivity. In other words, it establishes precisely how the response of the microphone changes with frequency but does not provide at any frequency an accurate calibration of the output voltage for a given sound pressure. However, the sensitivity at one frequency (typically 250 Hz) can be measured using a pistonphone, so the combination of calibrations with an actuator and a pistonphone will give the pressure sensitivity over the working frequency range.

The principles underlying the electrostatic actuator are as follows. It is a common observation that objects are drawn towards a charged surface by electrostatic attraction. Similarly, the plates of a condenser are attracted to each other by a force that is proportional to the square of the voltage between them. The actuator consists of a perforated metal plate mounted on insulating studs that stand on the tension ring on the rim of the microphone (Figure 6.17). The actuator plate is thereby held parallel to the diaphragm at some small distance from it (the gap is typically between 0.4 and 0.8 mm). The arrangement produces a parallel-plate condenser comprising the actuator plate and the outer surface of the diaphragm while at the same time the condenser of the microphone itself (formed by the inner surface of the diaphragm and the microphone backplate) is unaffected. The application of a sinusoidal alternating voltage between the actuator and the diaphragm results in an alternating force on the diaphragm that is virtually indistinguishable from the force that would be produced by a sound

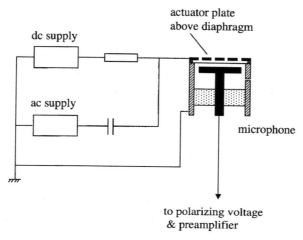

FIG. 6.17 Electrostatic actuator for measuring the pressure response of a condenser microphone.

pressure. The microphone can be operated in the normal way to record the response to this force.

A minor complication is that the electrostatic force is attractive regardless of the polarity of the applied voltage, whereas simulation of the sound pressure requires an alternating force corresponding to acoustic compression and rarefaction. The difficulty is overcome by adding a steady voltage to the alternating voltage. This is the electrical equivalent of the way in which sound pressure is an alternating pressure superimposed on the static atmospheric pressure. Because the electrostatic force depends on the *square* of the applied voltage, the alternating component of the force on the diaphragm is theoretically not a perfect sinusoid, and it can be shown that there is a component at twice the frequency of the applied voltage. But the steady-bias voltage is typically 800 V, and the alternating voltage is typically 30 V rms. In these circumstances the second harmonic is only 1·3%, that is, 38 dB, below the level of the fundamental and is therefore negligible. Absolute calibration is not possible with the electrostatic method because the gap between the actuator plate and the diaphragm is not known precisely.

Insert-Voltage Method

The open-circuit sensitivity of a microphone is, for unit sound pressure, the voltage that the microphone would produce if no electrical load were present at its terminals. This is a fundamental characteristic of the microphone that allows us to determine theoretically what output it will have when it is connected to the preamplifier. We also have to know the capacitance of the microphone and preamplifier and the stray capacitances in the microphone cartridge, connectors, and cables. Let M be the sensitivity of the microphone when it is connected to its preamplifier. This sensitivity tells us what voltage the microphone will produce at the input to the preamplifier when the diaphragm is exposed to unit sound pressure. Let M_0 be the corresponding voltage at the terminals of the microphone when no electrical load is present. Then

$$M = M_0 \frac{C_m}{C_m + C_L}$$

where, as before, C_m is the capacitance of the microphone and C_L is the input capacitance of the preamplifier plus any stray capacitance in the connectors and cables. If p is the sound pressure at the diaphragm and g is the voltage gain of the preamplifier, then the output voltage V of the preamplifier is given by

$$V = pgM_0 \frac{C_m}{C_m + C_L}$$

The microphone capacitance is typically greater than 15 pF, whereas C_L is likely to be much smaller, perhaps only 0·3 pF. The ratio of M to M_0 is therefore close to unity, so that C_m and C_L do not have to be known with great accuracy and the calibration is determined principally by the value of the open-circuit sensitivity, M_0.

The open-circuit sensitivity can be found using the insert-voltage method as follows. To start with, a known sound pressure p is applied to the microphone, and the output of the preamplifier is noted. Call this voltage V. It is equal to kV_0, where k is a constant depending on the gain of the preamplifier and on the capacitances just described, and where V_0 is the open-circuit voltage of the microphone. The sound source is now turned off and a known voltage V_1 is inserted in series with the microphone. This operation is made possible by means of electrical connections provided on a preamplifier specially designed for the insert-voltage measurements. The inserted voltage V_1 is adjusted so that the output of the preamplifier is again equal to V and therefore to kV_1. It follows that

$$V_0 = V_1 \quad \text{and} \quad M_0 = V_0/p$$

An alternative to the insert-voltage method is the *charge injection* method. As the name suggests, a known charge is inserted into the microphone rather than a known voltage. An advantage of this method is that mechanical or electrical faults affecting the capacitance of the microphone soon become apparent.

Reciprocity Method

It is improbable that an audiologist will ever find it necessary to carry out a reciprocity calibration of a microphone, but the method is of fundamental importance as a primary means of calibration and the brief description given here needs no further justification. The *reciprocity theorem* is one of the most delightful creations of science. Its application is extremely wide, taking in optics, acoustics, mechanics, heat transfer, electricity, and any part of physics where a direct analogy may be made with force and velocity or displacement. Before explaining the theorem, let us state the conditions in which it is valid. It applies in systems that are passive and linear and it describes forces and displacements or velocities (or their equivalents) that are harmonic. In other words, the forces and the corresponding displacements are sinusoidal functions of time. It is in principle limited to systems in which each of the component vibrations can be described as a variation about an equilibrium value, so, for example, it might not apply to the transmission of sound in a flowing medium. To say that a system is *passive* means that it does not contain any amplifiers or internal sources of energy; to say that it is *linear* means that a sinusoidal force gives rise to a sinusoidal displacement at the same frequency without overtones. Much has been written about the reciprocity principle, and its general description would require mathematical discourse well beyond the scope of this book. Instead, some simple examples will be given to illustrate the underlying ideas. It will then be shown in outline how these ideas can be put to use in calibrating acoustic transducers.

Suppose there is a room, divided partially by an internal wall (Figure 6.18). There are two tables, *A* and *B*, one on each side of the wall. On table *A* there is a lamp, the only source of light in the room, and on table *B* there is a light meter. It will be found that the reading on the meter will be the same if the meter and the lamp are interchanged. This will be true regardless of the presence of reflecting or absorbing obstacles in the room or the colour of the paint on the walls (for example, one side of the partition could be painted black and the other white). The same experiment can be made acoustically, replacing the lamp by a simple sound source and the light meter by a microphone. The only proviso is that in each case the source and receiver must be nondirectional; that is, the lamp or sound source must radiate uniformly in all directions and the light meter or microphone must respond uniformly.

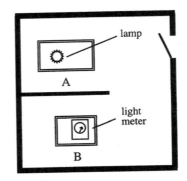

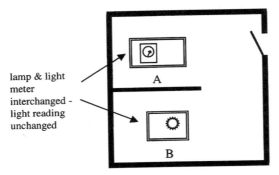

FIG. 6.18 Illustrating the reciprocity principle.

The acoustic experiment can be extended and expressed more formally as follows. Suppose that a simple source of strength Q_1 at a point A produces a sound pressure p_2 at a point B and that, alternatively, a simple source of strength Q_2 at B produces a sound pressure p_1 at A. Then

$$\frac{Q_1}{p_2} = \frac{Q_2}{p_1} = J \qquad\qquad 6.4$$

where J is a constant whose value depends on the relative positions of A and B and the acoustic environment in which they exist.

A simple demonstration of reciprocity in a mechanical system is provided by the following example. Suppose a metal bar is held horizontally by clamping it at one end while the other end remains free (Figure 6.19). If an oscillatory force F is applied at a point A on the bar and the displacement

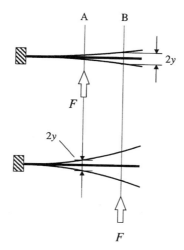

FIG. 6.19 The reciprocity principle applied to a mechanical system.

amplitude y is observed at a point B, then the same displacement will be observed at A if the same force F is applied at B.

Let us now consider what are called *reversible transducers*. These are transducers that can perform their conversion in either direction. A loudspeaker, for example, produces sound when current is passed through its speech coil. If the speaker is placed in a sound field, its diaphragm will vibrate in response to the acoustic pressure and a voltage will be generated across the ends of the coil. All microphones described in the foregoing paragraphs are reversible, with the exception of the hearing aid microphones that have built-in amplifiers. For the condenser microphone, we saw, when describing the electrostatic actuator, that an alternating electric field between the plates of a condenser gives rise to an alternating force. The condenser microphone therefore becomes a sound generator when an alternating voltage is added to its polarizing voltage. Condenser microphones are sometimes used in this way when the application requires a small and accurately reproducible source.

Reversible transducers have an interesting property that allows the reciprocity principle to be extended to include systems whose acoustic elements include microphones and speakers. As an example, suppose that a dynamic microphone is made into a sound source by passing a current through it. The current causes a force to be generated within the microphone which makes its diaphragm vibrate. Let the force be F when the current in the microphone coil is I. Because the microphone is a linear transducer, the force is directly proportional to the current and we can write

$$F = hI$$

where h is a constant called the *electromechanical coupling coefficient*. The really interesting thing is that this constant can be defined in an entirely different way and yet the same answer is obtained. Suppose the microphone is disconnected from its electrical supply and that in response to sound pressure or some other vibratory input to the diaphragm a sinusoidal motion is produced with velocity u. In this condition let E be the open-circuit voltage across the ends of the coil. Then

$$E = hu$$

and h has the same value as before. It follows from this that

$$\frac{EI}{Fu} = 1$$

This relationship is true for the dynamic microphone and for the other reversible transducers so far described. For such transducers the reciprocity principle can be expressed in the following way. Suppose that the transducer is operated as a sound receiver (as a microphone). Let p_A be the sound pressure at an arbitrary reference point A close to the transducer (Figure 6.20), and let E be the corresponding open-circuit voltage that appears at the terminals. Now suppose that the transducer is operated as a source supplied with a current I. At a point B whose distance from A is d, the sound pressure due to the operation of the transducer is p_B. The reciprocity theorem (in its electroacoustic form) then states that

$$\left(\frac{E}{p_A}\right) \div \left(\frac{p_B}{I}\right) = J \qquad\qquad 6.5$$

The first term is the 'microphone response' of the transducer, and the second term is its 'speaker response'. The ratio of these quantities is the constant J, called the *reciprocity constant* or *reciprocity coefficient*. If the transducer is operating in a free field and if it is small enough to be in effect a source of spherical waves when viewed from B, then J is given by

$$J = \frac{2d\lambda}{\rho_0 c} = \frac{2d}{\rho_0 f} \qquad\qquad 6.6$$

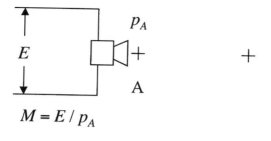

$$M = E / p_A$$

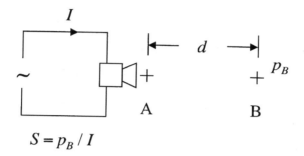

$$S = p_B / I$$

FIG. 6.20 Microphone response *M*, and speaker response *S*, of a reversible transducer.

where f is the frequency of the sound and ρ_0 is the density of air. Exactly the same result applies in Equation 6.4.[7]

Equation 6.5 will look a little tidier if we write the microphone response as M and the speaker response as S; that is,

$$M = E/p_A \quad \text{and} \quad S = p_B/I$$

Then

$$\frac{M}{S} = J = \frac{2d\lambda}{\rho_0 c}$$

In other words, the reciprocity constant of a reversible transducer is the ratio of its open-circuit voltage sensitivity to its sensitivity to current (its speaker response) at a specified distance.

[7]The reader may confirm this by using Equation 4.1 to obtain an expression for Q_1/p_2. Note that p_2 should be treated as a pressure amplitude rather than an rms quantity, in keeping with Q which is by definition a peak value. The root 2 in the denominator of Equation 4.1 is then not required. The radial distance r is replaced by the distance d between points A and B.

The relationships just described are the basis of reciprocity methods for calibrating microphones. Calibration can be done in various ways to give the free-field or the pressure sensitivity. There are, to be sure, practical difficulties and refinements to be considered according to the type of calibration and the type of microphone being calibrated, but we shall overlook these and present the method in its most basic form. We shall suppose that the free-field response is required and we shall not insist that the microphone being calibrated (the 'test' microphone) must itself be reversible. In addition to the test microphone, we require a sound source and a reversible transducer. The calibration is made in two steps as described next and illustrated in Figure 6.21.

STEP 1

    a. Place the reversible transducer in the sound field of the source and measure its open-circuit voltage E_R.

    b. Replace the reversible transducer with the test microphone and measure its open-circuit voltage E_1.

Because the reversible transducer and the test microphone were each exposed to the same sound field, the measured voltages are in the same ratio as the corresponding microphone sensitivities, here denoted by M_R and M, respectively. Accordingly,

$$\frac{E_R}{E_1} = \frac{M_R}{M}$$

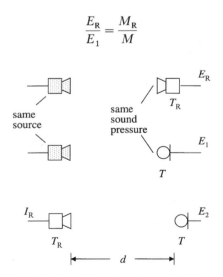

FIG. 6.21 Reciprocity calibration of a microphone. T_R is a reversible transducer. T is the microphone being calibrated.

Therefore

$$M = E_1 \times \frac{M_R}{E_R} \qquad 6.7$$

STEP 2. Place the test microphone at a distance d from the reversible transducer. Pass a current I_R through the latter and record the open-circuit voltage E_2 of the test microphone.

Let p_R be the sound pressure at the diaphragm of the test microphone. If S_R is the speaker response of the reversible transducer, then, by definition,

$$p_R = I_R S_R$$

But from the reciprocity theorem, $M_R/S_R = J$ and therefore

$$p_R = I_R M_R / J$$

But $E_2 = M p_R$ and therefore

$$E_2 = M \times \frac{I_R M_R}{J}$$

so that

$$M = \frac{E_2 J}{I_R M_R} \qquad 6.8$$

Multiplying Equations 6.7 and 6.8 gives M^2, so on taking the square root and substituting for J from Equation 6.6, we arrive at the desired result, namely,

$$M = \sqrt{\frac{E_1 E_2}{I_R E_R}} \times \frac{2d}{\rho_0 f}$$

The important thing to notice about this expression for the free-field sensitivity of the test microphone is that it contains only terms that can easily be measured. The difficult problem of measuring sound pressure directly is avoided. This is a particular merit of the reciprocity method.

One further point concerning the reciprocity principle is that the directivity pattern of a transducer (the way its response varies with the direction of the transmitted or received sound) is the same whether it is acting as a source or a microphone. The directional properties of a condenser microphone mounted in an infinite baffle would be the same as those of a similarly mounted circular piston. If two omnidirectional microphones are separated by a small distance and their outputs are connected so that one is subtracted from the other, the directional response of the combination would be that of the acoustic dipole described in Chapter 4.

Sound-Level Meters

A sound-level meter is an instrument for measuring sound pressure level. Its development has a long history. At first, efforts were made to produce an instrument that would give an objective indication directly related to the 'response of the human ear' or at least to the subjective sensation that we call loudness. As we shall see, weighting circuits (filters) were incorporated to shape the frequency response of the meter so that it matched the loudness response of the ear to pure tones. The weighting functions were therefore the inverse of the equal loudness curves. A major obstacle was that the shape of these curves is not independent of sound level, so that different frequency weightings had to be selected according to the overall sound pressure. Moreover, it was found that loudness is not a simple function of the frequency-weighted sound pressure but a quality that depends also on other physical attributes, including particularly the distribution of sound energy within the spectrum. While objective correlates of the subjective qualities of sound do exist — for example, in loudness and annoyance ratings — they are not usually derived from the result of a single measurement but are more often obtained following calculations from a number of separate observations. Some weighting functions have survived, and nearly all modern instruments can provide an A-weighted response, but their purpose is not to give a direct indication of loudness.

There are various ways in which sound-level measurements can be made, and there are many different sounds and many environmental and measurement variables to be considered. Standardization is therefore essential if observations from different origins are to be compared. The operating characteristics of sound-level meters are the subject of international agreement as specified in IEC 60651. The preamble to this standard contains the following paragraph:

> Owing to the complexity of operation of the human ear, it is not possible at present to design an objective noise measuring apparatus to give results which are absolutely comparable, for all types of noise, with those obtained by subjective methods. However, it is considered essential to standardize an apparatus by which sounds can be measured under closely defined conditions so that results obtained by users of such apparatus are always reproducible within stated tolerances.

In keeping with the standard, sound-level meters are classified according to the application and measurement accuracy required:

Type 0   Laboratory reference standard
Type 1   Laboratory use or use in the field,[8] where the acoustical environment can be closely specified or controlled

[8]Not to be confused with a *sound field*. Here it means a work area that is not a laboratory.

Type 2 General field applications
Type 3 Intended primarily for noise surveys to determine whether an established noise limit has been significantly violated

Type 2 meters are appropriate for general audiological applications such as measuring or monitoring the sound fields for audiometry or hearing aid tests; a type 1, or preferably type 0, meter is required for calibrating audiometers.

All sound-level meters provide the user with an output indication in decibels that is related in a standard way to the sound pressure at the microphone. The following components are needed:

- Microphone and preamplifier
- Input amplifier (variable gain)
- Weighting circuits and other filters
- Output amplifier (variable gain)
- Squaring and averaging circuit (and logarithmic converter)
- Analogue or digital display

These components are shown schematically in Figure 6.22. In addition to weighting filters, provision is often made for inserting external filters such

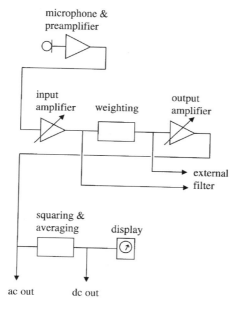

FIG. 6.22 Components of a sound-level meter.

as those needed for octave or third-octave analysis. Some sound-level meters are equipped to measure peak sound levels and impulse-weighted levels. These measurements will be described later. Many sound-level meters provide an amplified (and filtered) version of the microphone output ('ac output'), and some provide a rectified signal ('dc output') proportional to the supply to the display unit. The ac output can be taken to a frequency meter or an oscilloscope; the dc output can be used, for example, to supply sound level recording equipment. The components just listed will be described further in the following paragraphs.

Microphone and Preamplifier

The design of microphones has already been described. Laboratory sound-level meters (types 0 and 1) are very likely to have condenser microphones, so that a preamplifier is essential. This will normally be included within the measuring instrument or supplied as an accessory if the microphone is to be separated from it by a cable.

Sound-level meters are often portable instruments intended to be hand-held or mounted on a tripod. The microphone is usually attached to the front of the instrument case; the case is therefore an acoustically significant object that modifies the sound field. An even greater modification is inevitable if the observer is also present, as is necessary unless some arrangement is made to read the meter remotely. The influence of the instrument case on the indicated sound level is greatly reduced if the microphone is mounted separately using a cable or if it is separated from the instrument by an extension. In these circumstances a reasonably uniform directional response can be obtained by fitting a special attachment to the microphone known as a *random incidence corrector*. This cancels the directional effects caused by diffraction. It is useful when measurements are made in a diffuse field. The reflection of sound from the body of the observer is greatest for frequencies around 400 Hz. If the observer stands close to the microphone, the measured values are likely to be in error, possibly by as much as 6 dB. The arrangement shown in Figure 6.23 is recommended.

Input and Output Amplifiers

In simple instruments, only one main amplifier is used, but in better instruments the amplifier is split into two independently controlled stages on either side of the filter, as shown in Figure 6.22. An advantage of this

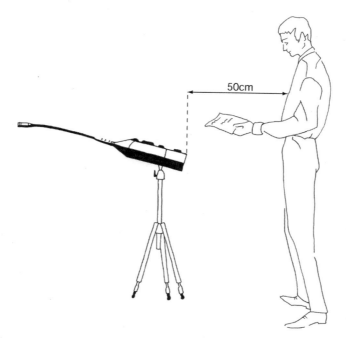

FIG. 6.23   Using a sound-level meter. (Redrawn from a sketch that appeared in Jens Broch, *Acoustic Noise Measurements*. Brüel & Kjaer, Denmark, 1973.)

arrangement is that each amplifier can be operated at a gain that gives the best performance (least internal noise) consistent with the input from the microphone. It also gives the user some control over the signal level at the filter, which may be important when an external filter is used, particularly if it has components that generate significant electronic noise. The gain of each amplifier can be changed in steps of exactly 10 (or sometimes 5) decibels. The overall gain (in decibels) is the sum of the gains of each amplifier; this has to be added to the indicated output. On some instruments the addition may be done automatically.

Good-quality sound-level meters have indicators to warn the user of overload in one or both amplifiers. This is important because an overload may not be obvious from the value of the displayed output. It is good practice to check, if possible, that the output remains unchanged when the gain of one amplifier is increased by 10 dB while the gain of the other is reduced by the same amount.

Weighting and Filters

The standard frequency weightings, designated as A, B, and C, are historically related to loudness functions that describe how the loudness of a pure tone varies with sound pressure and frequency. Loudness is measured in *phons*. The sound to be measured is compared with a 1-kHz tone, and the intensity of the tone is adjusted until it is judged to have the same loudness. The loudness of the measured sound in phons is then numerically equal to the sound pressure level of the tone. A formal definition is given in the last part of this chapter.

Curves showing the variation of sound pressure with frequency for constant loudness are called *loudness contours*. The A, B, and C weightings correspond approximately to the inverse of the 40-, 70-, and 100-phon contours. For example, the 40-phon contour passes through a point corresponding to a sound pressure of 51 dB SPL at 200 Hz. This means that the sound pressure for a pure tone must be 11 dB greater at 200 Hz than at 1 kHz for the same loudness, and the A-weighting filter should reduce the sensitivity of the sound-level meter by 11 dB at 200 Hz. The frequency response of filters as specified in the IEC standard is, however, only an approximate version of the inverse loudness contours. This is partly because the filters are specified in a way that allows them to be realized using simple, passive circuits whose branches contain only resistors and capacitors. Another reason is that the equal loudness contours were further investigated and redefined after the standard weighting characteristics had become established.

The standard A, B, and C weighting functions are given in Table 6.1 and shown graphically in Figure 6.24. The weightings require the use of three filters: filter number 1 gives the C-weighted response, filter number 2 added in series with filter number 1 gives the B-weighted response, and all three filters together give the A-weighted response. Consistent with the definition of the phon, all filters have unit gain (0 dB) at 1 kHz; that is, they do not affect the response of the sound-level meter at this frequency. Sound calibrators are sometimes operated at 1 kHz so that the calibration can be verified with the filters in place. When the filters are used, the weighted sound levels are expressed as dB(A), dB(B), or dB(C).

Given that sound-level meters are no longer intended to be instruments for measuring loudness, what virtue is there in retaining the standard weightings? So far as the B-weighting is concerned, the answer is very little. This weighting is now virtually obsolete. The C-weighting is useful as a means of limiting the operating bandwidth of a sound-level meter in a well-defined way by providing a nearly uniform response between 31·5 and

TABLE 6.1 IEC Frequency Weighting
Characteristics

Frequency (Hz)	A (dB)	B (dB)	C (dB)
10	−70·4	−38·2	−14.3
12·5	−63·4	−33·2	−11·2
16	−56·7	−28·5	−8·5
20	−50·5	−24·2	−6·2
25	−44·7	−20·4	−4·4
31·5	−39·4	−17·1	−3·0
40	−34·6	−14·2	−2·0
50	−30·2	−11·6	−1·3
63	−26·2	−9·3	−0·8
80	−22·5	−7·4	−0·5
100	−19·1	−5·6	−0·3
125	−16·1	−4·2	−0·2
160	−13·4	−3·0	−0·1
200	−10·9	−2·0	−0·0
250	−8·6	−1·3	−0·0
315	−6·6	−0·8	−0·0
400	−4·8	−0·5	−0·0
500	−3·2	−0·3	−0·0
630	−1·9	−0·1	−0·0
800	−0·8	−0·0	−0·0
1000	**0**	**0**	**0**
1250	+0·6	−0·0	−0·0
1600	+1·0	−0·0	−0·1
2000	+1·2	−0·1	−0·2
2500	+1·3	−0·2	−0·3
3150	+1·2	−0·4	−0·5
4000	+1·0	−0·7	−0·8
5000	+0·5	−1·2	−1·3
6300	−0·1	−1·9	−2·0
8000	−1·1	−2·9	−3·0
10000	−2·5	−4·3	−4·4
12500	−4·3	−6·1	−6·2
16000	−6·6	−8·4	−8·5
20000	−9·3	−11·1	−11·2

Note: The table shows the weighted response of the
sound-level meter according to frequency and the selected weighting characteristic. The tabulated values are
added to the unfiltered response. For example, if the
meter reads 60 dB without weighting at 400 Hz, its
A-weighted reading will be $60 - 4 \cdot 8 = 55 \cdot 2$ dB(A).

The frequencies shown in the table are the nominal
frequencies recommended by the IEC. They are rounded
values of frequencies given by the formula $f = 1000 \times 10^{n/10}$, where n is a positive or negative integer.

The information given in the table is taken from Table
IV in IEC 60651 (1979).

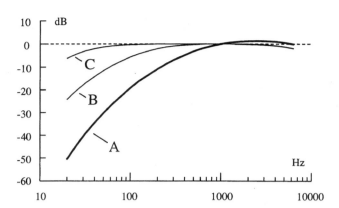

FIG. 6.24   IEC frequency weighting characteristics.

8000 Hz. The attenuation for frequencies outside this range is given in Table 6.1. If no filter is used, the bandwidth is of course limited, but it then depends on the response of the microphone and the various amplifiers and other electronic components. The nonfiltered (nonweighted) response of a sound-level meter is often described as 'linear', and the measured sound level is then given as 'dB LIN'. The 'linear' response is nonstandard and its specification is left to the manufacturer or user.

The A-weighting remains the most important response-shaping function because it gives a response that has some relevance to auditory physiology. Its relation to loudness has already been described. It is also a clinically useful approximation to the threshold of hearing. Below 5500 Hz the sound pressure in dB(A) is within a few decibels of the threshold of hearing for binaural listening in a free field (the minimum audible field;[9] see Figure 6.25). For this reason it is often considered acceptable in 'sound field' behavioural audiometry to treat threshold levels measured in dB(A) as though they were true hearing levels. A further merit of the A-weighted response is that it yields a result that is at least roughly proportional to the sound energy delivered to the cochlea in a given time. Again, the upper frequency limit is about 5500 Hz. In Figure 6.25 the crosses show estimated values of the free-field sound pressures that are needed for a constant sound pressure at the input to the cochlea. Measurement of exposure to noise and estimates of the likely harm so caused often rely on A-weighted sound levels. It is believed that damage to the cochlea caused by noise is related to the

---

[9]The *minimum audible field* (MAF) is defined as the threshold of audibility for pure tones heard binaurally when the listener is facing the source of plane progressive waves.

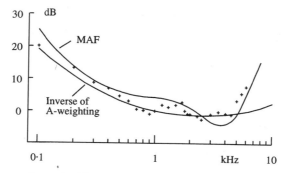

FIG. 6.25 Comparison of a binaural threshold of hearing in a free field (minimum audible field, MAF) and the inverse of the A-weighting functions. The crosses show estimated values of the free-field sound pressures that are needed for a constant sound pressure at the input to the cochlea. (Data are from J. Zwislocki in *The Nervous System*, Vol. 3, ed. D. B. Tower. Raven Press, NHew York, 1975, p. 53. The data points have been displaced vertically to pass through 0 dB at 1 kHz.)

total acoustic energy received by the cochlea. It is therefore logical that A-weighting should be applied. Noise dosimetry will be considered later in this chapter.

Squaring and Averaging

Sound level is, by definition, expressed in decibels and related directly to the acoustic power emitted by the source or sources of the measured sound. The quantities p_n and p_0 in Equation 6.2b are normally represented by their root mean square values, so that when these averages are squared the result is a quantity that is proportional to acoustic power. The electrical signal in the sound-level meter is proportional to the instantaneous sound pressure at the microphone; it is therefore necessary to square and average this signal. It is not, however, necessary to follow this process by finding the square root of the average, because when the result is expressed in decibels the number representing the root is simply half that representing the square. The difference is then merely a component of the scale factor in the final display. Squaring can be accomplished as a mathematical operation using digital electronics, but for simpler instruments various analogue methods are available, each with its particular merits and shortcomings. The simplest method is to rectify the signal and pass it through a circuit whose gain is proportional to the instantaneous value of the input. Ideally this would require the gain to be continuously variable, but in practice sufficient accuracy can be obtained with a series of constant-gain sections, the gain increasing in steps from one section to the next. Diodes are used to switch the gain to the next as the signal changes.

The way in which the squared signal is averaged is of fundamental importance because it determines the nature of the response of the sound-level meter to the various sound pressure waveforms that may be present at the input. In order that this important feature is understood, we should remind ourselves of the way statisticians define the *mean* of a set of values. Suppose we have N values $y_1, y_2, y_3, \ldots, y_N$ of some variable y. The mean of y, here denoted as $\langle y \rangle$, is given by

$$\langle y \rangle = \frac{1}{N} \sum y_1 + y_2 + y_3 + \cdots + y_N$$

If y is a continuous variable, then the summation must be replaced by integration, the result of which can be shown as the area under a curve, and instead of dividing by N we divide by the range of the variable over which the average is obtained. In the present context, we have a voltage v that corresponds to the sound pressure at the microphone and we require the average of the square of this voltage. The average is to be taken over time T. The true average of v^2 is then

$$\langle v^2 \rangle = A/T \qquad 6.9$$

where A is the area under the graph of v^2 against time for the interval T seconds (Figure 6.26). The output of the sound-level meter would ideally represent an average calculated in this way. It would be a moving average that included only the recent history of the signal from time T to the time of observation. Such an average can indeed be obtained electronically, especially if digital methods are used, but for most purposes a much simpler procedure is available. This is sometimes called *time-weighted exponential averaging*. Readers will be relieved to hear that this is not as fearsome as the name suggests. It is important to know something about this type of averaging because it is, at least historically, the basis of the way in which averaging times are specified.

Exponential averaging is possibly the simplest type of averaging that can be accomplished electronically. At its heart, the averager comprises just a resistor and a capacitor connected in series. The signal to be averaged is applied across both components, and the output — representing the average — is the voltage that appears across the capacitor (Figure 6.27). Suppose the capacitor is initially discharged and that a step voltage V_0 is applied to the input and maintained constant. As the capacitor acquires a charge, the

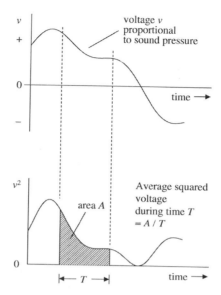

voltage *v*
proportional
to sound pressure

Average squared
voltage
during time *T*
= *A* / *T*

area *A*

FIG. 6.26   Definition of the average squared voltage during the averaging time *T*.

voltage on it rises, but as it does so the voltage across the resistor must diminish because the sum of the voltage across the resistor and the capacitor is constant. The charging current therefore decreases progressively, and the rate at which the output voltage rises is reduced. If the output voltage *V* is plotted against time, the result is an inverted exponential that approaches the value of input voltage asymptotically (Figure 6.28). We can write this formally as

$$V = V_0(1 - e^{-t/RC})$$

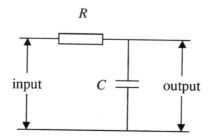

R

input

C

output

FIG. 6.27   Resistor-capacitor circuit for averaging.

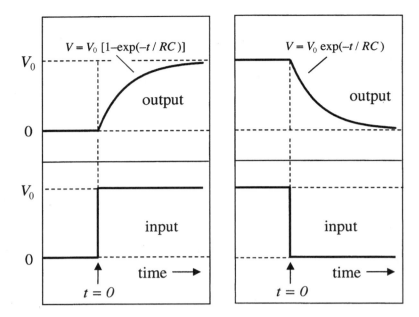

FIG. 6.28   The charge and discharge of a capacitor through a resistor.

The resistance R multiplied by the capacitance C is called the *time constant* of the integrator. If the resistance is measured in ohms and the capacitance in farads, the time constant will be in seconds. Now suppose that the capacitor was fully charged to start with and that the input was made equal to zero at time $t = 0$. As time passes the capacitor discharges through the resistor and the output voltage falls. It does so exponentially:

$$V = V_0 e^{-t/RC}$$

We can regard the resistor-capacitor (RC) averager as a simple low-pass filter. If its input is changing at a low frequency, the output on the capacitor follows the input very closely because there is sufficient time for the charge to adjust to the changing voltage. The opposite is true if the input varies at a high frequency. There is little time during each cycle for the capacitor to charge and discharge, and the voltage on it due to the high-frequency components of the input is small. The frequency response for the RC circuit is shown in Figure 6.29. It should be remembered that it is the *squared* value of the acoustic signal that is passed to the averager. If the frequency is high this value will fluctuate rapidly about its average (equal to half the square of the amplitude if the signal is a sinusoid). It is the fluctuation that is

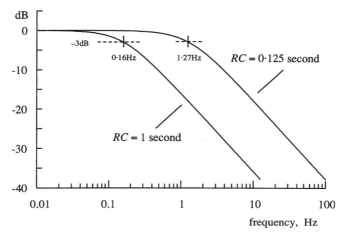

FIG. 6.29 The frequency response of the RC circuit shown in Figure 6.27.

suppressed by the RC filter. The condenser will charge or discharge until its voltage is equal to the average.

To understand how the RC averager works as part of the sound-level meter, we proceed as follows. We suppose that observations are made at time t. In other words, t is an arbitrary time at which a reading of the output of the sound-level meter is made. The level shown will depend on the history of the input insofar as it has contributed to the charge remaining on the capacitor. We shall use the symbol τ to denote the time of events preceding the moment of observation. The time that has passed between such events and the time of observation is therefore $t - \tau$. Suppose that at time τ the input to the averager is a brief rectangular pulse having a voltage V_τ and lasting for a time $\delta\tau$. It can easily be shown that the record of this pulse in the output of the averager at time t is the voltage δV given by

$$\delta V = \frac{V_\tau}{RC} e^{-(t-\tau)/RC} \delta\tau$$

When the sound-level meter is used, the voltage presented to the averager is a continuous function of time, proportional to the square of sound pressure at the microphone. This function can be treated mathematically as though it consisted of a sequence of rectangular pulses of the type just described, where the pulse height V_τ is equal to the voltage presented to the averager during the brief interval from τ to $(\tau + \delta\tau)$. The total voltage at the output of the averager is then found by adding the contributions of all the elementary pulses starting with those in the remote past (at $\tau = -\infty$) and ending just before the moment of observation (at $\tau = t$). This addition can be carried out mathematically by integration. The result is

$$V = A'/RC \qquad\qquad 6.10$$

where V is the output of the averager and A' is the area under a graph of $V_\tau \exp[-(t - \tau)/RC]$ plotted against τ from times past to the time of observation. We therefore see that the output of the averager is the sum of the contributions from past events weighted by the exponential factor according to the elapsed time $t - \tau$ (see Figure 6.30).

Equations 6.9 and 6.10 allow us to compare the output of the RC averager with the true moving average over time T up to the moment of observation. We should remember that the voltage derived from the microphone is squared prior to averaging, so that V is equivalent to $\langle v^2 \rangle$. It seems natural to choose values of R and C so that RC is equal to T. The difference between the two averages then resides in the difference between A and A'. One thing to note is that the ideal averager completely disregards all events that occurred before time T, whereas the RC averager has an output that, in principle, depends on the entire history of the signal. If significant sounds were recently present, the user must allow the meter to settle before making observations. This is not a problem if the time constant is short, but it can be troublesome if it is equal to several seconds. Another thing to note is that if the signal is unchanging (if it is a constant-amplitude pure tone, for example) the RC average is identical with the true moving average. The RC averager therefore behaves almost as an ideal averager provided that the signal amplitude remains constant for a time that is at least several times longer than the time constant. It is not, however, possible to make general statements about the average that will be obtained for varying signals except to say that because the properties of the RC averager are simple and clearly specified, the result can often be interpreted if enough is known about the signal waveform. For example, the response to tone bursts whose duration is greater or less than the integration time can be anticipated and used to test the performance of the averager.

This account is by no means the whole story. The best modern instruments use digital processing to calculate the root mean square of the signal. Simpler instruments rely on analogue circuits, but even here it is possible to improve on the simple squaring and RC averaging just described. In one very successful technique, part of the output of the averager is used as a feedback to modify the signal-squaring operation. The overall result is shown to be an output that is directly proportional to the root mean square of the original signal.

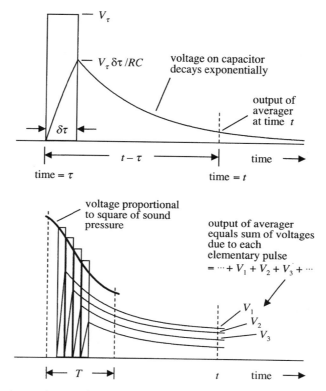

FIG. 6.30 Averaging a voltage proportional to the square of the sound pressure. The simplest method, which uses an RC circuit with a time constant equal to the required averaging time of the meter, is sometimes called *exponential time-weighted averaging*. In this illustration the input waveform (after squaring) is treated as if it were a series of rectangular pulses. The output is the sum of the responses from each pulse. The upper diagram shows the voltage rise during an elementary pulse of height V_τ and duration $\delta\tau$, and the subsequent decay of this voltage during time $t - \tau$. The lower diagram shows a portion of the squared waveform divided into a series of pulses, each of width $\delta\tau$.

The international standard[10] specifies two averaging times, which it designates as F (fast) and S (slow). They are 125 ms and 1000 ms, respectively. The use of exponential averaging is not an explicit requirement of the standard, but its authors clearly had this in mind. They provide tests

[10]At the time of printing (January 2002), IEC 60651 (1979) was being revised. The new version is very likely to retain the current specifications for the averaging time of the detector and its response to tone bursts.

using tone bursts that should be used to verify the response characteristics. These tests are based on the assumption that the squared signal is passed to an RC averager having a time constant equal to the specified S or F averaging time.

Output and Display

The sound-level meter must provide an indication in decibels. If its output is to be displayed digitally, or if it is to be displayed in analogue form using a linear scale, then it becomes necessary to find the logarithm of the signal leaving the averager. Logarithmic conversion is straightforward if done digitally. It may be more difficult to achieve in analogue circuits, but, as with so many electronic requirements, the designer has integrated circuits that perform such tasks to a high standard.

Many of the simpler sound-level meters (types 2 and 3) use a moving-coil meter as the indicator. The conversion of the output voltage, proportional to the mean square of the sound pressure, to a number of decibels can be done simply by giving the indicator a nonlinear scale graduated in decibels. The disadvantages to this are that the meter may be difficult to read accurately unless the selected amplification brings the indication to the central region of the scale, and that there is no dc output for connection to other equipment.

Other Features

As previously mentioned, indicators are often fitted to alert the user to amplifier overload that might otherwise go unnoticed.

Equipment that is intended for use in the field rather than the laboratory has to be battery powered. It should be provided with a means of checking that the battery voltage is adequate. It is good practice to check the battery condition (and sound-level calibration) before and after measurement.

Good-quality instruments usually have an internal reference oscillator running, probably, at 1 kHz. This provides a standard signal that can be applied to the input amplifier for self-testing. Connections should also be available for replacing the microphone output with an electrical signal from an external source. This allows the sound-level meter to be tested independently of its microphone. It also turns it into a very useful rms-indicating voltmeter.

Sound-level meters complying with the IEC standard must provide at least one of the specified A, B, or C weightings. Additional frequency responses that might be made available include D-weighting (for aircraft noise measurements) and the so-called linear response that was mentioned previously.

In addition to the indicator characteristics previously described, good sound-level meters often have circuitry for measuring the peak or the impulse-weighted response to transient signals. We will consider these in more detail presently. The option of a long averaging time (perhaps 10 or 20 seconds) is also helpful. It can be used to obtain a statistically representative sample of a fluctuating signal and is sometimes useful when correcting the measurement of weak signals for the influence of background noise. This correction will be described later.

Measuring Impulsive Sounds and Other Transient Signals

Transients such as acoustic clicks and tone bursts are often used as stimuli for auditory evoked potentials. Transient signals are also audiologically important when they are a potential source of harm. Impulse noise from gunfire is an example. A complete characterization of such signals requires a knowledge of the waveform, but usually one has to be content with a far simpler description. The following sections describe terms that are related to measurements that are often made.

Peak Sound Pressure Level

This is a level derived from the greatest sound pressure during the existence of the signal. Some sound-level meters have a circuit that captures the peak pressure and holds the peak level in the display until it is reset or overwritten by a greater value. The sign of the pressure is disregarded: it may be positive (compression) or negative (rarefaction). If p_m is the maximum pressure so defined, then the peak sound pressure level is given by

$$20 \log(p_m/p_0) \text{ dB SPL (peak)}$$

where unless otherwise specified, the reference sound pressure p_0 is equal to $20\,\mu\text{Pa}$. According to IEC 60050, the standard frequency weightings may apply to measured values of the peak level (in other words, the signal is frequency weighted before it arrives at the peak detector); if no weighting is specified, then A-weighting should be assumed. But one cannot rely on a universal adherence to this convention. When results are reported it is good practice to state explicitly how the measurements were made.

Peak-to-Peak Equivalent Sound Pressure Level

This measure is intended for use in audiometry, particularly in relation to evoked potentials. The following definition is based on that given in IEC 60645:

> *Peak-to-peak equivalent sound pressure level* The numerical value of the sound pressure level due to a long-duration sinusoidal signal which, when fed to the same transducer under the same conditions, produces the same peak-to-peak sound pressure as the short-duration signal. For clicks the long-duration signal should have a frequency of 1 kHz; for tone bursts it should have a frequency equal to the fundamental frequency of the tone.

The definition can be extended to include vibratory stimuli (measured as the force output of a bone vibrator). Frequency weighting is not implied. The name is often shortened to *peak equivalent sound pressure level* and written as 'dB peSPL'.

In this definition the expression *peak-to-peak* sound pressure means the maximum pressure range that encompasses the transient signal. 'Peak-to-valley' would be a more accurate (if unconventional) description. It is important to understand that the peak equivalent level refers a peak-to-peak range in the transient to the equivalent range in a sinusoid. The peak-to-peak range in a sinusoid is twice its amplitude and therefore $2\sqrt{2}$ times its rms value. If we remember that $20 \log \sqrt{2} = 3$ and $20 \log 2 = 6$, we can see that the peak sound pressure level in a transient lies between 3 and 9 dB above the peak equivalent level, depending on whether the transient is symmetrical about the baseline or whether it is entirely on one side of it. This is illustrated in Figure 6.31.

Sound-level meters do not show peak equivalent pressure levels directly, but if the ac output is taken to an oscilloscope, comparison can be made between the pressure waveform of the transient and the sinusoidal waveform generated by the long-duration tone. The peak-to-peak range in the oscilloscope display of the transient is noted. The transducer is then connected to a sinusoidal supply, which is adjusted until the same peak-to-peak range is displayed. The root mean square sound pressure of the tone as indicated by the sound-level meter is then numerically equal to the peak equivalent pressure of the transient. For further discussion, see Chapter 8.

Impulse Response

Some sound-level meters provide detectors that offer an impulse or 'I-characteristic'. This follows squaring and RC averaging similar to that used for the S and F responses, but the time constant is much smaller (35 ms). The maximum output of the averager is acquired by a peak detector. The

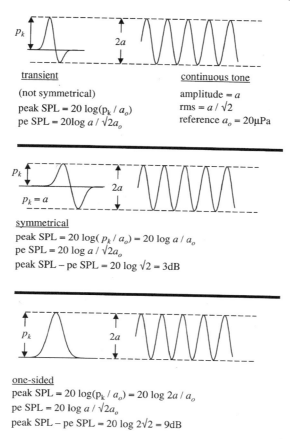

FIG. 6.31   Peak equivalent (pe) sound pressure level.

rise and fall times of the detector are designed to be unequal, so that its output follows closely that of the averager during the time that the averaged sound pressure is rising, but when the average is diminishing, the output falls slowly (2·9 dB/s) to give the user time to read the maximum value. It can be seen that the impulse response differs from the peak response described earlier because the maximum being measured corresponds to the greatest averaged sound pressure rather than the greatest instantaneous pressure. The averaging time (35 ms) is long compared with the duration of many impulsive sounds; for this reason, the impulse response increases with the duration of the impulse and with its acoustic energy. This may be an advantage in circumstances where energy has some significance. We therefore find that the impulse characteristic is used for noise assessment. It has also been used to express the level of single test items in speech audiometry.

Simple Frequency Analysis

It often happens that we want to know how the acoustic power in a sound field depends on frequency. Audiological applications include analysis of the masking noise produced by an audiometer and the measurement of background noise in test rooms used for audiometry. Another application is in tests that evaluate the listener's ability to hear speech against competing noise. Such tests are often used to assess the performance of hearing aids. The spectrum of the noise is a critical factor.

A full-spectrum analysis requires special and usually expensive equipment capable of narrow-band filtering or real time Fourier analysis. A relatively simple alternative is to divide the frequency scale into a series of bands using octave or third-octave filters, and to measure the sound pressure level in each band. The filters should ideally have a uniform response in their pass bands and an infinitely steep cut-off at the boundaries, and their bandwidths should form a contiguous series of frequency bands throughout the frequency range to be explored. In practice this ideal can be approached quite closely, and good-quality filter sets are usually available from manufacturers of sound-measuring equipment. Such filters are usually inserted between the input and output amplifiers of the sound-level meter, where they either replace or add to the standard frequency weightings. An obvious acoustical requirement is that the spectrum of the sound being investigated has to be reasonably constant during the time needed to complete the measurements.

Combining Measured Sound Levels

There are occasions when it is necessary to combine — to add or subtract — quantities related to acoustic power, given measured values expressed in decibels. To put it loosely, we want to 'add or subtract decibels'. We might, for example, have measured the sound pressure levels due to source A acting alone and source B acting alone and would like to know the level to be obtained if both sources are on at the same time. Again, we might, having obtained an octave band analysis of a sound, want to confirm that total sound level measured without filtering is equivalent to the sum of the measurements made with the octave band filters. As a further example, we might want to correct a measured sound level for the background noise that was present at the time of measurement.

The combination of sound levels requires three steps:

1. Convert the number of decibels in each measurement to a number proportional to the absolute power of the source.
2. Add (or subtract) the absolute values.
3. Convert the sum (or difference) of the absolute values to decibels.

Step 2 needs some justification, and there are times when it is not valid. We shall return to this later and deal first with the addition of sound power from sources that independently produce sound levels L_A and L_B. Let the corresponding sound pressures be p_A and p_B and let the reference pressure (20 μPa) be p_0. If both sources are on at the same time, the sound pressure is p_{AB}, for which the sound level is L_{AB}. The algebra that follows may seem rather messy because of the subscripts and exponents that are needed, but the end result is a very simple formula that is easily handled using a pocket calculator. We shall find that the reference pressure does not feature in the calculation, but it is included here to make the reasoning easier to understand.

By definition,

$$L_A = 10 \log \frac{p_A^2}{p_0^2} \qquad \text{and} \qquad L_B = 10 \log \frac{p_B^2}{p_0^2}$$

Therefore,

$$\frac{p_A^2}{p_0^2} = 10^{L_A/10} \qquad \text{and} \qquad \frac{p_B^2}{p_0^2} = 10^{L_B/10}$$

We shall assume that addition is on an energy basis; that is, the squares of the sound pressures may be added:

$$p_{AB}^2 = p_A^2 + p_B^2 \qquad \qquad 6.11$$

Therefore,

$$\frac{p_{AB}^2}{p_0^2} = \frac{p_A^2}{p_0^2} + \frac{p_B^2}{p_0^2}$$

$$= 10^{L_A/10} + 10^{L_B/10}$$

Therefore,

$$10 \log \frac{p_{AB}^2}{p_0^2} = 10 \log[10^{L_A/10} + 10^{L_B/10}]$$

so that

$$L_{AB} = 10 \log[10^{L_A/10} + 10^{L_B/10}] \qquad \qquad 6.12$$

The sound level L_{AB} is easily found using a pocket calculator because terms of the form 10^x can be evaluated using the inverse log function. Equation 6.12 can be extended to include the addition (or subtraction) of as many terms as are required. The following numerical example shows how this is done. Suppose that sources A, B, and C when acting alone produce the sound pressure levels 65, 70, and 72 dB SPL, respectively. What level should we expect when all sources are on at the same time? Substituting in Equation 6.12 gives

$$L_{ABC} = 10 \log[10^{6.5} + 10^{7.0} + 10^{7.2}]$$

$$= 74.6 \text{ dB SPL}$$

It will be noticed that each of the terms in the brackets is a large number. If the arithmetic is done with a calculator, these numbers can be added in memory before taking the logarithm. If it is done by hand, it is convenient to first take out a suitable factor, which in this example would be 10^6. Then

$$L_{ABC} = 10 \log[10^{6.5} + 10^{7.0} + 10^{7.2}]$$

$$= 10 \log\{10^6[10^{0.5} + 10 + 10^{1.2}]\}$$

$$= 10 \log 10^6 + 10 \log[10^{0.5} + 10 + 10^{1.2}]$$

$$= 60 + 10 \log[3.16 + 10 + 15.8]$$

$$= 74.6 \text{ dB SPL}$$

Correction for Background Noise

An important case involving the subtraction of sound energies is the correction of measurements for the presence of background noise. The noise may be present in the test environment or may be inherent within the measuring equipment. Two sound-level measurements are needed: one with the signal turned off and the other with the signal present. If the object of measurement (the signal) is the environmental sound itself, then the no-signal condition (with only the inherent noise in the equipment) can be obtained by covering the microphone with a noise-excluding cap or placing it in a relatively sound-free enclosure. Let the measured sound levels be L_{SN} when both signal and noise are present and L_N when only the noise is present. Then the corrected sound pressure level of the signal is L_S, where, from Equation 6.12,

$$L_S = 10 \log[10^{L_{SN}/10} - 10^{L_N/10}] \qquad 6.13$$

The correction given by Equation 6.13 can be evaluated using the arithmetic just described, but an alternative method is often suggested. This involves finding the value of a correction L_C that must be subtracted from L_{SN} to obtain the true level of the signal L_S. If D decibels is equal to the difference $L_{SN} - L_N$, then it is easily shown that the correction is given by

$$L_C = -10 \log(1 - 10^{-D/10}) \qquad\qquad 6.14$$

This leads to a positive value of L_C because the quantity in the brackets (the argument of the logarithm) is less than 1. The correction could be found by direct substitution into Equation 6.14, but the arithmetic is not simplified. The real value of this expression is that it can be represented graphically, allowing the correction term to be found without calculation (Figure 6.32). The final arithmetic is then merely the subtraction of one number from another. For example, suppose readings of 33 dB and 31 dB are obtained for the noise-plus-signal and noise-only conditions, respectively. The difference is 2 dB, and we see from the graph that the correction term is 4·3 dB. The true level of the signal is therefore $33 - 4·3 = 28·7$ dB.

It can be seen that when measurements are made in a noisy background, the estimates of the signal level become increasingly inaccurate as D becomes smaller; that is, a small variation or error in the measurement of D leads to a large variation in the correction factor. Ever more precise

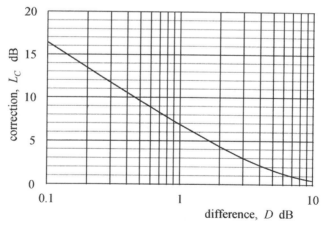

FIG. 6.32 Correcting sound-level measurements for background noise or inherent noise in the measuring equipment. $D = L_{SN} - L_N$ dB, the difference between readings of (signal + noise) and noise alone. Subtract the correction L_C from L_{SN} to find L_S.

measurements of L_{SN} and L_N are called for. If the noise and the signal are stable, the precision can be greatly improved by increasing the averaging time. The 'S' response or one with a longer time constant should be selected if it is available.

Combining Sound Levels: Statistical Independence

It was stated earlier that some justification is required for the assumption that sound levels can be combined by adding the squares of the absolute sound pressures (as in Equation 6.11, for example). The assumption is generally valid only if the vibrations to be combined are statistically independent. To see why this is necessary, suppose that a particular waveform is sampled to provide a set of values that taken in sequence represent the wave but are otherwise just a set of numbers of the sort that could be obtained when any physical quantity is measured repeatedly. The sampling can be imagined as a 'thought experiment', but it is a reality if the measuring equipment uses digital processing. The average of the sample is a number representing the average sound pressure during the sampling period. Now, sound pressure is a quantity that varies above and below atmospheric pressure, and, if atmospheric pressure, is disregarded, the mean of the sample must approach zero as the sample size is increased. The variance in the sample is defined as the mean square deviation from the sample mean, in this case the mean square deviation from zero. It is the mean square sound pressure during the sampling period. Imagine now that there are two variables, x and y, drawn from any two populations that may or may not have the same characteristics. Suppose that x_i and y_i are the values of randomly selected members from each of these populations. Adding x_i and y_i gives a new variable, $z_i = x_i + y_i$. Statistical theory tells us that if the samples of x and y are independent (meaning that the value of one is not influenced by the value of the other), then the variance of z is equal to the variance of x plus the variance of y.

The implication for acoustics is that sound levels can be combined by adding the mean squares of the corresponding sound pressures, providing that the sources of the sounds are independent. In many applications such independence can be virtually guaranteed. For example, the environmental noise in an audiometric test room is unrelated to the instrument noise within the sound-level meter. There are, however, occasions when independence is less certain. To give an instance of this, suppose we are attempting to measure the pure tone output of an audiometer at some level close to threshold. We find that the measured level falls by only a few dB when the tone switch is turned off, and we are tempted to make a correction for what appears to be background noise. But we might find on careful investigation that the noise observed with the tone turned off was just an attenuated version of the signal that had broken through the switching circuit. A correction using the methods just described would be invalid.

Measuring Subjective Qualities of Sound

We have scarcely mentioned the subjective attributes of sound, the sensations that it evokes, except in the reference to loudness earlier in this chapter. Anyone who is not a mystic would have to accept that all auditory sensations other than tinnitus are ultimately determined by the physical signal. Even the emotional quality that we call music has to be mediated by the physical events that precede the act of hearing. Yet despite the diversity of acoustic sensations, only two, pitch and loudness, have a simple relationship to the physical world. All others have a complicated and often barely understood connection with intensity and waveform, with the sound spectrum, and with the spatial and temporal distribution of physical occurrences. We still use the word *volume* to mean something akin to loudness, but not quite the same as loudness, yet how is such a sensation related to the concept of three-dimensional space or to the distribution of sound within such a space?

Loudness is the sensory concomitant of acoustic intensity, and pitch is likewise related to frequency, though neither is quite independent of the other. The pitch of a note depends on its loudness, and loudness is obviously related to frequency, if not to pitch itself. Because both intensity and frequency are measurable physical quantities, we can express the corresponding sensations in terms of their physical counterparts.

Loudness

We described earlier how the standard frequency weightings were originally based on loudness contours expressing loudness measured in *phons*. The official definition of this unit is as follows:

> *Loudness level*—of a sound, in phons, numerically equal to the median sound pressure level in decibels, re 20 μPa of a free progressive wave having a frequency of 1000 Hz presented to listeners having normal hearing facing the source that in a specified number of trials is judged equally as loud as the unknown sound (IEC 60050).

Loudness can also be expressed in *sones*. The sone scale is designed to give numbers that are approximately proportional to loudness. It is defined in terms of loudness level. If P is the number of phons and S is the number of sones, then

$$S = 2^{(P-40)/10}$$

or

$$P = 40 + \frac{10 \log_{10} S}{\log_{10} 2}$$

We see from these formulae that a loudness level of 40 phons corresponds to a loudness of 1 sone; doubling the loudness in sones increases its level by 10 phons. Methods are available for calculating loudness level from measurements of the sound spectrum in third-octave bands[11]. Methods similar to those for expressing loudness have been devised for expressing 'noisiness' in terms of the sound pressure level of a single band of noise or the pressure levels in series of third-octave bands.

Pitch

The pitch of a tone is almost entirely dependent on its frequency, but it is influenced to a small degree by intensity. The pitch of a complex tone depends on its spectral content or on the periodicity of the complex.[12] The unit of pitch is the *mel*, which is defined as follows:

> *Mel* Unit of pitch. A pure tone frontally presented, having a frequency of 1000 Hz and a sound pressure level of 40 dB, causes a pitch of 1000 mels. The pitch of a sound that is judged by the listener to be n times that of a 1000 mels tone is n thousand mels (IEC 60050).

Although judgements of pitch intervals (musical intervals) can be made with great precision, judgements of what might be called 'pitch extent' are considerably more variable, making the mel scale difficult to realize. In general it can be said that the pitch of a sound rises continuously with increasing frequency, but the relationship is nonlinear. Doubling the frequency of a sound produces a smaller relative change in pitch. For pure tones, the pitch in mels is said to be directly related to the position of

---

[11]Original articles by E. Zwicker, *Acoustica* 8: 237–258, 1958 and 10: 304–308, 1960. Also, ANSI S3.4-1980 (R1997), *Procedure for the computation of loudness of noise*. This is based on methods developed by S. S. Stevens.

[12]See, for example, B. Moore, *An Introduction to the Psychology of Hearing*, 4th ed. Academic Press, San Diego, 1997.

greatest vibration on the basilar membrane such that equal changes of pitch correspond to equal distances along the membrane.

Measuring Noise Exposure

Exposure to high sound levels for long periods damages the ear permanently. Although the harmful effects of noise, particularly in military and industrial contexts, have been known for a long time, it is only in the last 50 years that a quantitative relationship between noise exposure and hearing loss has been established. As might be expected, the relationship is highly variable from one individual to another, but statistical predictions of the hearing loss within a noise-exposed population are possible. There therefore exists a technical and statistical basis for setting limits on the allowable exposure to noise in the workplace.

If the noise is more or less continuous, as distinct from impulsive, its potential to harm the ear is related to what is called the *sound exposure*. In the restricted technical sense, this term describes an energy-like quantity based on the average squared sound pressure multiplied by the duration of exposure to noise. If the noise is impulsive, the hazard is more difficult to assess. Relevant characteristics include the peak level in the impulse, the energy, and the time during which a specified sound pressure is exceeded. No single descriptor appears to be capable of defining the likely consequences of exposure.

The protection of workers from the effects of noise has become an important element of occupational medicine and health and safety regulation. It has also led to considerable expansion in the manufacture of acoustic instruments, particularly those capable of measuring environmental noise and personal noise exposure (*noise dosimetry*). In parallel with this, there have been advances in methods for reducing noise at sources and in the working environment. Even when noise is not directly harmful to the ear, it can be a highly significant source of annoyance with psychological and social consequences that are far from trivial. There is now a large industry concerned with all aspects of noise control, noise measurement, and noise legislation.

Occupational health measures, designed to protect noise-exposed individuals, are almost universally based on the so-called *equal energy principle*. According to this, equal A-weighted sound exposures are equally noxious regardless of the particular exposure history, at least in the short term. A high-level sound present for a short time has the same potential to harm the ear as less intense sound present for a longer time, provided that the product of the squared sound pressure and the exposure time is the same in each case. It is, of course, highly unlikely that anyone will be exposed to noise at a constant level; it is therefore necessary to take into account the variation that occurs during the period for which the exposure is to be reckoned. To do this, it is necessary to divide the exposure time into indefinitely short intervals and add the sound exposures in each interval—an example of a process called *integration*. The sound exposure accumulated over a time T is therefore the area under a graph showing the square of the sound pressure plotted against time. As an example, Figure 6.33a shows a signal proportional to the square of the sound pressure in a sample of random noise. The shaded area is equal to the sound exposure E over the interval T. It has units of Pa^2s. In a practical instrument for measuring sound exposure (an integrating sound-level meter), it is likely that exposure would be derived from an average of the squared signal. A suitable method of averaging long-duration signals is to use the standard 'slow' response RC averaging described earlier. This provides a running average with a time constant of 1 second. When T is much greater than 1 second, the area under the averaged signal (Figure 6.33b) is almost the same as that under the raw signal.

It is usual to express sound exposure in decibels. The reference exposure E_0 is $(20 \ \mu Pa)^2 \times 1 \ s = 4 \times 10^{-10} \ Pa^2 \ s$, and the *sound exposure level L* is then given by

$$L = 10 \log(E/E_0)$$

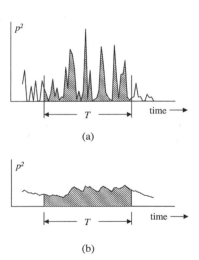

(a)

(b)

FIG. 6.33 Measuring sound exposure. (a) Raw signal proportional to the square of the sound pressure. (b) The waveform shown in the top part of the figure after RC averaging. The shaded areas, which are approximately the same in each case, are equal to the sound exposure E over the interval T.

The sound exposure level is a useful description of 'noise dose' for both long-duration exposures and single events. It has even been used to characterize impulsive sounds, notwithstanding that if the impulse is sufficiently brief and sufficiently intense, the equal energy principle is likely to be invalid. A-weighting is usually implied, but it can, if necessary, be made explicit by adding the subscript 'A', as in L_A. Other subscripts may be used to denote alternative frequency weightings or to distinguish sound exposure level from other decibel levels denoted by the same symbol. Unfortunately, the symbols used to describe exposure to noise are frequently decorated with numerous and complicated subscripts in an attempt to avoid duplication. Provided that the meaning is clearly stated according to context, it would be better to accept duplication as a reasonable concession to simplicity.

As an alternative to the reference exposure defined earlier, exposures are often expressed in terms of the noise energy present over an 8-hour working day or over a 24-hour period of which 8 hours are notionally taken to be potentially noisy. The duration for calculating E_0 is then the number of seconds in 8 hours, so that

$$E_0 = (20 \ \mu\text{Pa})^2 \times 8 \times 3600 \text{ s} = 1\cdot15 \times 10^{-5} \text{ Pa}^2 \text{ s}$$

With this reference, the exposure level is called the *noise exposure level*. For the same sound and duration, it is numerically 44·6 dB less than the *sound exposure level*. A-weighting is used. When the noise exposure level is used as a measure of an individual's exposure during a 24-hour period, the term *daily personal noise exposure level* is used. In principle, the noise is added over the 24 hours, but the 8-hour reference is applied. Symbols are $L_{p,d}$ or $L_{EX,8h}$.

It is often convenient to express noise exposure in terms of the sound pressure in a steady sound that would be associated with the same acoustic energy as that present in an actual exposure, in other words, an *equivalent* steady sound. If the shaded area in Figure 6.33 is divided by the duration *T*, the result is the average squared sound pressure in the interval. The average is equal to the squared sound pressure of the equivalent steady sound:

$$L_{eq} = 10 \log \left[\frac{\text{average squared sound pressure during time } T}{(20 \ \mu\text{Pa})^2} \right]$$

The interval is usually taken to be 8 hours; again, A-weighting is implied.

To give some idea of the significance of these measures, it may be said that a daily exposure of less than 80 dB is unlikely to be harmful. Exposures exceeding 85 dB, averaged over 8 hours, may be damaging; where they occur in the workplace, the employer may be required to take some action. In the United Kingdom, a personal 8-hour exposure of 85 dB is said to define 'the first action level', obliging the employer to assess the environmental noise, to identify the exposed personnel, and to advise them of the potential hazard.[13] Hearing protectors must be available on request. A 'second action level' is identified with an 8-hour exposure of 90 dB. When this is exceeded the employer must provide hearing protection. If the noise is impulsive, its harmful effects may not be adequately represented by an average exposure. In this case a 'peak action level' is defined for impulses in which the maximum sound pressure is 200 Pa. This has the same statutory significance as the 'second action level'.

The accumulated noise exposure during a lengthy period of work in a noisy occupation is sometimes expressed as a *noise immission level* (NIL). An important study undertaken in the United Kingdom in the 1960s demonstrated a statistical relationship between this measure of noise and the hearing loss acquired by employees.[14] The definition of noise immission level is similar to that of the noise exposure level except that the reference period is 1 year. It is assumed that the year comprises 1740 working hours and that noise exposure outside the workplace is negligible. The corresponding reference level is then $2 \cdot 51 \times 10^{-3}$ Pa^{2}s. Noise immission levels are usually calculated on the basis that the average A-weighted sound pressure level L_A in the workplace is constant. Accordingly, the immission level after employment for N years is given by

$$E_A = L_A + 10 \log N \text{ decibels}$$

Statistical distributions of hearing loss in an exposed population have been tabulated for values of E_A between 80 and 130 decibels.[15]

---

[13] *The Noise at Work Regulations.* Statutory Instrument 1989 Number 1790. The Stationery Office Ltd, London, 1989.

[14] *Immission* is not to be found in English-language dictionaries. The term was invented to denote this particular energy-related quantity. It is usually represented by the symbol E_A, where the subscript shows that A-weighting is required. The study referred to was undertaken by the Medical Research Council in collaboration with the National Physical Laboratory. From its findings, Douglas Robinson produced a simple mathematical model that showed how the noise immission level could be used to predict the distribution of noise-induced hearing loss in the noise-exposed population. See W. Burns and D. W. Robinson. *Hearing and Noise in Industry.* HMSO, London, 1970; W. Burns. *Noise and Man,* 2nd ed. John Murray, London, 1970.

[15] D. W. Robinson and M. S. Shipton. *Tables for the Estimation of Noise-Induced Hearing Loss.* National Physical Laboratory Acoustics Report Ac 61, 1977.

Questions and Exercises

Practical exercises are marked with an asterisk.

6.1 Explain the differences between *sound pressure level, hearing level,* and *sensation level.*

6.2 What is meant by the sensitivity of a microphone? At 1 kHz, a microphone produces an output of 0·20 mV for a sound pressure of 750 mPa, while at 2 kHz its output is 0·40 mV for a sound pressure of 1300 mPa. Calculate the sensitivities at these frequencies and express in decibels the frequency response at 2 kHz relative to the frequency response at 1 kHz.

6.3 Describe briefly the construction of a condenser microphone and the way in which the microphone works. When a positive sound pressure is applied to the microphone does its capacitance increase or decrease? Suppose that the polarizing voltage makes the backplate electrically positive relative to the diaphragm. In response to positive sound pressure, is the output voltage of the microphone then positive or negative? Justify your answers.

6.4 Why is it necessary to vent the capsule of a condenser microphone? What effect does this have on its frequency response?

6.5 The sensitivity of a condenser microphone can be increased by reducing the tension in the diaphragm. What are the disadvantages of doing this?

6.6 Why does a condenser microphone usually require a preamplifier?

6.7 What is *electret*? Why is the electret microphone particularly well suited to hearing aid applications?

6.8 What is *piezoelectricity*? Give an example of a commonplace nonacoustic application of the piezoelectric effect in the home.

6.9 What are the relative merits of *crystal* and *ceramic* microphones?

6.10 What is meant by a *cardioid* response? If a microphone has this response and its sensitivity to directly incident sound ($\theta = 0$) is 0·017 V/Pa, what is its sensitivity for sound incident (a) from one side ($\theta = 90°$) and (b) obliquely from behind ($\theta = 130°$)? Express the results in absolute terms and in decibels relative to the sensitivity for frontally incident sound.

6.11 Distinguish between the *free-field sensitivity* and the *pressure sensitivity* of a microphone. Why are they different?

6.12 What is a *pistonphone*? Why is this instrument suitable only for measurements at low frequencies? What are its particular merits?

6.13 Explain the principle of the *electrostatic actuator*. If a microphone is calibrated with an electrostatic actuator, what additional information is needed if it is to be used for (a) precision measurement of sound pressure in a closed cavity or (b) free-field measurements?

6.14 A microphone is calibrated with a pistonphone at 250 Hz. Its pressure response measured with an electrostatic actuator shows that its sensitivity is 1·7 dB greater at 6 kHz than it is at 250 Hz. When measuring the sound pressure produced by an earphone in a closed cavity, a level of 72·3 dB SPL is indicated at 6 kHz. What is the true sound pressure level in the cavity?

6.15 An actuator is operated with a dc bias of V_B and an alternating sinusoidal drive whose amplitude is V_0 volts. If F_1 and F_2 are the amplitudes of the electrostatic forces at the fundamental and second harmonic frequencies respectively, show that

$$\frac{F_2}{F_1} = \frac{V_0}{4V_B}$$

(A knowledge of elementary algebra and trigonometric functions is necessary. See Appendix A. Remember that the electrostatic force is proportional to the *square* of the applied voltage.) Use this formula (whether or not you have been able to derive it) to find the force level in the harmonic relative to that in the fundamental when the bias is 500 volts and the drive voltage is 25 volts rms (from the rms value, first find the amplitude V_0).

6.16 A condenser microphone has been calibrated using the insert-voltage method so that its open-circuit sensitivity M_0 is known accurately. If this sensitivity is used without correction as the basis of measurement, calculate the error in decibels if the preamplifier has an input capacitance of 0·5 pF and the microphone has a capacitance of 15 pF.

6.17* Attempt the photometric reciprocity experiment described in the text. It does work. The light source can be a 25-W lamp from a motor car (with suitable power supply), but a more powerful light might be needed unless you have a photometer that is able to operate at low light levels. Cover both tables with black cloth so that it will not matter if the photometer is unable to detect light from beneath it. Try the experiment acoustically. The source needs to be physically small so that its radiation is practically uniform. A small loudspeaker supplied with masking noise from an audiometer would be appropriate.

6.18 What is it about the reciprocity method that makes it particularly important as a method for calibrating microphones? Not all microphones can be calibrated by the reciprocity method. What special characteristic is essential?

6.19 List the principal components of a sound-level meter and draw a block diagram to show how they are combined.

6.20 Why is it necessary to square the electrical signal within a sound-level meter?

6.21 What is meant by *frequency weighting* in the context of sound-level measurements? Describe the features of the A-weighting function. Without weighting, a sound-level meter indicates 52·7 dB SPL when measuring the level of a narrow band of noise centred on 250 Hz. What reading is expected when A-weighting is included?

6.22* In this exercise we carry out arithmetically the operation of squaring and averaging a signal. A spreadsheet would be helpful but is not essential. We know that after amplification the signal voltage v in a sound-level meter is proportional to the sound pressure at the microphone. Suppose that the voltage varies with time t according to the equation

$$v = 2 \sin(3000t)$$

where v is in volts and t is in seconds. Choose a series of increasing values of t in the range $t = 1·5$ to $t = 2·5$ milliseconds. You need at least ten values, but more would be better. For each t, calculate v and v^2. Remember that the argument of the sine function is in radians. Make a table listing v and v^2 and t. Plot graphs of v and v^2 against t, drawing smooth lines through the data points. By 'counting squares', estimate the area between the graph of v^2 and the t-axis. The area should be expressed in units of volt^2 × seconds. Divide the area by the range of t (i.e., divide by 1×10^{-3} seconds). Notice that this gives the average value of v^2 over the chosen time interval. Compare your result with the mathematically correct value, which is 1·92 volt^2. For those who have a spreadsheet program, an extension or an alternative to this question is as follows. Create a column of values of t from 1.5 to 2.5 ms, in steps of s milliseconds. Tabulate v, v^2, and v^2s. Add all the numbers in the v^2s column except the last. Observe that the sum (divided by the range, 1 ms) tends to 1.92 volt^2 as the step size s is decreased.

6.23 During discharge through a resistance, the voltage V on a capacitor falls exponentially from its initial value V_0 according to the equation $V = V_0 e^{-t/\tau}$, where the time constant τ is equal to RC. If the initial voltage is 2·2 mV, calculate the voltage after a time equal to 3 time constants. If the time constant is equal to 125 ms, what time is required for the voltage to fall from 1·0 mV to 0·1 mV?

6.24 RC averaging is sometimes called 'exponential time-weighted averaging'. Why is this? If the IEC slow response is selected (time constant = 1 second), what weight is given to the squared sound pressure of an event that occurred 2·5 seconds ago compared with the weighting for events at the moment of observation?

6.25 What is loudness? Explain how loudness can be measured in phons. How is loudness on the sone scale related to loudness level in phons? Calculate the loudness in sones for a band of noise whose loudness level is (a) 40 phons and (b) 97 phons.

6.26 What loudness level in phons corresponds to a loudness of (a) 1 sone and (b) 20 sones?

6.27 What is the *equal energy principle* concerning the harmful consequences of exposure to noise?

6.28 Define *sound exposure level*. Calculate the A-weighted sound exposure level for a 30-second exposure, without protection, to noise at 103 dB(A). A task requires an employee to be exposed to this noise for 1 minute. Ear protectors are to be worn so that the sound exposure level at the ear will not exceed 100 dB. What is the least attenuation that the ear protectors must produce?

6.29 Explain what is meant by a *daily personal noise exposure level*. During an 8-hour shift an employee's task entails three exposures, each lasting 30 seconds, to noise at 112 dB(A), and a single 3-hour exposure to 92 dB(A). Calculate the personal noise exposure level for the day.

6.30 If a man works for 15 years in a factory where the noise level is 83 dB(A), express his noise exposure as a *noise immission level*. If the man subsequently works at another site where the noise level is 91 dB(A), what is the combined noise immission level for both periods of employment after 6 months in the new job?

7
Impedance

Impedance is a term and a concept widely used in the physical sciences, particularly in the electrical and electronic technologies, in mechanics, and in acoustics. Its special importance for audiology is in the description of middle ear and cochlear physiology and in the assessment of middle ear function. Those who have some knowledge of electronics will have little difficulty understanding the subject, but for many audiologists it is troublesome. The difficulty is probably that it involves mathematical ideas for which there are no obvious physical analogies. Attempts to explain impedance in terms of more familiar concepts such as 'flow of energy' are definitely misleading. The way to understand the subject is to approach it directly, accepting that it is somewhat technical and that it does require careful thinking. It is hoped that this chapter will provide a clear explanation of the ideas and terminology. The underlying principles will be demonstrated and the necessary mathematical ideas introduced as required.

Although it is *acoustic* impedance that is directly relevant to audiological practice, it should be understood that the same ideas are found almost identically in the theory of electrical circuits and in the description of mechanical systems (including the middle ear and cochlea!); indeed, electrical, mechanical, and acoustic impedance often appear together in the same problem, for example, in the design of electroacoustic transducers such as loudspeakers. Moreover, impedance has a sibling called *admittance*, and several cousins such as *resistance, reactance, inertance*, and so on; the entire family will be the subject of the following account. Before describing acoustic impedance, we will consider electrical impedance because this is the most widely known form and a general description is possible without the need to introduce an entirely new terminology. We will then consider mechanical impedance — the subject is important in its own right and it provides a good introduction to the acoustic case, particularly because mechanical systems are often easier to visualize than their acoustic counterparts. Finally, we will discuss how these ideas apply to acoustics, both as the specific acoustic impedance in an extended medium and as acoustic impedance within acoustic systems.

Resistance is a property of electrical conductors. It defines the relationship between the current carried by a conductor and the electrical potential difference (the voltage) across it. For most practical purposes the resistance of a given conductor is constant, that is, independent of the strength of the current. Its electrical effect is the same for both steady and varying currents. The Ohm's law relationship is

$$V = R \times I$$

If the current has a sinusoidal variation, then

$$I = I_p \sin \omega t$$

so that

$$V = RI_p \sin \omega t$$

The voltage waveform is therefore exactly the same as the current waveform (both are perfect sinusoids) and the voltage and current have *exactly the same phase*, so if an electrical circuit contains only resistors, we can find the voltage across its terminals simply by multiplying the current by the resistance of the circuit. The answer we obtain is the same for direct and alternating currents (Figure 7.1).

Now, although there are many occasions when such calculations are required, they exemplify a particular rather than a general case because most

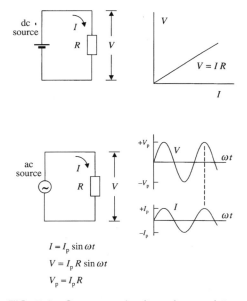

$$I = I_p \sin \omega t$$
$$V = I_p R \sin \omega t$$
$$V_p = I_p R$$

FIG. 7.1   Current and voltage in a resistor.

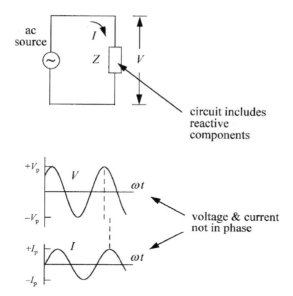

FIG. 7.2 Relationship between voltage and current for an arbitrary impedance, *Z*.

electrical circuits have components that are not simply resistances. Current and voltage are in phase only if the circuit is, or behaves as, a perfect resistance, but in general the phases differ (Figure 7.2). It is then impossible to calculate the voltage due to an alternating current, at any instant during the cycle, by a simple arithmetic multiplication. To understand this, consider that there are times in the current cycle when the current is momentarily zero but the voltage, which has a different phase, is not zero. Obviously, multiplying the current by a factor will not give the desired result.

The problem is solved by defining a property of the circuit called *impedance*. Impedance behaves rather like resistance — indeed, a component of it *is* resistance — but it is treated as a vector whose magnitude is the ratio of the peak voltage to the peak current and whose direction shows the relative phase of the voltage and the current. Fortunately, working with impedance is not the daunting prospect it might seem and we will show presently how it is done. There are, however, two conditions. The first is that we can deal only with sinusoidal variation (complex waveforms must be handled in terms of their component sinusoids), and the second is that systems must be linear. In the electrical case, for example, this means that if the current is a sinusoid then the voltage has also to be a sinusoid of the same frequency.

Impedance as a Vector

Suppose an alternating current is flowing in a circuit that is not simply a resistance. We observe that the voltage across the circuit is not in phase with the current. The task is to calculate the amplitude and phase of the voltage, knowing the current and the impedance of the circuit. Impedance has two parts, *resistance* and *reactance*, defined in such a way that when the alternating current is multiplied by resistance, we obtain an alternating voltage in phase with the current; and when it is multiplied by reactance, we obtain an alternating voltage whose phase differs from that of the current by one quarter of a cycle. Note that reactance behaves in the same way as resistance but additionally contains the 'instruction' for the quarter-cycle phase change. When the voltages derived from the resistive and reactive parts are added, the result is the alternating voltage that actually appears across the circuit.

To show how this process works, suppose that the current is given by

$$I = I_p \sin \omega t$$

where I_p is the peak current, that is, the amplitude of the current waveform. Let the resistive part of the impedance be R and the reactive part be X. Multiplying the current by the impedance produces the two voltages V_R and V_X where

$$V_R = I_p R \sin \omega t$$

and

$$V_X = I_p X \sin(\omega t + \pi/2) \quad \text{if } V_X \text{ leads the current}$$

or

$$V_X = I_p X \sin(\omega t - \pi/2) \quad \text{if } V_X \text{ lags the current}$$

It can be shown that when V_R and V_X are added, they produce the voltage

$$V = V_R + V_X = Z I_p \sin(\omega t + \phi)$$

where

$$Z = \sqrt{R^2 + X^2}$$

and

$$\tan \phi = \frac{X}{R}$$

or

$$\tan \phi = \frac{-X}{R}$$

The problem of finding V given I is therefore solved provided we know the components R and X of the impedance. The relationships just demonstrated suggest that we can represent these components as vectors and add them to obtain Z. This representation is shown in Figure 7.3, where, conventionally, resistance is drawn horizontally and reactance is drawn vertically upwards for a phase lead or downwards for a phase lag.

The *magnitude* of the vector impedance is the ratio of the peak voltage to the peak current, and therefore also the ratio of the root mean squares of these quantities. That is,

$$Z = \sqrt{R^2 + X^2} = \frac{V_p}{I_p} = \frac{V_{rms}}{I_{rms}}$$

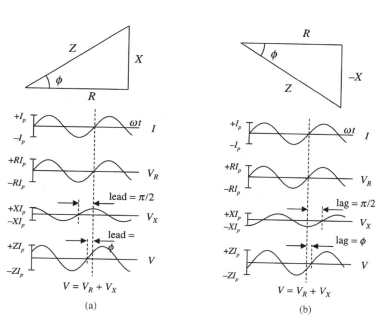

(a)   (b)

FIG. 7.3 (a) Vector representation of impedance. A current I flows through the circuit whose impedance is Z. The voltage across the circuit has a component V_R that is in phase with the current, and a voltage V_X that leads the current by a quarter cycle. The sum of these voltages is V. Its waveform leads that of the current by the angle ϕ. (b) Same as for part (a) except that the reactive component of the voltage lags the current by a quarter cycle. The voltage V across the circuit now lags the current by the angle ϕ.

Example

Suppose a circuit comprises a resistance of 400 ohms in series with an inductance that has a reactance of 300 ohms. To understand this example it is not necessary to know what is meant by inductance except that it is a circuit component that provides the stated reactance and that the voltage across it leads the current by $\pi/2$ (90°). If the rms current is 100 mA, what is the rms value of the voltage and what is the phase of the voltage relative to the current?

To answer this question we calculate the impedance Z and the phase angle ϕ either by scale drawing or by using trigonometric tables or an electronic calculator. The result is as follows:

$$Z = \sqrt{R^2 + X^2} = \sqrt{400^2 + 300^2} = 500 \text{ ohms}$$

and

$$\phi = \tan^{-1}\frac{300}{400} = 0.644 \text{ radians}$$

This angle is 36·9° or just over one-tenth of a cycle. The voltage across the circuit has therefore an rms value of $100 \times 500 = 50,000$ mV (i.e., 50 volts), and the voltage leads the current by 36·9° (Figure 7.4).

It is essential to understand the principle at work here. It is based on the fact that any sinusoidal vibration can be expressed as the sum of two other vibrations of the same frequency that differ in phase by $\pm\pi/2$. Multiplying the current by the impedance is an operation that goes beyond ordinary arithmetic because it produces, in effect, two such component vibrations. The process is exactly the same as that used in other applications, such as mechanics, to handle vector quantities. A single force, for example, can be replaced by two other forces acting in perpendicular directions. The importance of this is that if several forces act simultaneously, their components in the two directions can be added separately and then recombined to give the strength and direction of the resultant force. The same process is used to add impedances; that is, the resistive and reactive parts are added separately, and the combined impedance has the effect of the total resistance plus the total reactance.

Impedance as a Complex Number

It would be a great labour and a serious barrier to analysis if vector diagrams and trigonometric calculation were required for all terms involv-

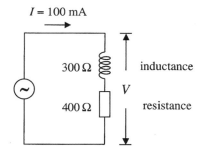

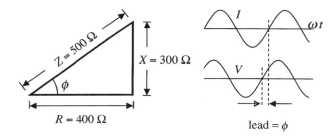

FIG. 7.4 Current and voltage relationships in the worked example.

ing impedance. It is one of the great joys of alternating-current theory (and acoustic theory likewise) to find that these difficulties are all but completely removed with the introduction of complex numbers. It is undoubtedly the role of complex numbers in solving practical problems in electronics that has made the teaching of this branch of mathematics an indispensable part of the technical curriculum. An introduction to the theory of complex numbers is given in Appendix A; if necessary, this should be consulted before proceeding further in this chapter.

Impedance can be represented by a complex number in which, by universal convention, the resistive part of the impedance is the real part of the number and the reactive part of impedance is the imaginary part of the number (Figure 7.5). Accordingly, this is written as

$$Z = R \pm jX \qquad\qquad 7.1$$

Reactance is multiplied by $+j$ for a phase lead and by $-j$ for a phase lag. Notice that Z is a complex number, whereas R and X are ordinary numbers. It will be immediately evident that the complex number expresses imped-

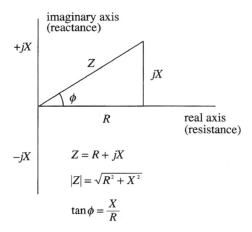

FIG. 7.5 Using complex numbers to represent impedance.

ance in the required vector form because the Argand diagram representing this number is identical to the vector diagram shown in Figure 7.3a. The advantage of the complex form is that Z can be treated as an ordinary algebraic variable. The value and simplicity of this will become apparent in the sections that follow.

The representation of Z as a complex number does, however, introduce a difficulty which needs some explanation. By definition, the ratio of the voltage V to the current I is equal to the impedance Z; that is,

$$\frac{V}{I} = Z$$

Now, if Z is a complex number, then V and I must also be complex; that is, they must have real and imaginary parts. But voltage and current are physical quantities and entirely real to anyone who has had to pay an electricity bill. We have to interpret the previous expressions as saying that V and I are vectors whose lengths are the real numbers that correspond to the amplitude of the alternating voltage across the terminals of the circuit and the amplitude of the current flowing in it. The ratio of the number of volts to the number of amperes is a real number equal to the magnitude of Z, and the phase of the voltage relative to that of the current is given by the direction of Z relative to the real axis. In some texts bold type is used whenever the variable is a complex number, so that voltage, current, sound pressure, particle velocity, and so on are printed in bold type. In other texts no distinction is made in the printed form, and the interpretation, which is usually obvious, must be inferred from the context. If it is necessary to distinguish a vector from its magnitude then magnitude signs can be used (for example, $|Z|$ is the magnitude of Z).

As already stated, X is a real number. It represents a physical quantity which we call reactance. When reactance is represented as part of a complex number, its full specification includes the factor $+j$ or $-j$ and we say that the imaginary part of the number is $+jX$ or $-jX$. It would be impossibly pedantic to make a verbal distinction between X and jX on all occasions, and the word *reactance* will be used to denote either the real number part or the vector as the need arises.

Finally, it is worth commenting that if a system has several components its reactance is likely to contain a number of separate terms, some preceded by $+j$ and some by $-j$. It is not unusual to find a $-j$ term described as a 'negative reactance'. Although not a hanging offence, this is arguably a misdemeanour. The reactance X is ordinarily positive, implying that a bigger current needs a bigger voltage to drive it regardless of relative phase or polarity. The converse is not impossible but is found only in unusual circumstances involving, for example, active components for which imped-ance principles are inapplicable. The point can be made by analogy. If a stone is thrown upwards its motion will be retarded by the downwardly directed gravitational acceleration. If the upward direction is (arbitrarily) taken to be positive, the gravitational force on the stone is $-mg$. It would be perverse to say in this case that the stone had a 'negative mass'.

Power Dissipation

Suppose that an electrical source of alternating current is connected to the terminals of a circuit that has reactive as well as resistive components. The impedance presented to the source at the terminals of the circuit is given by Equation 7.1, where R represents the combined effect of all parts of the circuit that contribute to the in-phase component of the current and X represents the combined effect of all parts that contribute to the component of current 'in quadrature', that is leading or lagging the voltage by a quarter cycle. When these components are added, the overall phase difference between current and voltage is ϕ radians, where

$$\tan \phi = \pm \frac{X}{R}$$

It is easily shown that the average power W supplied by the source in each cycle is given by

$$W = VI \cos \phi \qquad\qquad 7.2$$

where the voltage V and the current I are root mean square (rms) values. From Figure 7.3 we see that

$$\cos \phi = \frac{R}{\sqrt{R^2 + X^2}}$$

so that

$$W = VI \frac{R}{\sqrt{R^2 + X^2}}$$

and because $V/I = \sqrt{R^2 + X^2}$,

$$W = I^2 R = \frac{V^2 R}{R^2 + X^2}$$

The resistive part of the complex impedance is associated with energy dissipation, and the reactive part is associated with stored energy. If the impedance is entirely resistive, all the energy supplied by the source appears as heat generated in resistances within a circuit. At the other extreme, if the impedance is purely reactive, that is, if Z has only an imaginary part, then the phase angle is $\pm \pi/2$ and its cosine is zero. From Equation 7.2 we see that in these circumstances the power dissipation W is zero. Yet a voltmeter connected across the terminals of the circuit would register a finite rms voltage V, and an ammeter in series with the supply to the circuit would register a finite rms current I. This current is sometimes described as 'wattless'. In part of each cycle, energy supplied by the source is stored in reactive elements of the circuit, either through electric charge on capacitors or through magnetic fields around inductors; in other parts of the cycle, some of the stored energy is returned to the source. The same principle applies in mechanical and acoustic systems.

It should be understood that the resistance R and reactance X in the complex impedance are representative of the entire system to which that impedance relates. Although no energy is dissipated in the reactive components themselves, such components may contribute to the resistive part of the overall impedance. Similarly, the reactive part of the impedance may contain terms derived from individual resistances within the system even though resistances themselves are incapable of storing energy. This will be demonstrated later when a mechanical system is described.

Equivalent Circuits

In the remainder of this chapter we turn our attention to mechanical and acoustic systems. As already stated, the mechanical and acoustic elements of such systems have electrical counterparts that behave in exactly the same way so far as impedance is concerned. We will see that velocity is analogous to electric current, and mechanical force or acoustic pressure is analogous to the electrical potential difference across circuit components. It is therefore not surprising to find that mechanical and acoustic systems can be represented by their equivalent electrical circuits. This representation is nearly always helpful because electrical components can actually be joined together by pieces of wire to form a real circuit, whereas the corresponding mechanical or acoustic connections often have to be imagined. We will use equivalent circuits to assist the descriptions that follow. In doing so, elastic components will be shown by the symbol for a capacitor, mass components by the symbol for an inductor, and all forms of resistance by the usual electrical symbol. Inductance is analogous to mass, and capacitance is analogous to elastic compliance (the reciprocal of stiffness).

Mechanical Impedance

Mechanical impedance is used in the context of mechanical systems that undergo vibration. Just as electrical impedance is defined as the ratio of voltage to current, so mechanical impedance is the ratio of force to velocity. Mechanical impedance has a resistive component related to fluid friction and it has reactive components associated with mass and stiffness, which are the mechanical analogues of electrical inductance and capacitance. In understanding the origins of mechanical impedance it is necessary to remember that the velocity of any moving object is the rate of change of its position—that is, the rate of change of its 'displacement' or distance from a given position—and that its acceleration is the rate of change of velocity. For harmonic (sinusoidal) motion, the velocity leads the displacement by a quarter of a cycle and the acceleration leads the velocity by the same amount. The acceleration is therefore half a period ahead of the displacement, in antiphase with it. It must also be understood that when oscillations have an angular frequency ω, the velocity amplitude is ω times the displacement amplitude, and the acceleration amplitude is ω times the velocity amplitude.

Suppose that a vibrating object has an instantaneous velocity u given by

$$u = u_p \sin \omega t \qquad 7.3$$

The corresponding displacement ξ and acceleration a are therefore given by

$$\xi = \frac{u_p}{\omega} \sin(\omega t - \pi/2) \qquad 7.4$$

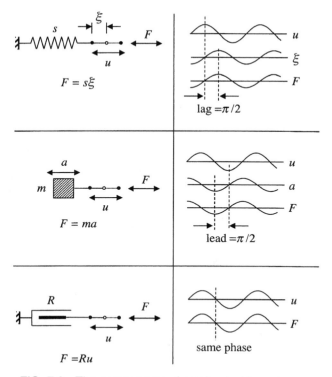

FIG. 7.6   The components of mechanical impedance.

and

$$a = \omega u_{\mathrm{p}} \sin(\omega t + \pi/2) \qquad\qquad 7.5$$

Imagine that the vibrating object is a spring whose mass is assumed to be negligible. We suppose that the spring is fixed at one end and that the vibration at the free end is maintained by a sinusoidal force F. This force must be equal to the stiffness s of the spring multiplied by the displacement ξ where the force is applied (Figure 7.6). The oscillatory force is therefore given by

$$F = \xi s = \frac{su_{\mathrm{p}}}{\omega} \sin(\omega t - \pi/2)$$

Comparing this with Equation 7.3, we see that the force is a sinusoid whose amplitude is s/ω times the amplitude of the velocity and whose phase lags the velocity by $\pi/2$. The mechanical impedance of the spring is therefore a pure reactance X_s which can be represented by the imaginary number

$$-jX_s = -js/\omega$$

At the other extreme, if the vibrating object is a mass m, then the force required to maintain its oscillation is the mass multiplied by the acceleration; that is,

$$F = ma = m\omega u_p \sin(\omega t + \pi/2)$$

Again, comparison with Equation 7.3 shows that the force waveform is a sinusoid whose amplitude is ωm times the velocity amplitude and whose phase leads the velocity by $\pi/2$. The mass therefore contributes the pure reactance

$$+jX_m = +j\omega m \tag{7.6}$$

The real part of the mechanical impedance is the *mechanical resistance*. It creates an oscillatory force in phase with the velocity and having an amplitude R times that of the velocity. The oscillatory force is therefore

$$F = Ru = Ru_p \sin \omega t$$

Mechanical resistance is often called 'friction', but the term needs to be used carefully. Friction is usually thought of as the force that opposes the sliding of one solid object over another, but in the definition of R it is fluid friction that matters. Solid friction is more or less independent of the relative speeds of the surfaces in contact, at least once some movement has commenced, but fluid friction generates a force that is directly proportional to velocity. We do in fact have everyday experience of the consequences of fluid friction: doors fitted with hydraulic dampers are easy to move slowly but much harder to move rapidly; a spoon in a jar of treacle falls slowly under just its own weight, but rapid movement requires great effort. Mechanical resistance is associated with friction in joints and bearings and with viscoelastic processes. It is also produced intentionally in devices such as hydraulic dampers. The symbol representing resistance in Figure 7.6 is supposed to suggest a fluid damper in which the movement of a central piston relative to a cylinder is opposed by the viscosity of the fluid that it contains. The fluid is not shown.

Units

Mechanical impedance and its components, resistance and reactance, express the ratio of force to velocity. The unit of measurement is therefore

Nsm^{-1}. This unit may be called a *mechanical ohm*, but the name is not widely used. Although the electrical ohm is unambiguously an MKS unit (one based on metres, kilograms, and seconds), the same is not true for mechanical and acoustic ohms because these names came into being at a time when the CGS (centimetre-gram-second) system was still in use. To prevent ambiguity, add the prefix 'MKS' or avoid the name altogether.

Forced Harmonic Oscillations (Series Configuration)

Chapter 2 described an elementary oscillator comprising a mass, a spring, and a damper. The response of this system to the action of a steady sinusoidal force can be calculated using the principles just described. The ease with which this is possible demonstrates the value of the 'impedance concept' and the advantage of using complex numbers.

Referring to Figure 7.7, we see that the oscillatory force F is applied at the connection between the free ends of the mass and the damper. It is opposed by the sum of the forces due to the inertia of the mass, the elastic

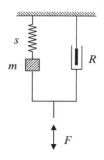

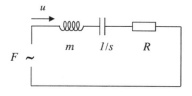

equivalent circuit

FIG. 7.7 Forced oscillation in a mechanical system, series configuration.

compression of the spring, and the fluid friction in the damper. The figure also shows the equivalent electrical circuit. Each of the mechanical elements undergoes an identical displacement and therefore moves with the same velocity. The oscillator is therefore the mechanical analogue of a series electrical circuit. The defining feature of a series circuit is that the same current flows in each component, and the voltage at the input terminals is the sum of the voltage differences across each of the elements. The total mechanical impedance[1] Z of the oscillator is the sum of the impedances of its components, so that

$$Z = R + j\omega m - js/\omega$$

Collecting the imaginary parts (the reactances of the mass and the spring) gives

$$Z = R + j(\omega m - s/\omega)$$

The magnitude of the impedance is therefore

$$|Z| = \sqrt{R^2 + (\omega m - s/\omega)^2}$$

and the phase angle ϕ is given by

$$\tan \phi = \frac{\omega m - s/\omega}{R}$$

The peak force in each cycle is $|Z|$ times the peak velocity, and the force leads the velocity by the angle ϕ.

The first thing to notice about this result is that the impedance changes with frequency, passing through a minimum value when $\omega m = s/\omega$ (Figure 7.8). The frequency at which the minimum occurs is the resonance frequency ω_0, so that

$$\omega_0 = \sqrt{\frac{s}{m}} \text{ radians per second}$$

or

$$f_0 = \frac{1}{2\pi} \sqrt{\frac{s}{m}} \text{ Hz}$$

Notice that the reactance vanishes at this frequency and that the oscillator then presents a pure resistance to the input force. Below the resonance frequency the phase angle is negative so that the force lags the velocity, whereas above resonance it leads the velocity. At low frequencies the acceleration and the velocity are low and the driving force is opposed

[1]To keep the notation simple, the same symbol Z will be used for all forms of impedance. Subscripts will be added where it is necessary to make a distinction.

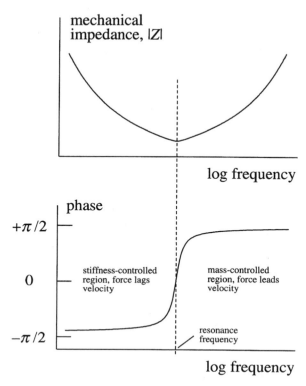

FIG. 7.8 Mechanical oscillator, series configuration: magnitude and phase of the mechanical impedance.

mainly by the action of the spring; at high frequencies the inertia of the mass is dominant. Going from a low to a high frequency, the system passes from springlike behaviour to masslike behaviour.

An important characteristic of the oscillator is its frequency response in the neighbourhood of the resonance frequency. Suppose that the amplitude F_p of the driving force is held constant while the frequency is varied. The change in velocity amplitude $F_p/|Z|$ with frequency is the velocity response of the oscillator, and resonance is revealed by a maximum in the response curve (Figure 7.9). The 'sharpness' of the peak is governed by the Q-factor, as described in Chapter 2. It is given by

$$Q = \frac{\omega_0 m}{R}$$

where $\omega_0 = \sqrt{s/m}$ is the angular frequency at resonance.

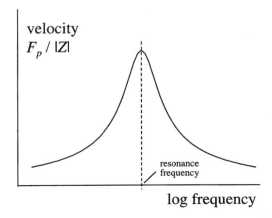

FIG. 7.9 Velocity response of a mechanical oscillator, series configuration.

To see how this formula is obtained, recall that Q can be defined as 2π times the mean stored energy in the oscillator divided by the energy that has to be supplied during each cycle to maintain the oscillation at resonance. The only components capable of storing energy are the spring and the mass that form the reactive part of the impedance. We have seen that the reactance vanishes at resonance and there is then no external force acting on the mass-spring combination other than the reaction at the fixed end of the spring. It follows that, overall, no energy is supplied or removed from the mass and the spring—they exchange energy with one another, but their total energy remains constant. At the moment when the velocity is greatest all the stored energy is kinetic, namely, $\frac{1}{2}mu_p^2$, where u_p is the velocity amplitude at resonance. The external force has only to overcome the friction in the damper. The average rate at which work is done on the damper is $(u_{\text{rms}})^2R$. [Compare this with the formula for the heat produced by an electric current in a resistor, $W = (I_{\text{rms}})^2R$.] The rms value of a sinusoidally varying quantity is equal to the amplitude divided by $\sqrt{2}$, so that in terms of the velocity amplitude, the power dissipation in the damper is $\frac{1}{2}u_p^2R$. The energy supplied during the cycle is the power multiplied by the period $2\pi/\omega_0$. Therefore,

$$Q = \frac{2\pi \times \text{stored energy}}{\text{energy supplied}} = \frac{2\pi \times \frac{1}{2}mu_p^2}{\frac{1}{2}u_p^2R \times 2\pi/\omega_0} = \frac{\omega_0 m}{R}$$

Forced Harmonic Oscillations (Parallel Configuration)

It is interesting to see what happens if the driving force is applied to the spring instead of the mass. The appropriate rearrangement of the mechanical system and the new equivalent circuit are shown in Figure 7.10. Although the mechanical parts seem to be in series, the equivalent circuit requires parallel connections. This is because the force applied to the free end of the spring is the same as the force exerted by the other end of the spring on the mass. This force is opposed by the inertia of the mass plus the friction in the damper. The mass and the damper have the same velocity, and this velocity added to that of the spring is the velocity at the input. It has to be understood that 'the velocity of the spring' means the rate at which one end of the spring moves relative to the other irrespective of any movement of the spring as a whole.

Let Z_1 be the impedance of the mass and damper and let Z_2 be the impedance of the spring. Then the impedance presented to the driving force is Z, where

$$\frac{1}{Z} = \frac{1}{Z_1} + \frac{1}{Z_2}$$

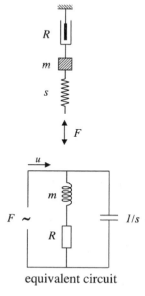

equivalent circuit

FIG. 7.10   Forced oscillation in a mechanical system, parallel configuration.

At this point we could, as before, write the impedances in terms of their real and imaginary parts and proceed to an expression for Z. We would soon discover that this required some long-winded algebra. It is tempting to look for an easier alternative. The easier way is to use *admittance* instead of impedance. Admittance is simply the reciprocal of impedance, that is, the ratio of velocity to force (expressed as a complex number). This quantity is usually represented by the symbol Y; the preceding expression then becomes

$$Y = Y_1 + Y_2$$

The simplification shows why parallel circuits are best dealt with in terms of admittance. Anticipating comments to be made later, it may be noticed that the term *otoadmittance* is sometimes used in connection with the acoustic measurement of middle ear function. At low frequencies the external ear and the eardrum are effectively in parallel.

To find Y, it is first necessary to express its constituents in terms of their real and imaginary parts. Accordingly,

$$Y_1 = \frac{1}{Z_1} = \frac{1}{R + j\omega m} \qquad \therefore Y_1 = \frac{R - j\omega m}{R^2 + \omega^2 m^2}$$

and

$$Y_2 = \frac{1}{Z_2} = \frac{1}{-js/\omega} \qquad \therefore Y_2 = j\omega/s$$

When Y_1 and Y_2 are added to obtain Y, the result, after some tidying up, is found to be

$$Y = \frac{R + j(\omega R^2/s + \omega^3 m^2/s - \omega m)}{R^2 + \omega^2 m^2}$$

The magnitude of the admittance is therefore

$$|Y| = \frac{\sqrt{R^2 + (\omega R^2/s + \omega^3 m^2/s - \omega m)^2}}{R^2 + \omega^2 m^2}$$

Notice that the real and imaginary parts of the complex admittance each contain terms in R and m; the mass contributes to the in-phase component of the admittance, and the resistance contributes to the quadrature component.

As the frequency of the driving force is increased, the admittance changes, passing through a minimum at a frequency very close to that for which the imaginary part vanishes (Figure 7.11). Because admittance is the ratio of velocity to force, the velocity will be a minimum at this frequency. If the amplitude of the driving force is kept constant, the response of the system

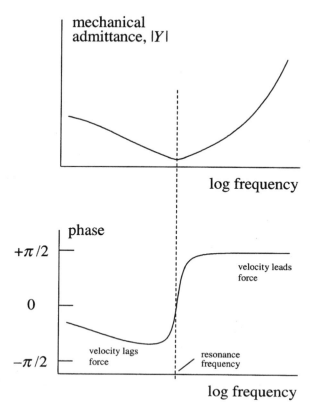

FIG. 7.11 Mechanical oscillator, parallel configuration: magnitude and phase of the mechanical admittance.

can be given in terms of velocity, and the existence of a minimum in the velocity is then an example of *antiresonance*. On the other hand, if the velocity amplitude at the free end of the spring is held constant, the force at this point becomes a measure of the response of the system. The force response passes through a maximum corresponding to a *resonance* (Figure 7.12). It can be shown that the Q-factor for this resonance is equal to $\omega_r m/R$, where ω_r is the resonance frequency whose value is given by

$$\omega_r = \sqrt{\frac{s}{m} - \frac{R^2}{m^2}}$$

We therefore see that in the parallel configuration the resonance frequency

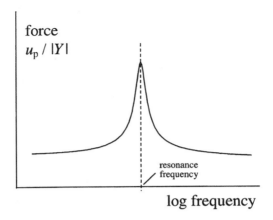

FIG. 7.12 Force response of a mechanical oscillator, parallel configuraton.

becomes slightly lower if the damping is increased, whereas in the series configuration it is independent of the damping.

Specific Acoustic Impedance

There are two kinds of acoustic impedance: the first, *specific acoustic impedance*, is helpful in understanding the propagation of sound in an extended medium and the reflection and transmission of sound as it passes from one medium to another; the second, known simply as *acoustic impedance*, is required when describing the radiation of sound from a vibrating surface and the propagation of sound in confining systems such as pipes and ducts. The latter include musical wind instruments, components attached to loudspeakers (horns, for example), acoustic attenuators in engine silencers, and, of course, the structures of the external ear. To begin the discussion we look briefly at specific impedance and give examples of its application.

Specific acoustic impedance is the ratio of sound pressure to particle velocity. It depends on the physical characteristics of the medium in which the sound is transmitted and on the type of wave. Expressing this symbolically, we write the specific impedance Z as

$$Z = \frac{p}{u}$$

where p and u are the sound pressure and particle velocity. As previously, we take Z to be a complex number with resistive and reactive parts R and X such that

$$Z = R \pm jX$$

It can be shown that the specific acoustic impedance in a plane progressive wave is given by

$$Z = \rho_0 c$$

where ρ_0 is the density of the medium and c is the velocity of sound. These are physical characteristics of a medium that depend entirely on its molecular composition and physical state (temperature and pressure); for this reason, the specific acoustic impedance for plane waves is also called the *characteristic impedance* of the medium. The characteristic impedance is an important acoustic property, especially in the context of sound transmission. In a plane wave, the sound pressure and the particle velocity have the same phase, and the specific acoustic impedance is therefore a pure acoustic resistance that depends only on the medium and not on the frequency of the sound. Resistance is associated with energy dissipation. In a sound wave, it relates to the energy needed to extend the sound field as the wave advances into the hitherto undisturbed medium. The rate at which this energy is supplied is proportional to the sound intensity, which, in a plane wave, is $(p_{rms})^2/\rho_0 c$, where p_{rms} is the root mean square sound pressure.[2]

If the wavefronts are not plane, then the specific acoustic impedance is not entirely resistive. For example, in spherical waves close to a simple source, the specific impedance is given by

$$Z = \frac{\rho_0 ckr(kr + j)}{1 + k^2 r^2} \qquad 7.7$$

where r is the distance from the source and $k = 2\pi/\lambda$. The reactive part of this impedance is preceded by $+j$, indicating that the sound pressure leads the particle velocity by the angle ϕ such that

$$\tan \phi = 1/kr \text{ radians}$$

Notice that if the distance from the source is increased to the point where $kr \gg 1$, then the impedance becomes resistive with the same value $\rho_0 c$ as exists for plane waves. The reactive component is, however, significant in regions close to the source. Its existence implies that the particle velocity can be thought of as having two components, one in phase with the sound pressure and one lagging the pressure by $\pi/2$. The kinetic energy associated

---

[2]The formula for acoustic intensity is analogous to the power dissipation V^2/R in an electrical resistor.

with the reactive part is not propagated with the wave and makes no contribution to the acoustic intensity.

Units of Measurement

Specific acoustic impedance has dimensions of pressure divided by velocity and can be measured in $Pa\,s\,m^{-1}$ or $N\,s\,m^{-3}$. Alternatively, we can express impedance in terms of the base units, mass $[M]$, length $[L]$, and time $[T]$, as follows:

$$[Z] = \frac{[\text{pressure}]}{[\text{velocity}]} = \frac{[\text{force/area}]}{[\text{velocity}]}$$

$$= \frac{[\text{mass} \times \text{acceleration/area}]}{[\text{velocity}]}$$

$$= \frac{[M] \times [LT^{-2}]/[L^2]}{[LT^{-1}]}$$

$$= [M][L^{-2}][T^{-1}]$$

In terms of base units, the unit of specific acoustic impedance is therefore $m^{-2}\,kg\,s^{-1}$. This unit is sometimes called a *rayl* or *MKS rayl* (after Lord Rayleigh).

Reflection and Transmission at a Boundary

In this context a *boundary* is an extended surface that is created when one substance is in contact with another. The surface of a lake, for example, is a boundary that separates the water from the air. Similarly, the surface of a window is a boundary between glass and air. When sound passes from one medium to another, there is an abrupt change (a discontinuity) in the specific acoustic impedance at the boundary. A general consequence of this is that sound is partially reflected and partially transmitted where the change of medium occurs.

   Suppose that a sound wave travelling in medium 1 is incident on a boundary with medium 2, as shown in Figure 7.13. The boundary is assumed to be perpendicular to the direction in which the waves are travelling. We can use subscripts 1 and 2 to distinguish properties of the two media, and subscripts i, r, and t to distinguish variables in the incident, reflected, and transmitted waves respectively. Now, although the impedance is not continuous across the boundary, quantities such as sound pressure

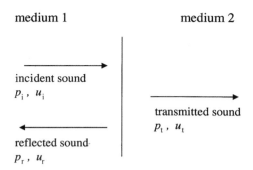

FIG. 7.13 Reflection and transmission of sound at a boundary between media of different specific impedance.

and particle velocity are continuous; that is, they do not undergo a sudden change. At points very close to the boundary, the sound pressure and particle velocity in medium 1 are the same as in medium 2. At first sight this may seem paradoxical and difficult to understand, but its truth is evident when one considers that an abrupt change in velocity would require infinite acceleration and the existence of infinite forces within the media. The sound pressure in the first medium is the sum of the pressures in the incident and the reflected waves (principle of superposition), and to preserve continuity this must equal the pressure in the transmitted wave, so that

$$p_t = p_i + p_r$$

By similar reasoning,

$$u_t = u_i + u_r$$

Also, from the definition of specific impedance,

$$Z_1 = p_i/u_i$$

$$Z_1 = -p_r/u_r$$

and

$$Z_2 = p_t/u_t$$

The minus sign in the second of these equations is needed because the reflected wave is propagated *towards* the source.

It only requires some simple algebra to show that these considerations lead to the following expressions for the sound pressures in the reflected and transmitted waves:

$$\frac{p_r}{p_i} = \frac{Z_2 - Z_1}{Z_2 + Z_1} \qquad\qquad 7.8$$

and

$$\frac{p_t}{p_i} = \frac{2Z_2}{Z_2 + Z_1} \qquad 7.9$$

These results show us immediately that if the specific impedances are equal, that is, if $Z_1 = Z_2$, then p_r is zero and $p_t = p_i$ so that the wave is transmitted from one medium to the other without reflection or change in amplitude. In all other circumstances, however, at least some reflection must occur. It is important once again to notice that although impedance is a vector represented by a complex number Z, it is treated as an ordinary algebraic quantity in the mathematics leading to Equations 7.8 and 7.9. The ratios of the sound pressures on the left-hand sides of these equations are also complex numbers having both real and imaginary parts. The physical meaning of this is that in general the reflection and transmission of sound involve a change in phase as well as amplitude.

The propagation of plane waves is, however, a special case for which a simpler result is obtained because the impedance in a plane wave is purely resistive. For a plane wave the complex impedance Z can be replaced by the characteristic impedance ρc, which is an ordinary number. The algebra will look tidier if we represent the ratio of the characteristic impedances in the two media by a single letter r, that is, if we let

$$r = \frac{\rho_2 c_2}{\rho_1 c_1}$$

With these substitutions in Equations 7.8 and 7.9 we find that

$$\frac{p_r}{p_i} = \frac{r - 1}{r + 1}$$

and

$$\frac{p_t}{p_i} = \frac{2r}{r + 1}$$

No phase change is involved except that if $r < 1$ the ratio p_r/p_i is negative. This means that if the impedance in the second medium is less than that in the first medium, the reflected wave is in antiphase with the incident wave. The transmitted wave, however, always has the same phase as the incident wave.

To know how much sound energy is reflected or transmitted, we need the sound intensities rather than the sound pressures. The ratio of the intensity in the reflected wave to that in the incident wave is called the *sound power reflection coefficient*, α_r. Similarly, the ratio of the intensity in the transmitted wave to that in the incident wave is the *sound power transmission coefficient*, α_t. For plane waves, as we have seen, sound intensity is equal to the square of the rms sound pressure divided by the characteristic impedance. The reflection and transmission coefficients can therefore be written as

$$\alpha_r = \frac{I_r}{I_i} = \frac{p_r^2}{p_i^2}$$

and

$$\alpha_t = \frac{I_t}{I_i} = \frac{p_t^2}{p_i^2} \times \frac{\rho_1 c_1}{\rho_2 c_2}$$

We can substitute ρc for Z in Equations 7.8 and 7.9 as before to obtain

$$\alpha_r = \frac{(r-1)^2}{(r+1)^2} \qquad\qquad 7.10$$

and

$$\alpha_t = \frac{4r}{(r+1)^2} \qquad\qquad 7.11$$

Because it has been tacitly assumed that no energy is absorbed at the boundary, the fraction of the incident energy that is reflected added to the fraction that is transmitted must equal unity, that is,

$$\alpha_r + \alpha_t = 1$$

Equations 7.10 and 7.11 are, of course, consistent with this requirement.

In the foregoing text we have seen how the reflection and transmission of sound at the interface between two media depend on the characteristic impedances. There are circumstances where it is necessary to reduce reflection (and enhance transmission) as far as possible. This aim can be achieved by matching the impedance values on either side of the boundary. Examples are the use of 'rho-c' rubber as a sealing compound in hydrophones and, in medical ultrasound, the application of oil to the patient's skin to form a nonreflecting layer where contact is made with the transducer.

Acoustic Impedance

Whereas specific acoustic impedance describes the relationship between sound pressure and the oscillatory movement of sound particles, ordinary acoustic impedance is concerned with the oscillatory flow of air at a surface. Acoustic impedance is therefore helpful in describing the interaction between sound fields and vibrating surfaces that are sources or receivers of sound. The tympanic membrane is a prime example, and measurement of acoustic impedance at the membrane is an important part of diagnostic audiology. Acoustic impedance is also useful in the treatment of sound transmission in closed systems. A simple cylindrical pipe is an example; for greater complexity, openings, side branches, and other attachments may be added. The chief example in audiology is the external ear, which can be thought of acoustically as a pipe of somewhat variable cross section terminating in the tympanic membrane and opening into an acoustically complex system involving the auricle, the head, and the torso. In general, the use of acoustic impedance is limited to circumstances where lateral dimensions (diameters of pipes) are small compared with the wavelength of sound, but there are no restrictions on the axial dimensions. Taking a pipe as an example, we would say that the sound is distributed along its length and that physical characteristics such as resistance, compliance, and acoustic mass (to be described presently) are likewise distributed. The transmission of sound in a 'distributed' system is then the acoustic equivalent of the transmission of high-frequency electric current in transmission lines and aerials. Systems in which the axial dimensions are short compared with the wavelength constitute a special case. A short pipe, for example, can be treated as a discrete acoustic element. Acoustic systems comprising such elements are equivalent to ordinary 'lumped' electrical circuits. Unfortunately for audiologists, the length of the external ear canal is such that it behaves as a discrete element only at low frequencies. This is a complication in middle ear measurement, which, accordingly, is usually undertaken at 226 or 660 Hz. It is also a problem in some hearing aid 'real-ear' measurement because higher frequencies cannot be ignored and the microphone tube must therefore be placed close to the eardrum.

Acoustic impedance involves a quantity called *volume velocity*. To understand what this is, imagine that sound is incident directly on the surface of an object whose vibrations are either caused by the sound (as in a microphone) or are the source of it (as in a loudspeaker or earphone). The particle velocity near to the surface must be the same as that of the surface itself. As the sound particles vibrate, their corporate movement is associated with an alternating flow of air perpendicular to the surface, so that at any instant the volume U flowing per second at the surface must be the product of the particle velocity u and the surface area S; that is,

$$U = uS$$

provided that u is uniform across the area.[3] The volume flow U is the

[3]In more general and rigorous definitions to be found in textbooks and standard glossaries, the particle velocity u is not assumed to be uniform or necessarily directed perpendicular to the surface. Instead, it is replaced by its component normal to each element of the surface, and the product of the normal component and elementary area are summed (integrated) to obtain the volume velocity over the specified surface.

volume velocity. It has the same sinusoidal variation (and of course the same phase) as the particle velocity. Volume velocity can also be defined with respect to an imaginary surface rather than the actual surface of some material object. To see this, suppose that a plane sinusoidal wave is propagated inside a pipe whose cross-sectional area is S. As the wave proceeds, the sound particles vibrate to and fro along the axis of the pipe to produce an alternating flow of air through any cross section in the path of the wave. The volume U flowing per second is then the particle velocity u multiplied by the area of the cross section.

The *acoustic impedance* Z at a specified surface can now be defined as the ratio of sound pressure to volume velocity; that is,

$$Z = \frac{p}{U}$$

The definition applies to the acoustic radiation at the interface between a vibrating surface and the surrounding air and to radiation transmitted through a specified section of a pipe.

To continue the description of acoustic impedance, it is necessary to define the terms that make up its real and imaginary parts and explain their origins. The real part of the acoustic impedance is called *acoustic resistance*. It is the ratio of the sound pressure to the component of the volume velocity that has the same phase as the pressure. For an example, suppose that sound is incident on the diaphragm of a microphone. To suppress resonances, the motion of the diaphragm will be damped, either electrically or by some internal mechanical or acoustic means. The damping is manifest as acoustic resistance at the surface of the diaphragm. If we look to the human ear for another example we will find that the acoustic impedance at the tympanic membrane is resistive over a substantial part of the auditory frequency range (Figure 7.14). This resistance has its origin in the cochlea, where the compliance of the basilar membrane interacts with the mass of the cochlear fluid to create a pressure at the oval window that is in phase with the velocity of the stapes. The stapes is therefore presented with a resistive impedance, and this resistance is transferred by the ossicular mechanism to the tympanic membrane, which appears as a resistance to the incoming sound.

Acoustic reactance is the ratio of the sound pressure to the component of volume velocity that leads or lags the pressure by a quarter cycle $(\pm \pi/2)$. When describing mechanical reactance, we considered the reactance of two basic elements, a spring and a mass. The reactance of the spring was shown to be s/ω (preceded by $-j$ in the complex number representation), and the

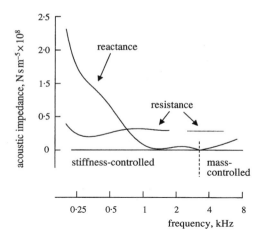

FIG. 7.14 Acoustic impedance measured at the eardrum. The solid lines show average values for men and women. The acoustic resistance is difficult to measure at frequencies above 2 kHz. Estimated values are shown by the dotted line. (Based on data from J. Zwislocki in *The Nervous System*, Vol. 3, ed. D. B. Tower. Raven Press, New York, 1975.)

reactance of the mass was shown to be ωm (preceded by $+j$). Let us now look for their acoustic counterparts, namely, *acoustic compliance* and *acoustic mass*.

Acoustic Compliance

In the mechanical case, we chose, arbitrarily, to express the elastic nature of a spring through its stiffness, s. Although acoustic stiffness exists, it is more usual to take its reciprocal, acoustic compliance, as the expression of elasticity in the acoustic sense. Mechanical compliance is the ratio of the displacement (at the end of a spring) to the applied force; correspondingly, acoustic compliance (C) is the ratio of the volume displacement to the sound pressure in a purely elastic system. The acoustic compliance (or just *compliance*) at the eardrum is a particularly important measure of middle ear function, and the term is widely used in clinical audiology. It is therefore important that its definition should be clearly understood.

To illustrate the meaning of compliance, imagine that a small tube is closed at one end by a piston and that the piston is connected to a

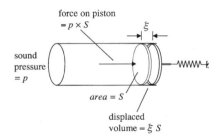

force on piston
$= p \times S$

sound
pressure
$= p$

$area = S$

displaced
volume $= \xi\, S$

FIG. 7.15   Diagram illustrating the meaning of acoustic compliance.

spring—an arrangement which has some similarity with the external ear (Figure 7.15). If a pressure p is applied to the piston, it will move outwards until the force pS due to the pressure is balanced by an equal force due to the elastic compression of the spring. If in the process the piston moves through a distance ξ, then the volume it displaces will be $S\xi$ and the acoustic compliance at its surface will be $S\xi/p$. In other words, the acoustic compliance of an acoustic element is defined as the volume displacement that is produced by the application of unit pressure.[4] There are other definitions that amount to the same thing. For example, acoustic compliance is the reciprocal of *acoustic stiffness*, which may be defined thus:

> In a system in which friction and inertia are negligible, quotient of sound pressure by the resulting in-phase volume displacement during sinusoidal motion.[5]

It is not meaningful to ascribe an acoustic reactance to a particular elastic element unless the motion is sinusoidal, and the requirement for sinusoidal motion is made explicit in the second of the definitions just presented. Acoustic compliance is analogous to mechanical compliance and is the exact analogue of electrical capacitance, acoustic pressure being analogous to voltage and volume displacement being analogous to electric charge. We usually expect the compliance of a spring or the capacitance of a condenser to be independent of the waveform of the applied force or voltage, and the same is true for acoustic compliance. In principle, then, the acoustic compliance of an element could be measured at any frequency or at no frequency at all, that is, as the volume change in response to the application of a steady pressure. One should be mindful, however, to choose the

[4]L. E. Kinsler and A. R. Frey. *Fundamentals of Acoustics*, 2nd ed. John Wiley & Sons, New York, 1962.
[5]IEC 60050-801 (1994), *International Electrotechnical Vocabulary—Chapter 801: Acoustics and electroacoustics.*

appropriate elastic constants when compliance is calculated: a greater compression of the air is produced by a steady pressure than by a rapidly alternating one because the former process is isothermal whereas the latter is likely to be adiabatic. As will be seen, the compliance of an air-filled cavity is usually calculated for the adiabatic rather than the isothermal condition.

To find an expression for the acoustic reactance of a compliant element, we use the same reasoning as we did when finding the mechanical reactance of a spring. We suppose that the element is in sinusoidal motion in response to the sound pressure. The particle velocity at the surface of the element is given by

$$u = u_p \sin \omega t$$

Correspondingly, the volume velocity is

$$U = U_p \sin \omega t$$

The particle displacement is

$$\xi = \frac{u_p}{\omega} \sin(\omega t - \pi/2)$$

as given previously in Equations 7.3 and 7.4.

From the definition of acoustic compliance C it follows that the sound pressure at the surface of the element is

$$p = \frac{\xi S}{C} = \frac{u_p S}{\omega C} \sin(\omega t - \pi/2)$$

But $u_p S$ is the amplitude of the volume velocity, so that

$$p = \frac{U_p}{\omega C} \sin(\omega t - \pi/2)$$

The sound pressure is therefore a sinusoid whose amplitude is $1/\omega C$ times that of the volume velocity and whose phase lags that of the volume velocity by $\pi/2$. The acoustic impedance of the element is therefore a pure reactance X_C which can be represented by the imaginary number

$$-jX_C = -j/\omega C$$

Acoustic Mass (Inertance)

The acoustic equivalent of mass or electrical inductance is called *acoustic mass* or *inertance*. It is the ratio of sound pressure to the rate of change of volume velocity (that is, volume acceleration). In other words, the acoustic mass M of an element is given by

$$M = \frac{\text{sound pressure}}{\text{volume acceleration}} = \frac{p}{aS} \qquad 7.12$$

where a is the particle acceleration at the surface S of the element. To gain some understanding of this concept, suppose that the spring in Figure 7.15 is replaced by a mass m so that the movement of the piston is now controlled by forces due to inertia rather than elasticity (Figure 7.16). The force required to produce an acceleration a is ma, and this force is equal to the sound pressure acting over the area S. The pressure is therefore equal to ma/S, so that

$$M = m/S^2 \qquad 7.13$$

To find the acoustic reactance of the element, we again start with the assumption that the element is in sinusoidal motion with velocity u and that this is also the velocity of the sound particles at the surface. Proceeding as before (Equations 7.3 and 7.5), we write the velocity and acceleration as

$$u = u_p \sin \omega t$$

and

$$a = \omega u_p \sin(\omega t + \pi/2)$$

But $u_p S$ is the amplitude U_p of the volume velocity, so that

$$a = \frac{\omega U_p}{S} \sin(\omega t + \pi/2)$$

From the definition of acoustic mass (Equation 7.12), the sound pressure is equal to MaS, so that

$$p = \omega M U_p \sin(\omega t + \pi/2)$$

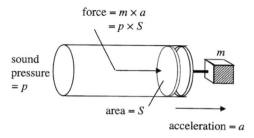

force = $m \times a$
   = $p \times S$

sound
pressure
= p

area = S

m

acceleration = a

FIG. 7.16   Acoustic mass.

We therefore see that the sound pressure is a sinusoid whose amplitude is ωM times that of the volume velocity and whose phase leads the velocity by a quarter of a cycle. The acoustic impedance of the element is therefore a pure reactance X_M which can be represented by the imaginary number

$$+jX_M = +j\omega M$$

Comparison of Acoustic and Mechanical Impedance

Mechanical impedance is the ratio of force to velocity, and acoustic impedance is the ratio of sound pressure to volume velocity. Both forms of impedance are helpful in solving problems concerning the production and reception of sound by mechanical systems. The mass element described in the previous section provides an example in which the driving force arising from sound pressure at the surface is opposed by mechanical inertia. If we substitute for M from Equation 7.13, we have

$$Z_M = jX_M = j\omega m/S^2 \quad \text{for the acoustic case}$$

and

$$Z_m = jXm = j\omega m \quad \text{for the mechanical case (Equation 7.6)}$$

So we find that the acoustic impedance of the element is equal to its mechanical impedance divided by the square of the surface area exposed to the sound field; that is,

$$Z_{\text{acoustic}} = Z_{\text{mechanical}}/S^2$$

This relationship is true generally, not just for the mass impedance. The squared term occurs because S is involved twice in the change from mechanical impedance to acoustic impedance — once in the change from force to pressure and again in the change from velocity to volume velocity. For example, the acoustic impedance presented to incoming sound at the eardrum is equal to the mechanical impedance of the middle ear divided by the square of the effective area of the tympanic membrane.

The mechanical impedance of a dynamic system such as a loudspeaker is made up of several components, including the interaction with the acoustic environment. That part of the mechanical impedance that is due to sound pressure is called the *radiation impedance*. It is equal to the acoustic impedance at the radiating surface multiplied by the square of the surface area.

Acoustic Compliance of an Air Volume

Suppose that air is trapped within a hard-walled container or cavity. We will allow a small opening for the passage of sound. The cavity can have any shape, but we stipulate that its dimensions in any direction must be small compared with the wavelength of sound. As a consequence of this, the sound pressure is the same at all points within the cavity. Seen through the opening, the enclosed air behaves as an almost perfect spring whose compliance is proportional to the contained volume. To calculate the acoustic compliance of the air volume, we recall that compliance is defined as the volume displacement brought about by the application of unit pressure. In context this may at first seem rather puzzling because the volume of the cavity is apparently constant, but it has to be understood that the compliance is required in circumstances where the cavity is connected to some other acoustic entity. The most readily visualized connection would be with a vibrating piston whose movement would produce a small oscillatory change in the enclosed volume. Alternatively, as in the Helmholtz resonator described later in this chapter, the connection is with a moving column of air that can be treated as a discrete element in much the same way as a mechanical piston. In middle ear measurement (tympanometry), the enclosure created by the external ear canal is similarly coupled to a moving column of air contained in the pipe that connects the ear to the sound source. The oscillatory changes in volume associated with such connections are of course exceedingly small compared with the volume of the cavity itself.

Referring to Figure 7.17, let V be the volume of air in the cavity and let P be the barometric pressure. For an adiabatic compression, the bulk elastic

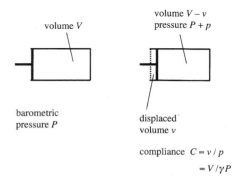

FIG. 7.17   Acoustic compliance of an air volume.

modulus κ of a gas is γP, where γ is the ratio of the principal specific heats (approximately 1·40 for air — see Chapter 1) and by definition

$$\kappa = \gamma P = V \times \frac{\text{increase in pressure}}{\text{decrease in volume}}$$

If the piston shown in the diagram moves to compress the air, the displaced volume can be regarded as a decrease in the volume of the cavity. It is accompanied by an increase in pressure. The ratio of the volume change to the pressure change is the acoustic compliance C of the air in the cavity, which is therefore given by

$$C = V/\gamma P$$

The velocity of sound is given by

$$c = \sqrt{\frac{\gamma P}{\rho_0}}$$

where ρ_0 is the density of air, so that we can write $\rho_0 c^2$ in place of γP to express the compliance in what are perhaps more familiar terms, namely,

$$C = V/\rho_0 c^2 \qquad\qquad 7.14$$

The acoustic impedance of an air volume is therefore the pure elastic reactance

$$-j\rho_0 c^2/\omega V$$

A small air cavity of known volume is a useful laboratory standard for calibrating impedance-measuring equipment, either in absolute units or directly as an *equivalent volume* as explained in the next subsection.

Units of Measurement

Acoustic impedance and its components — acoustic resistance and acoustic reactance — have dimensions of pressure divided by volume velocity. By analogy with electrical impedance, the former CGS unit corresponding to unit pressure divided by unit volume per second was called the *acoustic ohm*. Although some clinical instruments for measuring the impedance of the ear are calibrated in ohms, and much of the early data on middle aural impedance were expressed in this way, the use of the acoustic ohm is now in decline. The name *MKS ohm* can be used, but the standard MKS expression is $Pa\,s\,m^{-3}$ or $N\,s\,m^{-5}$. This is the same as $m^{-4}\,kg\,s^{-1}$ in base units. An acoustic impedance of $10^5\ N\,s\,m^{-5}$ is equal to 1 acoustic ohm (CGS).

Acoustic mass (inertance) can be expressed in $\mathrm{Pa\,s^2\,m^{-3}}$, $\mathrm{N\,s^2\,m^{-5}}$, or $\mathrm{kg^{-1}\,m^{-4}}$, and acoustic compliance has the units $\mathrm{m^3\,Pa^{-1}}$, $\mathrm{m\,N^{-1}}$, or $\mathrm{kg^{-1}\,s^2}$.

It is sometimes convenient to express acoustic impedance as an equivalent air volume. The equivalent volume is that volume of an air-filled cavity that has an impedance of the same magnitude as the actual impedance being described. If the impedance has magnitude $|Z|$, then its equivalent air volume V_e is given by

$$V_e = \rho_0 c^2 / \omega |Z|$$

At 226 Hz, an air volume of 1 cc has an acoustic impedance of 1×10^8 $\mathrm{N\,s\,m^{-5}}$. This is the same as 1×10^8 MKS acoustic ohms or 1×10^3 CGS acoustic ohms.[7]

Equivalent air volume is used widely in diagnostic audiology. It is useful for two reasons. The first is that the actual volume of a cavity is numerically identical to its impedance expressed as an equivalent volume. So expressed, the measured acoustic impedance of the external ear canal is the actual volume of the ear canal. The second useful feature is that the equivalence does not have to be restricted to acoustic impedance. It can apply, according to context, to its reciprocal (acoustic admittance) or even to acoustic compliance. Suppose a measurement X (of impedance, admittance, compliance, etc.) is made on an ear and the result displayed as an equivalent air volume. The same result would be recorded if the same measurement X were made on a cavity whose actual volume was equal to the equivalent volume.

The Helmholtz Resonator

The Helmholtz resonator is a passive oscillator comprising the simplest of acoustic elements, yet it can be finely tuned to provide a highly selective frequency response. Although it has long been superseded as an analytical tool, it provides a good illustration of the way in which some of the ideas described in this chapter can be applied to systems comprising discrete acoustic elements.

The resonator has a container of volume V with a small opening formed by a short tube or neck (Figure 7.18). Allowing for 'end effects', the neck has an effective length L and a cross-sectional area S. The dimensions of the tube

---

[6]Audiological instruments often indicate the reciprocal of impedance (admittance), and the CGS unit, the *mho*, is still in use. At 226 Hz, an air volume of 1 cc has an admittance of 1×10^{-3} mho (1 mmho [millimho]).

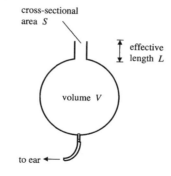

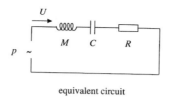

FIG. 7.18 The Helmholtz resonator.

and of the container are small compared with the wavelength of sound at the resonance frequency, so that (a) wave transmission in the neck can be ignored and (b) the sound pressure is practically uniform throughout the container. With these conditions, the air in the neck can be considered to oscillate as a single entity and we need only consider its mass. The contained volume of air acts as a spring and we need only consider its elastic behaviour. The mass of air in the neck is $\rho_0 LS$, and its acoustic inertance M is this mass divided by S^2; that is,

$$M = \rho_0 L/S$$

The acoustic compliance of the air volume is

$$C = V/\rho_0 c^2$$

The equivalent circuit of the resonator is the series arrangement shown in the figure. The airflow in the neck is equal to the flow in and out of the main vessel, and the pressure at the entrance to the neck is the sum of the acoustic pressure in the vessel and the acoustic pressure derived from the inertia of the air in the neck. The oscillator is damped by the resistance R, mainly because it radiates sound from the open end of the neck. At the angular frequency ω, the inertance M has an acoustic impedance $j\omega M$, and the compliance C has an impedance $-j/\omega C$. Sound from the external field arriving at the opening is presented with the combined impedance Z, where

$$Z = R + j\omega M - j/\omega C$$

so that

$$Z = R + j(\omega M - 1/\omega C)$$

The impedance is the acoustic equivalent of the mechanical impedance in the mass-spring-damper system described previously. It has the form $Z = R + jX$, where R is the resistance and X (which is equal to $\omega M - 1/\omega C$) is the reactance. As with the mechanical system, we find that the impedance is least at the frequency f_0 for which the reactive part is zero. Resonance therefore occurs at the angular frequency ω_0 for which $\omega_0 M = 1/\omega_0 C$. Remembering that $\omega = 2\pi f$, we find that the resonance frequency is

$$f_0 = \frac{1}{2\pi}\sqrt{\frac{1}{MC}}$$

Substituting for M and C gives

$$f_0 = \frac{c}{2\pi}\sqrt{\frac{S}{LV}}\ \text{Hz}$$

At resonance, the impedance as 'seen' from the external sound field is just the acoustic resistance R which is due principally to radiation of sound from the opening back into the field. Because, as stipulated, the diameter of the opening is small compared with the wavelength of sound at the resonance frequency, the resonator is a relatively feeble source. The resistance is small and the velocity amplitude of the vibration in the neck is typically several hundred times greater than it would be for the same acoustic pressure in a free plane wave. As a consequence, the sound pressure in the cavity is much greater than the pressure in the external field. To make practical use of the resonator, it is necessary to have some means of detecting the sound within it. One way is simply to listen to it through an ear tube attached to a small opening in the cavity. The pressure magnification enhances the audibility of weak sounds at the resonance frequency. It can be shown that the magnification is equal to the Q-factor $\omega_0 M/R$, which is typically[7] between 50 and 80. The resonator therefore has excellent gain and frequency selectivity, enabling sounds at the resonance frequency to be heard even in the presence of competing sounds at other frequencies.

[7]Data from Kinsler and Frey, *Fundamentals of Acoustics*, p. 194.

Acoustic Admittance

It has already been suggested that the behaviour of elements connected in parallel is more easily described in terms of their admittance than their impedance. A simple example will make this clear. The formula for calculating the combined resistance R of two resistors connected in parallel is

$$\frac{1}{R} = \frac{1}{R_1} + \frac{1}{R_2}$$

To obtain R we have to find the reciprocals of the component resistances, add the results, and then calculate the reciprocal of this sum. If, on the other hand, we are content to work with conductance G (the reciprocal of resistance), then the corresponding formula for the combined conductance of the two elements is simply

$$G = G_1 + G_2$$

Although this may seem a rather trivial advantage when only arithmetical operations are involved, it can lead to a useful simplification when dealing with complex numbers. Audiologists will be aware that middle ear measurements almost invariably involve admittance. As explained earlier, clinical measurements are made at low frequencies such that the sound pressure is virtually constant throughout the external ear canal and equal to the sound pressure at the eardrum. The volume velocity at the entrance to the ear canal is the sum of the velocity due to the elastic compliance of the air in the canal and the velocity due to the response of the eardrum. The ear canal and the middle ear are therefore acoustically in parallel even though the anatomical layout suggests a series arrangement.

Acoustic admittance is defined as the ratio of volume velocity to sound pressure. It can be treated in exactly the same way as impedance, using complex numbers to represent the in-phase and quadrature components. The real part G of the admittance is called *acoustic conductance*, and the imaginary part $\pm jB$ is called *acoustic susceptance*.[8] These ideas are expressed formally as follows:

$$Y = \frac{1}{Z}$$

$$|Y| = \frac{1}{|Z|}$$

---

[8]These terms are not often used except in the audiological context, and they are not mentioned in standard texts and glossaries. IEC 1027 (1991), *Instruments for the measurement of aural acoustic impedance/admittance*, defines acoustic conductance and acoustic susceptance as, respectively, the real and imaginary component of the complex acoustic admittance.

and

$$Y = \frac{U}{p} = G \pm jB$$

Notice here that if $+j$ is required in the specification of impedance, $-j$ is required in the specification of admittance, that is, if $Z = R + jX$ then $Y = G - jB$. Conversely, if $Z = R - jX$ then $Y = G + jB$.

To explore the relationship between acoustic impedance and admittance, imagine that a particular arrangement has an impedance Z and an admittance Y where, by definition, $Y = 1/Z$. Suppose for the sake of argument that the volume velocity has amplitude U_p and the sinusoidal variation

$$U = U_p \sin \omega t$$

The corresponding sound pressure is then

$$p = |Z| U_p \sin(\omega t + \phi)$$

Taking ϕ to be positive, we interpret this result as showing that the pressure leads the velocity by ϕ radians. The origin of the velocity-time curve is quite arbitrary, however, and we could equally well have stipulated that

$$p = p_0 \sin \omega t$$

where p_0 is the pressure amplitude. We would then have found the corresponding volume velocity to be

$$U = |Y| p_0 \sin(\omega t - \phi)$$

showing that the velocity lags the pressure by ϕ radians? This is exactly the same as saying that the pressure leads the velocity by the same angle. The point being made here is that the phase relationships for impedance and admittance are complementary: a pure mass element has reactance $+ j\omega M$ and susceptance $- j/\omega M$; a purely elastic element has reactance $-j/\omega C$ and susceptance $+ j\omega C$. Figure 7.19 provides further illustration.

When examining data from different sources, it may be necessary to compare resistance and reactance with conductance and susceptance. If an acoustic impedance is a pure resistance R, then the corresponding admit-

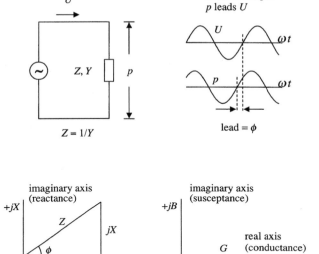

FIG. 7.19 Comparing acoustic impedance and admittance.

tance is $1/R$; if it is a pure reactance X, the admittance is $1/X$. In all other cases the relationship is slightly less simple, and the following formulae should be applied. Their derivation is given in Appendix A.

$$G = \frac{R}{R^2 + X^2} = \frac{R}{|Z|^2}$$

$$B = \frac{X}{R^2 + X^2} = \frac{X}{|Z|^2}$$

$$R = \frac{G}{G^2 + B^2} = \frac{G}{|Y|^2}$$

$$X = \frac{B}{G^2 + B^2} = \frac{B}{|Y|^2}$$

Clinical Measurement of Acoustic Admittance: Tympanometry

Measurement of the acoustic admittance of the ear is a well-established part of clinical practice. Measurement is, however, far from easy except when it is done at low frequencies; the clinician has usually to be content with results obtained at 226 Hz, or occasionally 660 Hz. Although measurements are of necessity made through the external ear, the object of interest is usually the middle ear. The only middle ear structure that is normally accessible is, of course, the tympanic membrane, and measurement of admittance of the middle ear is therefore measurement at the membrane, that is, *tympanometry*. The development of methods for this kind of measurement has a long history, and in the past it was usual to present the results as impedance rather than admittance. The equipment was called an impedance meter or impedance 'bridge' (by analogy with equipment used by electrical engineers). Not unreasonably, *impedance* has persisted as a generic term in middle ear measurement, although other names have been introduced.[9] The word *tympanometry* has also come to denote a class of measurements rather than a single specific one. It generally signifies the measurement of aural acoustic impedance or admittance, or the measurement of some related quantity such as acoustic compliance, as a function of the static pressure in the external ear. The graphical representation of the result is called a *tympanogram*. A qualifier can be added to create names such as *impedance tympanogram*, *admittance tympanogram*, *susceptance tympanogram*, and so on, but unless otherwise stated, we shall assume that the measured quantity is acoustic admittance.

The physical principles underlying tympanometry are as follows. The measuring equipment has a small probe that is inserted into the external ear canal. A soft rubber plug is used to make an airtight seal (Figure 7.20). Three fine-bore tubes pass through the plug to connect the ear to a microphone, a sound source (miniature earphone), and a pneumatic system that controls the static pressure in the ear canal. Although in the measurement process the volume velocity in the ear canal is *not* constant, the sound source can be described as a 'constant velocity source'. What this means is that the impedance of the source is relatively high, making it the dominant factor in determining the flow at the entrance to the ear canal. If, then, the amplitude of the voltage across the terminals of the source is held constant, the amplitude of the volume velocity in the sound entering the ear is virtually unaffected by any change in the ear's impedance. Within design

---

[9]*Immittance* has been suggested as a catch-all, but the term is deprecated (IEC 60050-801).

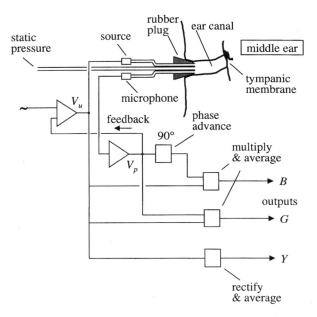

FIG. 7.20 Clinical measurement of the acoustic admittance of the ear. (Adapted from Figure 15 in *Otoadmittance Handbook 2*, Grason-Stadler, 1973, with permission from Grason-Stadler, Inc., to whom the copyright of the original diagram belongs.)

limits, a given voltage produces the same volume velocity regardless of the impedance of the ear or cavity to which the probe is fitted.

The volume velocity U at the entrance to the ear canal is therefore proportional to, and in phase with, the voltage at the source, or at least at some suitably chosen reference point in the circuit supplying the source. Similarly, a voltage derived from the output of the microphone is proportional to the sound pressure and in phase with it, and this voltage is used to supply a control circuit that adjusts the output of the earphone so as to keep the sound pressure p constant. By definition, the admittance Y at the probe is equal to U/p, so that if p is constant then U is proportional to Y. This means that the voltage at the earphone is also proportional to Y. Call this voltage V_u. If only the magnitude $|Y|$ of the admittance is required, all that is necessary is to rectify and average V_u and provide a display proportional to the average. The displayed output can then be made to show admittance directly after calibration with a standard cavity. If, however, the components G and B are required, some simple processing is needed. One method is as follows.

Let the sound pressure have the sinusoidal variation $p = p_0 \sin \omega t$. The volume velocity then has two components, one proportional to $G \sin \omega t$ and the other proportional to $B \sin(\omega t \pm \pi/2)$. The latter is more conveniently written as $\pm B \cos \omega t$. If the voltage V_u representing the volume velocity is multiplied by a voltage V_p representing the sound pressure (the voltage derived from the output of the microphone), the product will contain two terms, one proportional to $G \sin^2 \omega t$ and the other proportional to $B \cos \omega t \sin \omega t$. The average value of $\sin^2 \omega t$ taken over a whole number of cycles is $\frac{1}{2}$, whereas the average of $\cos \omega t \sin \omega t$ is zero. Averaging the product of the source and microphone voltages therefore gives a result proportional to the acoustic conductance. To obtain the susceptance, all that is needed is to advance the phase of the pressure-dependant voltage by one quarter of a cycle. This is easily done using a phase-shift circuit. Multiplication by V_u will then give terms proportional to $B \cos^2 \omega t$ and $G \sin \omega t \cos \omega t$ so that it is now the susceptance that remains after averaging.

As already stated, the ear canal can be treated as a 'discrete' acoustic element in parallel with the middle ear (Figure 7.21). If Y_c and Y_d are respectively the acoustic admittances of the ear canal and the eardrum (at

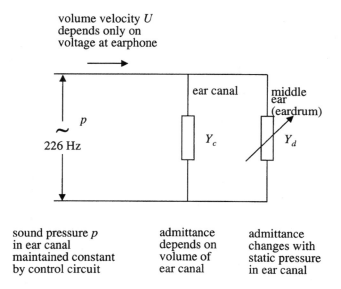

volume velocity U
depends only on
voltage at earphone

p

226 Hz

ear canal

middle ear
(eardrum)

Y_c

Y_d

sound pressure p
in ear canal
maintained constant
by control circuit

admittance
depends on
volume of
ear canal

admittance
changes with
static pressure
in ear canal

FIG. 7.21   The ear canal and middle ear as discrete acoustic elements.

the tympanic membrane), then the admittance presented to the probe is Y such that

$$Y = Y_c + Y_d$$

In tympanometry the static pressure in the external ear canal is changed progressively, usually from $+200$ daPa to -200 daPa or less, relative to the ambient atmospheric pressure. The pressure change is usually accompanied by a small change in Y_c and a very much greater change in Y_d. The change in canal admittance is usually negligible. It is caused by small changes in canal volume and changes in the density of the air. The admittance at the tympanic membrane is normally greatest when the pressure of the air in the external ear is equal to the pressure of the air in the middle ear space. If a pressure difference exists, the tympanic membrane is either forced inwards or drawn out, depending on whether the external ear pressure is greater or less than the middle ear pressure. The corresponding reduction in eardrum admittance is probably caused by an increase in the stiffness of the membrane, but the acoustic consequences for the middle ear are unknown except that the overall result is generally a slight impairment of the sound transmission. It is usually found that both the susceptance and the conductance of the tympanic membrane are lower when the membrane is stressed.

The result is that admittance and its components are all likely to have a maximum value when the pressures on the membrane are balanced. The location of this point on the pressure scale of a tympanogram is therefore a diagnostically useful indication of middle ear pressure (Figure 7.22).

The shape of a tympanogram can give useful diagnostic information. The simplest and most widely used interpretation is based on two crude but acceptable assumptions. These are that the low-frequency admittance of a normal middle ear is practically that of a pure compliance (Figure 7.14) and that the external ear is acoustically nothing more than an air-filled volume that also behaves as a pure compliance. The middle and external ears are acoustically in parallel and their admittances are directly additive. Because each has the same phase, the addition can be made arithmetically. All that is necessary is to subtract the admittance at a pressure of, say, $+200$ daPa from peak admittance (at middle ear pressure). The result is the admittance at the eardrum. It will usually be expressed as an 'admittance' or a 'compliance' in equivalent volume units. An abnormally low or high admittance or the absence of a peak in the tympanogram is an indication of middle ear pathology (Table 7.1).

A more refined analysis is possible if the components of the admittance are available separately. The measured admittance at $+200$ daPa is, hypothetically, the admittance of the external ear canal itself on the assumption that the application of a static pressure makes the admittance

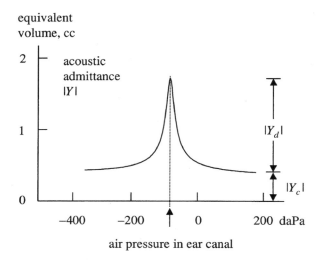

FIG. 7.22 Admittance tympanogram showing slightly low middle ear pressure (arrow).

at the tympanum negligible. The admittance of the ear canal is then

$$Y_c = G_c + jB_c$$

If the ear canal between the probe and the tympanic membrane behaves as the simple compliance of an air volume V, we expect the conductance G_c to be negligible and the susceptance B_c to be equal to $\omega V/\rho_0 c^2$. This, then, is the susceptance measured at $+200$ daPa or at some other static pressure for which the eardrum admittance is effectively zero. Measurements at any other pressure then yield the admittance Y_d at the eardrum once the canal susceptance has been subtracted:

$$Y_d = G + j(B - B_c)$$

so that the magnitude of the eardrum admittance is

$$|Y_d| = \sqrt{G^2 + (B - B_c)^2}$$

The volume velocity leads the sound pressure by the angle ϕ radians, where

$$\tan \phi = \frac{B - B_c}{G}$$

Tympanograms showing the admittance at the tympanic membrane after allowance for the volume of the ear canal are said to be *meatus compensated*.

TABLE 7.1 Tympanometric Findings Associated with Middle Ear Abnormalities

Disorder	Tympanometric indication	Explanation
Eustachian tube dysfunction	Normal admittance; low middle ear pressure (tympanogram displaced to left)	Selective absorption of oxygen in middle ear space
Fluid (liquid) in the middle ear	Low admittance or no peak; low middle ear pressure	Air in middle ear replaced by relatively incompressible and dense liquid; middle ear compliance much reduced by loss of air volume in middle ear space; possible mass loading of tympanic membrane
Middle ear scarring/adhesions	Low admittance or no peak; middle ear pressure normal	Middle ear appears as a stiff and dense termination of the external ear canal
Healed perforations	Admittance may be high; peaks in tympanogram sharply defined; middle ear pressure normal	Flaccid areas in tympanic membrane offer little acoustic impedance except when static pressure is applied; middle ear admittance approximately that of the air in the middle ear space without the normal loading of the ossicles and other structures
Perforation	High admittance, unchanged by static pressure ('flat' tympanogram)	Middle ear admittance equal to that of the air volume of the middle ear space 'seen' through the perforation, static pressures are identical on both sides of the membrane
Otosclerosis	Admittance often lower than normal; middle ear pressure normal	Fixation of the stapes increases the stiffness of the middle ear
Ossicular discontinuity (rare)	Similar to healed perforation; very high admittance	Mechanical loading of tympanic membrane reduced; admittance is mainly that of the air volume of the middle ear space

Tympanograms are described as *uncompensated* if no such allowance has been made.

The assumption has been that the measured conductance G_c at the probe is negligible when the tympanic membrane is sufficiently stressed by a static pressure. If G_c is not negligible, then either the ear canal is not behaving as a simple cavity (a pure compliance) or the residual conductance of the middle ear is significant. The results are then difficult to interpret.

Distributed Acoustic Impedance

When the wavelength of sound is not so long that elements can be treated as discrete, the essential wave nature of sound has to be recognized when considering its transmission in confining systems. As previously stated, the treatment of radio-frequency electric currents in transmission lines provides a direct analogy. The physics is considerably more challenging than for the discrete systems, and only some very general ideas will be presented here. The relevance for audiology is that even though formal analysis is rarely attempted, it is necessary to be aware of the wave nature of sound when considering its transmission in the external ear and in hearing aid components at frequencies above about 2 kHz.

Suppose sound is transmitted in a circular pipe whose cross-sectional area is S. This area is numerically equal to the volume of unit length of the pipe so that, by Equations 7.13 and 7.14, the inertance M_L and compliance C_L per unit length of the pipe are given by

$$M_L = \rho_0/S$$

and

$$C_L = S/\rho_0 c^2$$

Unless the pipe has a very small diameter, the acoustic resistance (per unit length) associated with viscosity and heat exchange through the walls is negligible. Nevertheless, some losses must occur in transmission, so that if the pipe is very long only the transmitted wave will be observed, any wave reflected from the far end having died out. It can be shown that in these circumstances the acoustic impedance at any cross section of the pipe is given by

$$Z = \sqrt{\frac{M_L}{C_L}} = \frac{\rho_0 c}{S}$$

This is simply the characteristic impedance divided by the area of the pipe. The pipe therefore behaves as a perfect acoustic resistance, and the ratio of sound pressure to particle velocity is the same as that found in a plane progressive wave. This result is to be expected because, by definition, a plane wave is not divergent and the confinement offered by an open pipe running perpendicular to the wavefront has no influence on the motion of the air except at the boundaries (assuming the wall of the pipe to be extremely thin). Boundary effects have been discounted because viscosity and heat exchange have been assumed to be negligible.

Suppose, however, that the pipe has a finite length and sound is introduced at one end. Let the other end of the pipe be terminated by some impedance that differs from the characteristic impedance. Sound will be reflected at the termination, and interference will occur between the incident and reflected waves. The impedance of the pipe will then vary from one place to another along the pipe. It will depend on the length of the pipe, the frequency of the sound, and the nature of the terminating impedance. A general treatment will not be attempted here, but one special case will be mentioned. This is where the pipe is closed by a solid cap so that the terminating impedance is infinite. The impedance Z_o at the open end of the pipe is then found to be

$$Z_o = -j \frac{\rho_0 c}{S} \cot kl$$

where l is the length of the pipe and $k = 2\pi/\lambda$. The cotangent (cot kl) is defined as $1/\tan kl$; that is,

$$\cot kl = \cos kl/\sin kl$$

We see, therefore, that the impedance varies between $+j \times \infty$ and $-j \times \infty$ depending on the wavelength and the length of the pipe (Figure 7.23). The pipe therefore presents a pure reactance, either of the mass type or the elastic type. For certain wavelengths, the reactance is zero and the corresponding frequencies are the resonance frequencies for the closed pipe. They occur when $\cos kl$ is zero, that is, when kl is an odd multiple of $\pi/2$. Accordingly,

$$kl = (2n - 1)\pi/2$$

where $n = 1, 2, 3, \ldots$ for resonance. But $f = ck/2\pi$, so that the resonance frequencies are the series of odd harmonics

$$f_1 = c/4l, \; f_2 = 3c/4l, \; f_3 = 5c/4l \ldots$$

previously described in Chapter 3.

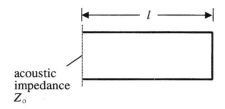

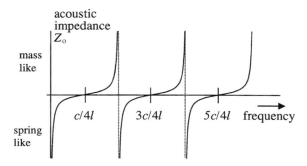

FIG. 7.23   The acoustic impedance at the entrance to a closed pipe.

Acoustic Impedance of the Human Ear

A knowledge of the impedance of the ear is important in the specification and evaluation of the performance of audiometric earphones and hearing aids. Ear simulators have been developed that present an earphone or hearing aid with the same acoustic load as the average human ear and which produce a sound pressure at the microphone corresponding to that at the tympanic membrane. Further information about ear simulators can be found in Chapter 8.

Audiometric Earphones

Figure 7.24 shows the magnitude and phase of the acoustic impedance of the ear as viewed through the aperture of an earcap (earphone cushion). The data are taken from the international standard[10] specifying the perfomance of an artificial ear to be used for calibrating supra-aural earphones. The

[10]IEC 60318-1 (1998), *Ear simulator for the calibration of supra-aural earphones.*

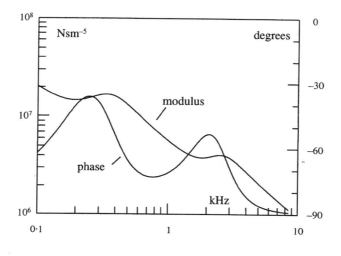

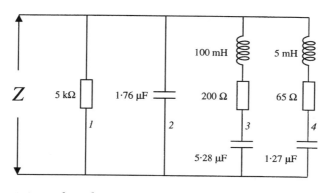

1 ohm ≡ 10^5 Nsm^{-5}

FIG. 7.24 Acoustic impedance of the ear under the cushion of an earphone.

equivalent electrical circuit has four parallel branches. The first of these is a high resistance (not less than 5 kΩ) to represent the acoustic effect of a small leak needed to remove any excess static pressure at the microphone. The second branch represents the volume of air between the ear cushion and the pinna. The third branch influences the observed impedance below about 500 Hz. Its effect is to reduce the acoustic reactance at low frequencies. The physiological basis for the low reactance observed in real ears is not known for certain, but the phenomenon remains even when the meatus is plugged.

The compliance of the pinna may be a factor because the reactance depends on the force between the earcap and the ear. The fourth branch represents the influence of the ear canal and middle ear. The impedance simulated by this branch is not needed if the ear canal is blocked.

Hearing Aids

Hearing aids are usually coupled to the ear by an earmould which occupies about half the length of the ear canal. The distance between the tip of the earmould and the tympanic membrane is then about 14 mm. The acoustic impedance of this short part of the ear canal, together with the termination provided by the middle ear, has an important influence on the performance of hearing aids. Most hearing aid earphones behave as high-impedance sources insofar as the deflection of the diaphragm is virtually unaffected by the acoustic load normally encountered. This means that the sound pressure at the tip of the earmould will be proportional to the acoustic impedance presented to it. At any given frequency the sound pressure at the tympanic membrane is proportional to the sound pressure at the earmould, so it is desirable that the impedance of the ear is correctly replicated. An appropriate acoustic load is provided by the standard occluded-ear simulator.[11]

The impedance of the simulator, which is close to that of the average human ear (suitably occluded) is illustrated in a simplified way in Figure 7.25. Impedance is represented as if measured at a reference plane in the ear canal, 14 mm from the tympanic membrane. The salient features are as follows:

(A) The impedance is that of the air volume between the earmould and the tympanic membrane, together with the stiffness-controlled impedance of the eardrum (as described earlier concerning tympanometry).

(B) In this region the reactive part of the eardrum impedance changes from stiffness controlled to mass controlled, with a middle ear resonance at about 1300 Hz.

(C) Here, and at higher frequencies, the impedance at the reference plane is virtually independent of the middle ear. It is the impedance of a closed tube.

(D) The minimum at about 7 kHz is produced by resonance in which the length of the ear canal is a quarter of a wavelength. There is a velocity node at the eardrum and a velocity antinode at the end of the earmould (reference plane). The velocity is maintained by the output

[11]IEC 60711 (1981), *Occluded-ear simulator for the measurement of earphones coupled to the ear by ear inserts.*

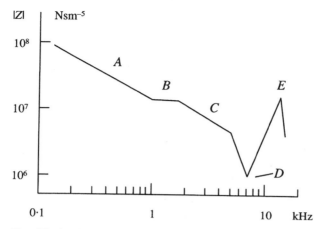

FIG. 7.25 Simplified representation of the acoustic impedance in an occluded ear. (Adapted from Figure 12 in *Impedance of Real and Artificial Ears*. Brüel & Kjaer.)

from the hearing aid. There is a pressure antinode at the eardrum and a pressure node at the earmould. The low sound pressure and high velocity at the earmould produce a low impedance at this point.

(E) The maximum at about 14 kHz (a frequency above the normal working range) corresponds to resonance in which the length of the ear canal is half a wavelength. There are velocity nodes at the eardrum and at the end of the earmould. There are corresponding pressure antinodes at these points so that the sound pressure is high at the earmould. The high pressure and low velocity give a high impedance at this point.

Questions and Exercises

7.1 A steady current of 15·9 mA is flowing in a pure resistance of 7·1 ohms. What is the potential difference across the resistor? How much power is dissipated in it?

7.2 What current will flow through a 22-kΩ resistor when it is connected across a 6-V supply?

7.3 Three 10-ohm resistors are connected in series with a 12-V battery. What current will flow? How much heat is generated in each resistor?

7.4 Two 133-kΩ resistors are connected in parallel across a 48-V supply. What current is drawn from the supply? How much heat is generated in each resistor?

7.5 A 14·3-kΩ resistance is to be made from the parallel combination of a 15·0-kΩ resistance and a resistance R. What, to the nearest 1000 ohms, is the required value of R?

7.6 A 1500-ohm resistor is carrying a sinusoidal current whose root mean square value is 9·1 mA. What is the peak voltage across the resistor?

7.7 A sinusoidal voltage has a frequency of 330 Hz. What is the corresponding angular frequency?

7.8 Express the following angles in degrees: (a) π, (b) $\pi/2$, (c) $3\pi/4$, (d) $-\pi/2$, (e) $\pi/3$.

7.9 What is meant by *reactance*? What can be said about the phase relationship between the current and the voltage in a reactive part of a circuit?

7.10 The hypotenuse of a triangle has a length of 7·7 cm. The base has a length of 5·6 cm. Find the length of the remaining side and the angle in radians between the base and the hypotenuse.

7.11 A sinusoid can be expressed as the sum of two other sinusoids of the same frequency, differing in phase by $\pi/2$ (90°). For example,

$$5 \sin(x + 0·927) = 3 \sin x + 4 \sin(x + \pi/2)$$

Draw a right-angled triangle with a base of 3 units and a height of 4 units. Verify that the hypotenuse has a length of 5 units $[\sqrt{(3^2 + 4^2)}]$. Convert 0·927 radians to degrees and, using a protractor, confirm that this is the angle between the base and the hypotenuse. Using a calculator or tables, confirm that this angle is equal to arctan(4/3). Chose arbitrary values of x and substitute into the given equation to show that it is correct.

7.12 An impedance has a resistive component of 2 ohms and a reactive component of 3 ohms. The voltage across the impedance leads the current. If the current at the angular frequency ω is represented at time t by the equation $I = 2 \sin \omega t$, write the corresponding equations for the in-phase and quadrature components V_R and V_X of the voltage. Sketch the current waveform over one complete cycle and, with the same time axis, sketch the waveforms of the two voltage components. Calculate the magnitude Z of the impedance and the phase angle ϕ. Using the same time axis, sketch the current waveform and the voltage waveform that would be observed across the impedance. What is the peak value of the voltage?

7.13 Make scale drawings to represent the complex numbers listed below. The numbers have the form $Z = A \pm jB$. In each case measure the length of the line representing Z and confirm that $Z = \sqrt{(A^2 + B^2)}$. Measure the angle between the vector and the real axis and confirm that it is equal to $\arctan(B/A)$ or $\arctan(-B/A)$ depending on the sign preceding j.

$$1 + j \qquad 1 - j \qquad 2 + 3j \qquad 5 + 7j \qquad 7 - 5j \qquad (2 + 2j + 3 + j)$$

7.14 What is meant by the *complex conjugate* of a complex number? Write the conjugate of the number $3 + 4j$. Multiply $2 - 5j$ by its conjugate. Note that the result is a real number. Find its square root.

7.15
$$Z = \frac{3 + 2j}{4 - j}$$

Multiply the top and bottom of the fraction by the conjugate of the denominator and hence express the result in the form $Z = A + jB$. What is the magnitude of Z and the angle between Z and the real axis?

7.16 Define *mechanical impedance*. When mechanical elements are represented by an equivalent electrical circuit, what is the electrical analogue of mechanical force? What mechanical property is represented by electric current?

7.17 A mechanical oscillator vibrates with an angular frequency of 471 rad s^{-1}. If the peak velocity in the cycle is 2×10^{-3} ms^{-1}, find the peak displacement and the peak acceleration. Calculate the frequency of oscillation in hertz and the rms acceleration in decibels relative to 1 ms^{-2}.

7.18 Calculate the magnitude of the mechanical reactance of a mass of 3 gm at 550 Hz. What is the peak force in the cycle at this frequency if the peak velocity is 1.7×10^{-5} ms^{-1}?

7.19 Define mechanical *stiffness*. What name is given to the reciprocal of this quantity? If the application of a compressive force of 0.028 N reduces the length of a spring by 0.01 mm, what is the stiffness of the spring? What is the magnitude of its mechanical reactance at 50 Hz?

7.20 Distinguish between the characteristics of solid friction and fluid friction. What type of friction is usually associated with mechanical impedance?

7.21 A mechanical system comprising a spring, a mass, and a damper, effectively in series, has a mechanical impedance Z given by

$$Z = R + j(\omega m - s/\omega)$$

Given $m = 2$ gm, $s = 3000$ N m^{-1}, and $R = 0.12$ N sm^{-1}, calculate the resonance frequency $\omega/2\pi$ Hz. If the system is driven sinusoidally by a force whose amplitude is 0.01 N, calculate the velocity amplitude at resonance and at twice the resonance frequency.

7.22 What is meant by the Q-factor of a resonant system? If ω_0 is the resonance frequency, then

$$Q = \frac{\omega_0 m}{R} = \frac{\sqrt{sm}}{R}$$

Substitute the values of m, s, and R given in the previous question to calculate the Q-factor of the system.

7.23 Explain the difference between *acoustic impedance* and *specific acoustic impedance*. What is meant by the *characteristic* impedance of a medium? If the temperature is constant, how do (a) the speed of sound and (b) the characteristic impedance of the air depend on atmospheric pressure? (See also Chapter 3.)

7.24 If, in a plane sound wave, the rms particle velocity is 8.7×10^{-7} ms^{-1} when the rms sound pressure is 3.6×10^{-4} Pa, what is the value of the specific acoustic impedance?

7.25 What is the sound intensity in a plane wave when the rms sound pressure is 2.4×10^{-3} Pa if the velocity of sound is 340 ms^{-1} and the density of air is 1.2 kg m^{-3}?

7.26 Explain what is meant by the *sound power reflection coefficient*. Plane sound waves travelling in medium 1 are reflected at a boundary with medium 2. The boundary is perpendicular to the incident sound. Find, in decibels, the sound pressure level in the reflected sound relative to that in the incident sound if the specific acoustic impedance in the second medium is 1.8 times that in the first medium.

7.27 Explain what is meant by *volume velocity*. A circular diaphragm has a radius of 1 cm. If the diaphragm vibrates with an rms velocity of 2×10^{-6} ms^{-1}, what is the rms volume velocity at its surface? What is the peak volume velocity if the diaphragm is driven at 1 kHz with a displacement amplitude of 50 μm?

7.28 Define *acoustic compliance*. If the middle ear behaves as a pure acoustic compliance equal to $5 \times 10^{-12}\,m^3\,Pa^{-1}$, what volume displacement is expected at the eardrum on the application of a pressure of 3 Pa? If the tympanic membrane is treated as a simple piston having an area of 60 mm^2, what is the rms velocity of the membrane at 220 Hz corresponding to an rms sound pressure of 1 Pa at the eardrum?

7.29 Calculate the compliance of the air in a cavity whose volume is 0·7 cc. Assume that the dimensions of the cavity are small compared with the wavelength of sound. The density of air is 1·21 kgm^{-3} and the speed of sound is 343 ms^{-1}. What is the acoustic impedance of the cavity at 220 Hz?

7.30 Refer to Table 1.1. Calculate the percentage change in the acoustic compliance of a small air-filled cavity when it is taken from sea level to 2000 m. What implication does this have for the calibration of tympanometers?

7.31 A Helmholtz resonator comprises a flask whose volume is $5·75 \times 10^{-4}\,m^3$ opening through a pipe having a diameter of 13·0 mm and an effective length of 11·0 mm. Calculate the resonance frequency if the speed of sound is 343 ms^{-1}.

7.32 What is *acoustic admittance*? Why is it more convenient to use admittance than impedance when describing acoustic elements that are effectively in parallel? Under what circumstances can the ear canal and the middle ear be thought of as a parallel combination of discrete elements?

7.33 You have been reading an old publication that gives the acoustic resistance and reactance of the ear as, respectively, 355 and 237 CGS ohms at 660 Hz. Find the corresponding values of the acoustic conductance and susceptance in CGS mmho. Find the admittance and express its value in SI units. (A 'mho' is the reciprocal of an acoustic ohm, and a 'mmho' is one thousandth of a mho.)

7.34 Find in absolute MKS units the impedance, admittance, and compliance at 226 Hz of the air in a 1-cc cavity if the density of air is 1·225 kgm^{-3} and the speed of sound is 340·3 ms^{-1}.

7.35 As applied to tympanometry, explain what is meant by *equivalent air volume* as a unit of acoustic admittance or compliance. Suppose the acoustic admittance of an ear is found to be 0·68 cc at 226 Hz. Express this result in absolute SI (MKS) units.

7.36 If the density of water is 1000 kgm^{-3} and the gravitational acceleration is 9·81 ms^{-2}, find the pressure in mm water corresponding to 100 daPa [1 daPa = 10 Pa].

7.37 Explain why the tympanogram of a normal ear shows a maximum acoustic admittance when the static pressure in the external ear canal is equal to that in the middle ear.

7.38 Explain the terms *distributed* and *discrete* (*lumped*) when describing the impedance of acoustic elements. If the external ear canal is to be treated as a discrete acoustic component of the ear, what can be said about the distribution of sound pressure in the ear canal and the wavelength of sound compared with the length of the canal?

7.39 Calculate the magnitude of the acoustic impedance at the entrance to a tube that is closed at one end, if the length of the tube is two-thirds of a wavelength and the cross-sectional area of the tube is 1·5 cm^2. The characteristic impedance of air is 415 Nsm^{-3}. Is this impedance 'springlike' or 'masslike'?

7.40 Referring to Figures 7.24 and 7.25, compare the acoustic impedance underneath the cushion of an audiometric earphone with the impedance at the exit of a hearing aid earmould for frequencies below 1 kHz. Why is the earmould impedance greater than the earphone impedance?

8

The Calibration and Testing of Audiometric Equipment

It is generally thought self-evident that audiometric equipment needs regular inspection to ensure that its performance meets the manufacturer's specifications and the demands of the appropriate national or international standards. Performance specifications exist to guide design and to inform the user. They do not constitute a list of items to be checked on each occasion. Even for basic equipment, such as a two-channel audiometer, the number of laboratory tests needed to ensure complete compliance with the specifications in all modes of operation would be impossibly large and it has to be assumed that the manufacturer's work has been done correctly. This raises questions as to what aspects of performance need to be checked and how often such checks should be made. The answers depend on the likelihood and consequences of equipment malfunction and on the resources available for inspection. An audiometer that is used for research involving the precise measurement of hearing thresholds would need more frequent inspection than one used in a community screening programme. For audiometers in general clinical use, it is usually thought appropriate to make a daily inspection of the external wiring and mechanical parts, to listen at least daily for evidence of gross malfunction, and to arrange an annual laboratory test of the signal output level and frequency. Other features of performance may be tested at longer intervals.[1] This three-tier approach is arbitrary in the sense that it is not based on a statistical appraisal of malfunction and its consequences, but it is the kind of approach in lieu of genuine risk analysis that is often applied elsewhere. In the United Kingdom, for example, all motor vehicles more than 3 years old are required to undergo an annual Ministry of Transport inspection. As a consequence, acoustic laboratories and garages are guaranteed a steady supply of work, though whether auditory diagnosis or road accident statistics are significantly improved remains a moot point. In favour of the annual event, it can be said that the duration of any discoverable fault will not exceed 12 months.

---

[1]Subjective stage A checks (daily checks) are listed in ISO 8253-1. This standard also lists laboratory checks (stage B) and refers to basic calibration (stage C).

The observations made in this chapter are intended to be general so that they apply in part to various kinds of audiological equipment, including tympanometers, speech equipment, and equipment used for evoked potentials, as well as audiometers. We shall, however, start with pure tone audiometers.

A distinction may be made between inspection or testing and calibration. A test for the absence of distortion is not calibration; adjustment of the output level of a tone or the sweep speed of a timebase is. The chief purpose of audiometer calibration is to ensure that signals are generated with the correct frequency and hearing level. Calibration is therefore the practical realization of the hearing level scale, and although this may seem a perfectly straightforward physical measurement, there are underlying difficulties which even now have not been fully resolved. We shall defer consideration of these until the end of this chapter and describe first the methods for calibrating pure tone audiometers.

It should be remembered that the hearing level scale for pure tones is a universal scale and thus a pure tone audiometer should deliver the same stimulus regardless of where it was calibrated. The standardization that makes this possible is coordinated by the International Organization for Standardization (ISO) and the International Electrotechnical Commission (IEC). Most national standards for audiometer calibration are now identical in all material respects to those published by the ISO and IEC. The performance of pure tone audiometers is covered in IEC 60645-1. The standard gives considerable detail that is not included in this chapter. Readers who intend calibrating their own equipment should consult the standard or its national equivalent. It will be found that standards usually contain numerous references to other standards. Unfortunately, this leads to something of a paper chase for anyone attempting faithful adherence to the prescribed methods for audiometer calibration and testing. A list of relevant international standards is included at the end of this chapter (Table 8.6).

Pure Tones: Frequency, Sound Level, and Force Level

To measure the sound output of an earphone, it is necessary to connect it acoustically to a microphone. This is done using a standard *acoustic coupler* or *ear simulator*. Most audiometers have supra-aural[2] earphones, for which

---

[2]A *supra-aural* earphone is one that rests against the pinna, as distinct from a *circumaural* earphone, which encloses it. An *insert* earphone is one that fits directly into the external ear canal.

there are currently two forms of coupler. The simplest of these, described in IEC 60318-3, is little more than a cylindrical cavity having a volume of approximately 6 cc. It is identical to the American NBS (National Bureau of Standards) type 9A coupler. This coupler now is used only to calibrate audiometers fitted with Telephonics TDH-39 or Beyer DT-48 earphones. Audiometers having any other type of supra-aural earphone should be calibrated using an IEC 60318-1 ear simulator. The latter is also called an *artificial ear*; it replicates the acoustic load presented to the earphone (see Chapter 7). The sound pressure at the microphone of the artificial ear is the pressure that would appear at the entrance to the ear canal of an 'average' human ear. Ear simulators and couplers are illustrated in Figure 8.1. Audiometers for paediatric work are sometimes fitted with insert earphones. They may be calibrated using hearing aid couplers, namely, the 2-cc coupler (IEC 60126) or the occluded-ear simulator (IEC 60711). The 2-cc coupler is merely a cylindrical cavity, but the occluded ear simulator presents the impedance of the occluded ear, and its microphone records a pressure equivalent to that at the tympanic membrane.

When calibrating an audiometer for bone conduction audiometry, a *mechanical coupler* is needed to connect the vibrator to a force transducer. Currently, only one mechanical coupler is specified (IEC 60373). It presents the vibrator with approximately the same mechanical impedance as that measured at the mastoid of the average human head. The transducer records the vibratory force that would be generated at the contact between the vibrator and the real human mastoid. The device is often called an *artificial mastoid* (Figure 8.1e).

Standards in the ISO 389 series give values of the sound pressure level to be obtained in the appropriate coupling device when the audiometer is operated at 0 dB HL. Collectively, these levels are called the *reference equivalent threshold sound pressure levels* (RETSPL). The corresponding force levels for bone conduction measurements are called *reference equivalent threshold force levels* (RETFL). The word *reference* implies that the values relate to a standard to which the calibration is referred; the word *equivalent* means that the output of the transducer at the reference level is the same as the output it would have at the normal threshold of hearing. Notice that the concept of 'normal hearing' is embodied in the definition of the reference levels. We shall return to this in the discussion at the end of the chapter. The reference levels are given in basic physical units: for sound pressures they are in decibels relative to 20 μPa; for force levels they are in decibels relative to 1 μN. They depend on frequency and, in some cases, on the particular combination of earphone and coupler that is employed. Examples are given in Table 8.1. The practical significance of the reference levels is this: if the audiometer is set to produce a tone at N dB HL, then

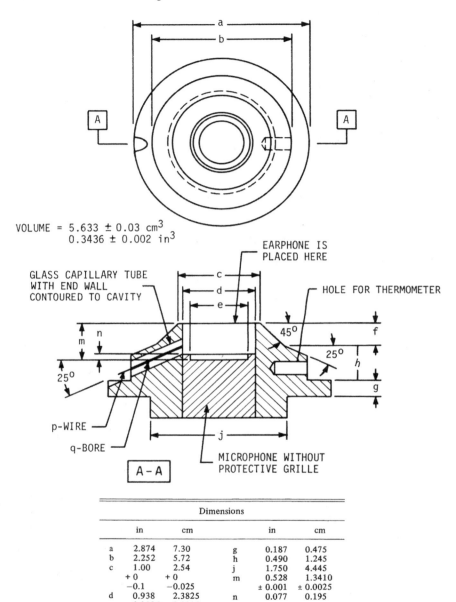

VOLUME = 5.633 ± 0.03 cm³
0.3436 ± 0.002 in³

	Dimensions				
	in	cm		in	cm
a	2.874	7.30	g	0.187	0.475
b	2.252	5.72	h	0.490	1.245
c	1.00	2.54	j	1.750	4.445
	+0	+0	m	0.528	1.3410
	−0.1	−0.025		± 0.001	± 0.0025
d	0.938	2.3825	n	0.077	0.195
	± 0.0006	± 0.0015			
e	0.728	1.85	p (dia)	0.016	0.041
f	0.295	0.75	q (dia)	0.024	0.061

FIG. 8.1   (a) Acoustic coupler for calibrating audiometric earphones (IEC 60318-3, formerly IEC 303/NBS 9A). (Reproduced with permission of the American Institute of Physics.).

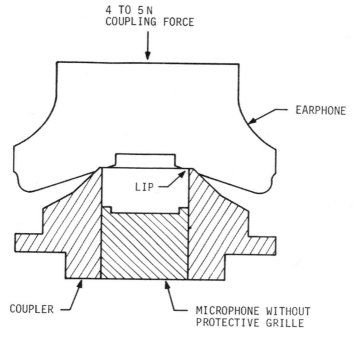

4 TO 5 N
COUPLING FORCE

EARPHONE

LIP

COUPLER

MICROPHONE WITHOUT
PROTECTIVE GRILLE

FIG. 8.1 (a) *continued.*

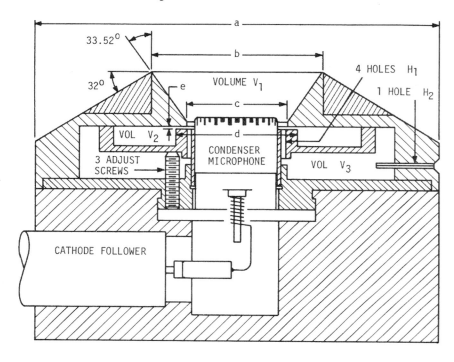

Dimensions					
	in	cm	in³	cm³	Comment
a	2.360	5.995	–	–	Diameter
b	0.984	2.5	–	–	Diameter
c	0.587	1.49	–	–	Diameter
d	0.689	1.75	–	–	Diameter
e	0.0039	0.01	–	–	Diameter
V_1	–	–	0.153	2.5	Volume
V_2	–	–	0.110	1.8	Volume
V_3	–	–	0.458	7.5	Volume
H_1 (hole)	0.018	0.045	–	–	Diameter
	0.15	0.38	–	–	Length
H_2 (hole)	0.012	0.03	–	–	Diameter
	0.35	0.90	–	–	Length

FIG. 8.1   (b) Artificial ear for calibrating audiometric earphones (IEC 60318-1). (Reproduced with permission of the American Institute of Physics.)

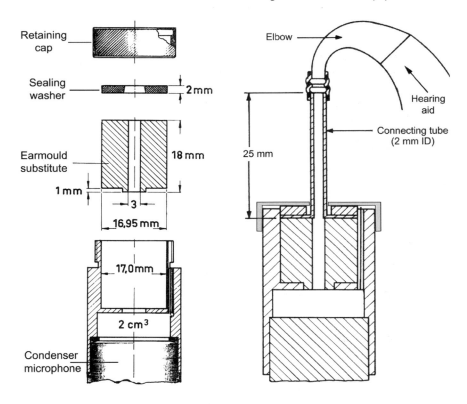

Fig. 8.1 (c) IEC reference coupler for the measurement of hearing aids using earphones coupled to the ear by means of ear inserts (IEC 60126). This simple 2-cc coupler can also be used to calibrate audiometers fitted with insert earphones. The coupler can be adapted for connection to different types of earphones. The diagram shows its basic components and an example of its use with a behind-the-ear hearing aid. (Reproduced with permission of the American Institute of Physics.)

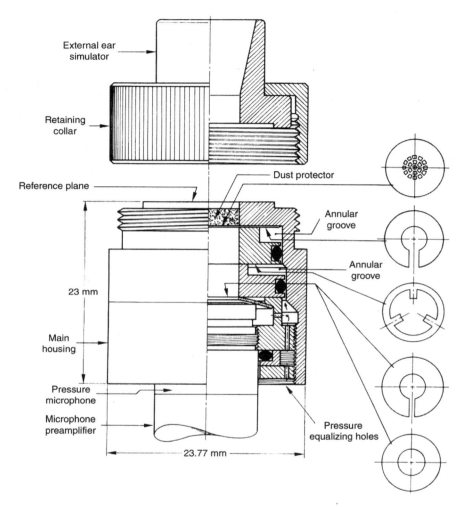

Fig. 8.1 (d) Occluded-ear simulator for the measurement of earphones coupled to the ear by means of ear inserts (IEC 60711). (Reproduced with permission of the British Standards Institution.)

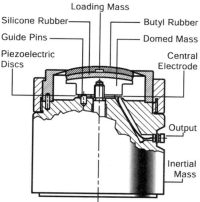

FIG 8.1 (e) Mechanical coupler for measurements on bone vibrators (IEC 60373). The labels in the photograph show the loading weight (1) and the bone vibrator under test (2). (Reproduced with permission of Brüel & Kjaer Sound & Vibration Measurements A/S, to whom the copyright belongs.)

TABLE 8.1   Reference Equivalent Threshold Sound Pressure Levels and Force Levels

Frequency (Hz)	RETSPL (dB re 20 μPa)			RETFL (dB re 1 μN)
	TDH-39 with IEC 60318-3 coupler	Beyer DT 48 with IEC 60318-3 coupler	Supra-aural earphone with IEC 60318-1 simulator ear	Bone vibrator with IEC 60373 mechanical coupler
125	45·0	47·5	45·0	—
250	25·5	28.5	27·0	67·0
500	11·5	14·5	13·5	58.0
750	7·5	9·5	9·0	48·5
1000	7·0	8·0	7·5	42·5
1500	6·5	7·5	7·5	36·5
2000	9·0	8·0	9·0	31·0
3000	10·0	6·0	11·5	30·0
4000	9·5	5·5	12·0	35·5
6000	15·5	8·0	16·0	40·0
8000	13·0	14·5	15·5	40·0

Note: The data in this table are taken from ISO 389 part 1 (1998) and part 3 (1994). Values are rounded to the nearest 0·5 dB.

its output measured with an appropriate acoustic or mechanical coupler should be $N + R$ decibels, where R is the specified reference level at the operating frequency. Calibration ensures that this requirement is fulfilled within a specified tolerance.

An example of equipment set up for audiometer calibration is shown in Figure 8.2. Some or all of this equipment is needed depending on the extent of the work to be undertaken. Essential components comprise the couplers, the sound-level meter, and the frequency meter. A good-quality sound-level meter is required. It should be a type 0 or type 1 as described in the previous chapter. A set of filters (preferably third-octave) or a spectrum analyser is needed to measure harmonic distortion and to measure the characteristics of masking noise. The oscilloscope provides a direct visual indication of distortion and any contamination of the signal by environmental and instrument noise. Ancillary equipment includes a pistonphone or other form of acoustic calibrator and a reference alternating-voltage source (or an oscillator and a voltmeter) to set the voltage sensitivity of the sound-level

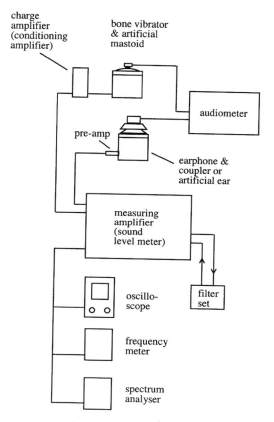

FIG. 8.2 Equipment for testing and calibrating audiometers.

meter when it is used with a direct electrical input. The direct input is needed for bone conduction calibration. It replaces the input from the microphone. Couplers or artificial ears nearly always have condenser microphones, so that a microphone preamplifier is essential. The preamplifier may be part of the sound-level meter itself, in which case the acoustic coupler has to be connected directly to it. The mechanical coupler can be used without a preamplifier or conditioning amplifier (charge amplifier), but it is then important to take particular care to minimise the capacitance in the cable that connects it to the sound-level meter. A suitable cable can be obtained from the manufacturer.

Practical Considerations

Audiometers are usually more reliable and more stable than their users believe. The most common sources of error are faulty connectors and damaged headphone cables; mechanical switches are also prone to failure. Battery-operated audiometers are usually received for calibration in want of a recharge or a new battery. Portable audiometers are seldom strong enough to withstand user abuse, but their earphones are surprisingly resistant to maltreatment. Bone vibrators have a bad reputation, but it is probably not deserved. Calibration should always be preceded by a thorough mechanical and electrical inspection, after which any significant faults should be dealt with. At this time it is also worth checking serial numbers to ensure that the audiometer has been sent with the correct transducers. Users often, and perhaps unknowingly, interchange headsets between audiometers.

Audiometric earphones and bone vibrators are usually of the moving-coil variety, so that the sound pressure or vibratory force that they generate depends on the current in the coil. The current is equal to the open-circuit output voltage of the audiometer divided by the combined impedance of the transducer, its leads and connectors, and the source impedance of the audiometer. The impedance of the leads and connectors is the most variable of these elements, and particular care should be taken to confirm the integrity of the wiring to the transducers to ensure that connectors make good electrical contact and to check that connecting screws (particularly those in the earphone) are tight.

Electrical safety should also be tested.[3] It may be observed in passing that there may be mechanical as well as electrical hazards. The coupling between the headband and the earphone or bone vibrator is often barely strong enough; in particular, the rivets that secure the stirrups are likely to fail at some time. The greatest mechanical stress occurs when the headband is expanded at the moment of fitting it to the patient, when failure could cause injury to the head of face or, more seriously, to an eye. For this reason it is good practice in audiometry to fit the headband from behind the patient's head, and it is good practice to examine headbands frequently for mechanical defects.

Calibration is best undertaken in a quiet room; otherwise, the signal from the audiometer can only be examined when it is at a high level. The workbench has to be reasonably vibration free. It may be helpful to place the artificial ear on an antivibration table. A suitable table can easily be made from a heavy block of material mounted on some soft springs; a

---

[3]National guidelines apply, for example, the Medical Devices Agency (MDA) DB9801 in the United Kingdom. The international safety specification is IEC 60601-1.

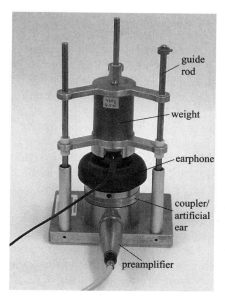

FIG. 8.3 Apparatus for applying an earphone to an acoustic coupler or ear simulator.

paving stone placed on the softly inflated inner tube of a tyre from a motor scooter is an excellent alternative. Vibration from the workbench is seldom a problem for bone conduction calibration because the mechanical coupler is itself quite massive (4·3 kg) and is supplied ready-mounted on soft springs.

It is necessary to apply the earphone to the coupler or artificial ear centrally and with the prescribed force, namely, 4·5 ± 0·5 N not including the weight of the earphone itself.[4] The manufacturer may supply a suitable loading device, but a 460-gm weight placed on top of the earphone is an adequate alternative. A more elaborate arrangement that ensures correct alignment[5] and provides the required weight is shown in Figure 8.3. Before placing an earphone on the coupler, care should be taken to see that it is free of debris that could fall on the diaphragm of the microphone. Consider-

---

[4]The phrase 'not including the weight of the earphone' is a curious anomaly that has been reproduced in many versions of the standards. It implies that the weight of the earphone is an additional force between the earphone and the coupler, but a similar requirement is not found in the specification for calibrating bone vibrators. The specified nominal force between the earphone and the pinna is also 4·5 N. This does not involve the weight of the earphone because the headband force is directed horizontally.

[5]P. M. Haughton, *B. J. Audiol.* 29: 188–191, 1995.

ation should be given to obtaining a good acoustic seal between the earphone and the coupler; otherwise, there is a likelihood of significant error, particular at low frequencies. Similar considerations apply to bone vibrators, except that the alignment of the vibrator and the mechanical coupler is less critical. A deadweight is probably more convenient than the loading spring provided by the manufacturer[6] (Figure 8.1e). The required weight is 550 gm (5·4 ± 0·5 N).

Although the reference equivalent levels specify the output of the audiometer when it is operated at 0 dB HL, it is not normal practice to calibrate directly at this level because the acoustic signal would be too weak to measure easily. Manufacturers may suggest levels at which calibration should be made (the levels are displayed on some audiometers when they are operated in the 'calibration mode'). So long as the signal level is well above that of the environmental and instrument noise level it is not critical. Somewhere in the middle of the available hearing level range, perhaps 60 dB HL, seems sensible. The signal level should be recorded because later, when the signal attenuator is checked, attenuator error will be zero at this level. Because bone vibrators are prone to distortion, it is usual to calibrate the bone conduction output at more modest levels, for example, 20 dB HL at 250 Hz, and 40 dB HL at higher frequencies.

It is good practice to record the output of the audiometer at all frequencies, for both air and bone conduction, before making adjustments. This is a somewhat tedious process, and computer assistance in acquiring and storing the information is helpful. The preadjustment record is important because any errors that may have affected the outcome of audiometric tests will remain traceable. Adjustment should then follow if permitted tolerances are exceeded. The signal level should be accurate to ±3 dB in the frequency range 125 to 4000 Hz and ±5 dB at 6000 and 8000 Hz (IEC 60645-1). Frequencies have a tolerance of between 1% and 3% according to the type of audiometer.[7] It should be remembered that these tolerances are the maximum allowed according to the standard. Most laboratories would impose stricter limits. The guiding principle is that errors in calibration should have a negligible influence on the outcome of audiometry. A 5-dB error is not negligible (though its effect on diagnosis probably is), but a 1-dB error would be undetectable in a clinical setting.

---

[6]At present the principal manufacturer is Brüel & Kjaer Ltd. Their type 4930 artificial mastoid has a loading arm that holds the vibrator in place by means of a spring. A simple and effective modification is to remove the spring and to attach a weight to the arm at a point just above the vibrator. The weight should be 5·4 N. (A small deduction can be made for the weight of the vibrator itself and the load on the vibrator due to the weight of the arm.)

[7]The types are designated 1 to 5 in IEC 60645-1. Types 1 and 2 require a frequency accuracy of 1% and 2%, respectively; for types 3 to 5, the accuracy is 3%.

Although ambient conditions (temperature, pressure, and humidity) have little effect on earphone calibration, the performance of the mechanical coupler is significantly dependent on temperature. The standard temperature for calibration is $23 \pm 1°C$. Unless the temperature is controlled, it is often likely to lie outside this range. Corrections for temperature changes are difficult to make because the mechanical impedance of the coupler as well as its force sensitivity are affected and the overall effect depends on the mechanical impedance of the bone vibrator. With the B71 vibrator, changes in the hearing level calibration exceeding 0·5 dB/°C can be expected at some frequencies.[8] It is therefore necessary to control the temperature or to accept that some error is inevitable.

Little has been said so far about measuring frequencies. This is seldom a difficult task because it is easily accomplished with the kind of frequency meter that is standard equipment in any electronics workshop. Care should be taken to see that the signal delivered to the frequency meter is noise free and of an adequate level. If there is any doubt, it is worth checking that the frequency read-out remains unchanged when the signal level is altered.

Equipment Fitted with Insert Earphones

Insert earphones have received relatively little attention as a standard source for audiometry. They do, however, offer some valuable advantages over supra-aural earphones in at least two applications, namely, as a source of masking noise generally and as a source of pure tones for hearing tests on small children. The advantage for masking is that the listener's head is unencumbered by the headband that would otherwise be needed to support one (or often both) external earphones; the advantage when testing children is again that no headband is needed, with the result that the child is likely to be more comfortable and more cooperative. For a given electrical supply, the sound pressure created in the ear canal by an insert phone depends to a significant extent on the depth of insertion and on the size of the ear canal itself. It is probable that this leads to a more variable stimulus than would be obtained with conventional earphones. In principle, it would be possible to define the threshold of hearing in terms of the sound pressure that is actually present in the ear canal. Such a definition would then apply equally in other contexts, such as tympanometry and other measurements where the sound source is coupled to the ear by a probe. In tympanometry, the level

---

[8]S. Dowson and H. McNeil. The effect of temperature on the performance of audiometric mechanical coupler systems. *B. J. Audiol.* 26: 275–281, 1992.

of the probe tone is often controlled so that the sound pressure in the ear canal is the same regardless of the canal volume; in otoacoustic emissions work, it is usual to monitor the sound pressure in the ear canal and to modify the electrical supply to the earphone to obtain the requisite stimulus level. It is also the case that real-ear measurements are frequently used to assess hearing aid performance, and for some digital hearing aids, fitting involves the measurement of hearing thresholds with the aid itself acting as the source of the stimulating tone. However, formal audiometry seldom, if ever, relies on direct sound measurement in the ear canal. There is, moreover, no reference standard that specifies the real-ear sound pressure for audiometry; instead, the usual procedure of referencing to an equivalent threshold sound pressure in a coupling device is adopted.

Reference equivalent thresholds have so far been established for only one insert earphone, the Etymotic Research ER-3A, coupled to the ear by inserts of type ER-3-14 (Table 8.2). This earphone is acoustically identical to the Eartone 3A. The inserts may have the name Earlink 3A. The methods for attaching the earphone to the standard 2-cc coupler (IEC 60126) or occluded-ear simulator (IEC 60711) and the corresponding reference equivalent thresholds are described in ISO 389-2. The earphone is normally coupled to the ear by a flexible sound tube having a length of 240 mm and an internal diameter of 2 mm. The sound tube ends in a small plastic

TABLE 8.2 Reference Equivalent Sound Pressure Levels in dB re 20 μPa for Etymotic ER-3A Insert Earphones and IEC 60126 Coupler and IEC 60711 Occluded-Ear Simulator

Frequency (Hz)	Acoustic coupler (dB)	Ear simulator (dB)
250	14.0	17.5
500	5.5	9.5
750	2.0	6.0
1000	0.0	5.5
1500	2.0	9.5
2000	3.0	11.5
3000	3.5	13.0
4000	5.5	15.0
6000	2.0	16.0
8000	0.0	15.5

Note: Data at selected frequencies are from ISO 389-2 (1997). For further information, refer to the standard.

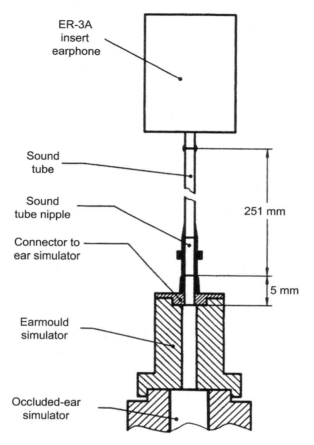

FIG. 8.4 Method for connecting an insert earphone to an occluded-ear simulator (IEC 60711). The same method is used for connection to a 2-cc coupler (IEC 60126). (Reproduced with permission of the British Standards Institution.)

connector that is attached by a short tube to a cylindrical foam eartip. The tip is normally inserted so that its outer end is 2 to 3 mm inside the entrance to the ear canal. For calibration, the tip is removed and the sound tube from the earphone is connected to the coupler or artificial ear by means of a plastic tube to make the outlet of the connector flush with the inlet nipple of the coupler. The total length of pipework, including the plastic connector, between the exit of the earphone and the entrance of the coupler is then 251 mm (Figure 8.4).

Signal Attenuator

Over much of its range, the signal from the audiometer is controlled by an attenuator whose output is connected directly to the earphone or bone vibrator. The level may also be controlled in part by changing the signal at the input to the attenuator, for example, by introducing additional amplification to extend the range to higher hearing levels. The comments that follow apply regardless of the way in which the change in output is accomplished, and for simplicity we shall use the word *attenuator* to denote the controlling mechanism. Attenuators work by switching the output connections between different points in a chain of resistors. The switching may be done mechanically or electrically. Mechanical switches are inherently noise free, at least when new, because they have no active components, but the switch contacts are likely to deteriorate and may eventually become unreliable. Nowadays most audiometers have electronic attenuators. A properly designed attenuator provides the audiometer with the same output impedance at all hearing levels. The impedance is generally low compared with that of the earphone or vibrator at the operating frequency, so that the audiometer behaves, at least approximately, as a 'voltage source'. Nevertheless, if the transducer is replaced by a simple electrical load (for test purposes), care should be taken to see that its impedance is approximately the same as that of the transducer.

The IEC standard specifies the accuracy of the output control in the following way:

> The measured difference in output between two successive indications of hearing level which are not more than 5 dB apart shall not differ from the indicated difference on the indicator by more than three-tenths of the indicated interval measured in decibels or by more than 1 dB, whichever is the smaller.

It should also be remembered that, overall, the output of the audiometer must always lie within the ± 3 or ± 5 dB tolerance mentioned previously. Any systematic error in the attenuator must therefore be such that the cumulative error does not take the output beyond these limits even if calibration at the nominated calibration point (e.g., 60 dB HL) is satisfactory.

Practical Considerations

Ideally, attenuators should be tested by examining the acoustic or vibratory output from the transducer. The difficulty is that environmental and instrument noise may preclude measurement at low signal levels. In practice it is best to supplement acoustic measurements with measurements of the voltage developed across a dummy load. A 10-Ω resistor is a satisfactory substitute for most earphones, but for some a higher value is needed. It is generally considered adequate to test the attenuator for only one mode of output and only one frequency. Usually this will be for air conduction at 1 kHz. A case can be made, however, for testing the output in the bone conduction mode at least once during the lifetime of the audiometer. Earphones have approximately the same impedance at all frequencies, but the load presented by a bone vibrator is likely to be small at low frequencies and large at high frequencies (typically 4·5 Ω at 250 Hz and 25 Ω at 4000 Hz). Some attenuators whose performance is irreproachable in air conduction are found wanting when bone conduction is selected, particularly if the frequency is low.

When testing the attenuator, it is essential to maintain a good signal-to-noise ratio at all indicated hearing levels. There are several ways of doing this. An obvious ploy is to choose a frequency for which the reference equivalent pressure or force level is relatively high. This leads to a correspondingly high sound pressure or force level for a given hearing level. It is particularly successful for bone conduction measurements at low frequencies (250 or 500 Hz). Another method is to work at frequencies where environmental noise is least troublesome. A third-octave or narrow-band filter is distinctly helpful and, for measurements at 1 kHz, A-weighting may be added for good measure. It is also possible to take advantage of the fact that the output of the audiometer has a much lower impedance than the input of the sound-level meter or measuring amplifier. When purely electrical measurements are made, that is, when the transducer is replaced by a dummy load, a step-up transformer can be inserted to raise the signal voltage, thereby taking it well above the inherent noise in the measuring equipment. A microphone transformer is ideal for this task (Figure 8.5).

In view of the need to be mindful of both the change in output between successive attenuator settings and the overall output of the audiometer, it is helpful to tabulate results in a way that shows the measured step size and the cumulative error. Computer spreadsheets make this easy. An example is shown in Figure 8.6.

The attenuator that controls the masking level can be tested in the same way as the signal attenuator, either using masking noise itself or, for two-channel audiometers, using a pure tone source. The tolerance on step size is the same as for the signal attenuator, but the overall accuracy required is from -3 to $+5$ dB relative to the indicated hearing level.

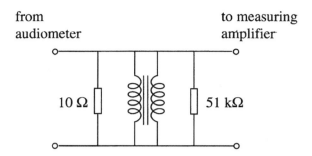

microphone transformer
primary 600 Ω
secondary 50 kΩ
voltage gain 9·1

FIG. 8.5 Step-up transformer as a substitute for a 10-ohm earphone.

Unwanted Signals

Ideally, the input to the earphone or vibrator should contain the pure tone signal (the test signal) and nothing else; the tone should cease altogether when the tone switch is turned off. Contamination of the test signal by masking noise or other signals from a second channel is often called *cross-talk* or *breakthrough*. It is of course essential that the test signal should not appear in the nontest earphone when the tone is presented. It is possible to check for the presence of unwanted signals by listening to the audiometer, and the IEC standard describes how this should be done. The acoustic environment in the laboratory is, however, unlikely to be sufficiently quiet for such methods, and direct electrical measurements are more appropriate.

The standard makes the following recommendations:

(a) At a hearing level control setting of 60 dB and with the tone 'OFF', the electrical signal in each one-third octave band within the range 125 Hz to 8 kHz shall be at least 10 dB below the signal corresponding to the reference equivalent threshold level for the centre frequency of the one-third octave band.

(b) With the tone 'ON', the unwanted signal in the non-test earphone or a substitute dummy load shall be at least 70 dB below the test tone as measured when the hearing level is set to 70 dB or greater.

Paragraph (a) refers to the test earphone. The filters are needed not merely

	A	B	C	D	E
	nominal hearing level dB HL	indicated level dB	measured hearing level dB HL	step dB	cumulative error dB
1	80	109·0	76·7	4·8	-3·3
2	75	104·2	71·9	3·9	-3·1
3	70	100·3	68·0	3·4	-2·0
4	65	96·9	64·6	4·6	-0·4
5	60	92·3	60·0	0	0
6	55	86·2	53·9	6·1	-1·1
7	50	81·3	49·0	4·9	-1·0
	45	76·3	44·0	5·0	
	35	67·1	34·8	4·9	-0·2
	5	40·8	8·5	4·8	3·5
	0	37·1	4·8	3·7	4·8
	-5	33·5	1·2	3·6	6·2
	-10	32·6	0·3	0·9	10·3

Row 5   Audiometer calibrated at 60 dB (in this example).
Column A   Selected audiometer output level.
Column B   Level indicated on measuring amplifier / sound level meter.
Column C   Actual hearing level *excluding* any calibration error at 60 dB HL.
Column D   Actual step size, working *from* row 5,
Column E   Cumulative error relative to row 5. For example, if the audiometer is perfectly calibrated when an output of 60 dBHL is selected, its output will be 4·8 dB high when 0 dBHL is selected.

Formulas (in rows 1 & 6 for example)
Above row 5: $C1 = \$A\$5 - \$B\$5 + B1$; $D1 = B1 - B2$; $E1 = \$A\$5 - \$B\$5 - A1 + B1$
Below row 5: $C6 = \$A\$5 - \$B\$5 + B6$; $D6 = B5 - B6$; $E6 = \$A\$5 - \$B\$5 + A6 - B6$

FIG. 8.6   Tabulating the results of an attenuator test.

to improve the signal-to-noise ratio in the measurements but also to tie the measurements to a specified reference equivalent threshold. It may be remarked that the standard offers no enlightenment as to the cross-channel interference that can be accepted in the test earphone when the tone is on. The guiding principle should be that the separation between channels should readily exceed the so-called interaural attenuation for air conduction audiometry. By convention, the least attenuation is taken to be 40 dB, but we should expect the channel separation to be much greater, perhaps 70 dB, in line with the other specified requirements. It is also possible that the test signal may be contaminated by signals from a second channel or other electrical source even when there is no intended output to the nontest earphone. The important point being made here is that the signal in the test earphone must not be contaminated with signals from other sources (notably from a second channel delivering masking noise to the nontest earphone) and it must not contain spurious signals that become audible when the test tone is turned on. It would, for example, be disastrous if mains hum became audible in the test earphone each time the tone switch was turned on.

Channel separation requires careful circuit design and layout. Cross-talk can lead to serious inaccuracies in audiometry, but fortunately it can often be detected in simple listening tests. The difficulty with objective tests is that they involve the measurement of low-level signals. They are nevertheless worth doing, at least on some occasions, particularly if the audiometer is one that has not been tested previously. They are best done electrically (that is, with dummy loads) as an adjunct to the attenuator checks.

Distortion

It was explained in Chapter 5 that a periodic, nonsinusoidal signal can be expressed as the sum of a series of sine and cosine terms whose frequencies are harmonics of the frequency corresponding to the fundamental period of the signal. These terms constitute the Fourier series representing the signal. If the output of an audiometer (or other oscillator) is a true pure tone, then, by definition, it has just one frequency (the fundamental) and no harmonics. If it is distorted, it will contain harmonic frequencies in addition to the fundamental. The amount of distortion, expressed in terms of the harmonics, is called *harmonic distortion*. It is particularly important that signals from an audiometer should be free of distortion. This is because the threshold of hearing changes considerably with frequency, so that in unfavourable circumstances harmonics might be audible even when the fundamental is below threshold. If the listener responds to the harmonics, the audiogram will show better hearing at the test frequency than actually exists.

Harmonic distortion is usually expressed as a percentage: it is 100 times the ratio of the rms sound pressure in the harmonic to the rms sound pressure in the signal as a whole. Quoting again the formulae given in Chapter 5,

$$100 \times \sqrt{\frac{p_2^2 + p_3^2 + p_4^2 + \cdots}{p_1^2 + p_2^2 + p_3^2 + p_4^2 + \cdots}}$$

expresses the total harmonic distortion as a percentage, and

$$100 \times \sqrt{\frac{p_n^2}{p_1^2 + p_2^2 + p_3^2 + p_4^2 + \cdots}}$$

gives the percentage distortion due to the nth harmonic alone. The same expressions apply to the distortion in bone conduction signals if the rms values of the sound pressures are replaced by the corresponding values of vibratory force. The sound pressures are those measured in an artificial ear or coupler, and the forces are those measured on a mechanical coupler. Corrections have therefore to be made for the frequency response of the microphone and, more important, for the frequency response of the mechanical coupler when it is treated as a force-measuring instrument. The maximum permitted distortion depends on the type of signal (air or bone) and on frequency (Table 8.3).

Harmonic distortion could be measured using a spectrum analyser, but a less expensive method is to measure the output of the audiometer after inserting appropriate narrow-band filters that pass only the required harmonic frequencies. One-third octave filters are satisfactory for this task. As with other performance tests, it is not practicable to measure harmonic distortion at all frequencies on all occasions, but it is worth checking at a low frequency (250 Hz, for example) where the transducer is operating at a relatively high output for a given hearing level. Bone vibrators are particularly prone to distortion at low frequencies, to the extent that they are of limited value in audiometry below 500 Hz. The appearance of the waveform

TABLE 8.3 Maximum Permissible Harmonic Distortion as a Percentage of Sound Pressure or Vibratory Force (IEC 60645-1)

	Air conduction			Bone conduction		
Frequency range (Hz)	125–250	315–400	500–5000	250–400	500–800	1000–4000
Hearing level (dB HL)	75	90	110	20	50	60
Total harmonic distortion	2·5	2·5	2·5	5·5	5·5	5·5

Note: The hearing level should be as specified in the table or be the level at the maximum output of the audiometer, whichever is lower; for circumaural and insert earphones, the hearing level will be 10 dB less than the levels specified in the table.

of the signal as seen on an oscilloscope is a useful guide to the presence of distortion. If the waveform is visibly nonsinusoidal, then it is likely that the permitted limits have been exceeded, particularly in earphone measurements. When testing at 500 Hz or at lower frequencies, the visible distortion can be enhanced by inserting the A-weighting filter to emphasise the high-frequency components. If significant distortion is observed, the transducer should be replaced by a dummy load to determine whether the problem is in the supply to the transducer or in the transducer itself.

Examples

Suppose that the audiometer is operating at 250 Hz and that the sound pressure level in the artificial ear is 102 dB SPL. When a filter is inserted that passes only a narrow band of frequencies centred on 500 Hz, the measured output is 70 dB SPL. How great is the second-harmonic distortion? To answer this question, let us start by simplifying the formula for percentage distortion by letting p_T represent the root mean square sound pressure in the unfiltered signal. Then

$$p_T^2 = p_1^2 + p_2^2 + p_3^2 + \cdots$$

so that the distortion in the second harmonic is

$$100 \times \sqrt{\frac{p_2^2}{p_1^2 + p_2^2 + p_3^2 + p_4^2 + \cdots}} = 100 \times \sqrt{\frac{p_2^2}{p_T^2}} = 100 \times \frac{p_2}{p_T} \text{ percent}$$

The fall in sound pressure level on inserting the filter allows us to calculate p_2/p_T as follows.

$$70 - 102 = 20 \log \frac{p_2}{p_T}$$

$$\frac{p_2}{p_T} = 10^{-32/20} = 0{\cdot}025$$

The distortion due to the second harmonic is therefore 2·5%.

Continuing with the same audiometer and earphone, suppose that the 500-Hz filter is replaced by one that passes frequencies centred on 1 kHz (the third harmonic of 250 Hz) and that a sound pressure of 76 dB SPL is then observed. Also suppose that measurements with higher-frequency filters show that components above the third harmonic are negligible. What is the distortion in the third harmonic, and what is the total distortion?

Proceeding as before,

$$76 - 102 = 20 \log \frac{p_3}{p_T}$$

$$\frac{p_3}{p_T} = 10^{-26/20} = 0.050$$

The distortion due to the third harmonic is therefore 5.0%. The total harmonic distortion is

$$100 \times \sqrt{\frac{p_2^2}{p_T^2} + \frac{p_3^2}{p_T^2} + \text{negligible terms}}$$

$$= 100 \times \sqrt{0.025^2 + 0.050^2}$$

$$= 5.6\%$$

Notice that this is less than the sum of the percentages for the individual components.

Table 6.1 shows that the A-weighting filter would increase the levels of the second and third harmonics by 5.4 and 8.6 dB, respectively, relative to the level of the fundamental. It will be found that the distortion in the A-weighted waveform is then 4.7% in the second harmonic, 13.5% in the third harmonic, and 14.3% in total. The waveform of the distorted signal is shown in Figure 8.7. It can be seen that distortion at 5.6% would be just visible to someone familiar with the appearance of sinusoidal signals. The A-weighting reveals a clearly nonsinusoidal form.

Harmonic Distortion: Physical and Audiological Meaning

As already explained, harmonic distortion is usually expressed in terms of the root mean squares of the sound pressures measured in an artificial ear or coupler, or in terms of the forces produced by the action of a bone vibrator at the input to a mechanical coupler. If sound measurements were made in a free field, the acoustic intensity associated with each harmonic relative to the intensity of the whole signal would be completely specified by the particular value of $(p_n/p_T)^2$ because the acoustic impedance in free sound waves is virtually independent of frequency so long as the wavefronts are approximately plane surfaces. Impedance is practically constant except in standing waves or at points close to a source or obstacle. Suppose that harmonic distortion is measured in the field of a loudspeaker. The formulas we have quoted provide a physically meaningful statement of the sound

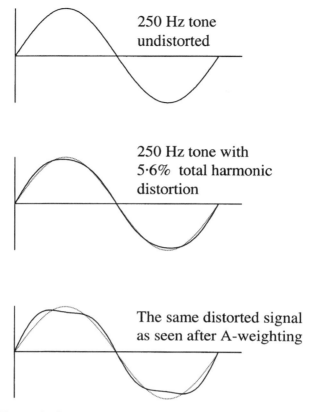

250 Hz tone
undistorted

250 Hz tone with
5·6% total harmonic
distortion

The same distorted signal
as seen after A-weighting

FIG. 8.7 Harmonic distortion in the output of an audiometer. The dotted lines in the lower two graphs show the undistorted waveform.

energy in each of the harmonics and in the harmonic content as a whole. In contrast, when measurements are made with an artificial ear or mastoid, the energy in the harmonics is not simply related to the energy in the fundamental because the acoustic or mechanical impedance varies greatly with frequency. Audiologists will not lose much sleep worrying over this complication, but they should be aware that the simple physical interpretation has been lost, though the method for calculating the harmonic distortion still provides a useful basis for description and specification.

A further difficulty, already alluded to, is that the threshold of hearing is also frequency dependent when it is expressed as a sound pressure or force level. This is reflected in the way the equivalent threshold levels change with frequency (Table 8.1). Although the calculation of harmonic distortion ignores this problem, it has been recognized in setting performance limits.

Signals are required to be almost free of the higher harmonics. With earphones, for example, the permitted distortion in the fourth and higher harmonics is only 0·3%.

Masking Noise

All audiometers, with the exception of those intended for simple screening tests (IEC types 4 and 5), provide noise sources for masking pure tones. Some audiometers also provide alternative noise sources for masking speech and for special tests. The ideal noise for masking tones is a narrow band whose width is equal to the psychoacoustic critical band.[9] This type of noise has the least loudness for a given degree of masking. The specification in IEC 60645-1 is for bands that are similar to the psychoacoustic bands while having a simple engineering definition, namely, a width of between one-third and one-half of an octave geometrically centred on the frequency of the tone. The bands are said to be slightly wider than the critical bands and to have a less tonal quality. From this definition it can be seen that the cut-off frequencies for bands of masking noise are given by

$$f \times 2^{-1/4} \text{ to } f \times 2^{-1/6} \quad \text{(lower limit)}$$

$$f \times 2^{1/4} \text{ to } f \times 2^{1/6} \quad \text{(upper limit)}$$

where f is the centre frequency. Beyond these limits, the spectrum density of the noise should fall by at least 12 dB per octave.

The noise spectra are intended as a specification of the sound produced by an audiometric earphone in a coupler or artificial ear. Because of the limitations of these devices, measurements at frequencies above 5 kHz should be made electrically across the terminals of the earphone or across a dummy load.

If the audiometer provides uniform broadband noise (white noise), the sound spectrum measured in a coupler or artificial ear should be uniform within ±5 dB between 250 and 4000 Hz. Noise for masking speech will be described later.

Noise spectra can be measured using a spectrum analyser or suitable narrow-band filters, but it is only on rare occasions that laboratories will be asked to make such measurements and, except where a fault exists, there is little that can be done if the spectra do not conform to the IEC specification.

[9]The ear behaves as though it possesses a contiguous series of band-pass filters, each having a bandwidth known as a *critical band*, defined empirically as a range of frequencies within which subjective responses are constant. The term was originally used by Fletcher and Munsen to denote a range of frequencies within a band of noise that alone contribute to the masking of a tone at the centre frequency. See also Chapter 9.

TABLE 8.4   Reference Levels for Narrow-Band Masking
Noise Derived from Wide-Band Noise Using Third-Octave or
Half-Octave Filters and Measured in an Artificial Ear
or Coupler

Centre frequency (Hz)	Third octave (dB)	Half octave (dB)
125	4	4
250	4	4
500	4	6
750	5	7
1000	6	7
1500	6	8
2000	6	8
3000	6	7
4000	5	7
6000	5	7
8000	5	6

Note: These levels are to be added to the corresponding reference
equivalent threshold sound pressure levels. The data have been
taken from ISO 389-4 (1994) at selected frequencies. For further
information, refer to the standard.

Fortunately, this characteristic is not critical and audiometers may be
considered perfectly serviceable even if they do not comply entirely with the
standard. It is desirable, however, that the level of the noise should be
consistent with the masking level indicated on the audiometer. In an early
version of the standard,[10] the requirement for masking pure tones was
simply that the sound pressure level of the noise should be 3 dB greater than
the level of the tone itself where the level of the noise is measured in a
third-octave band centred on the frequency of the tone. This meant that the
reference levels for narrow-band masking noise were 3 dB greater than the
corresponding reference equivalent threshold sound pressure levels. These
levels have now been refined[11] on the basis that the narrow-band noise is
derived from uniform wide-band (white) noise using third-octave or half-
octave band-pass filters with the cut-off frequencies given previously. Reference levels are then specified according to frequency and bandwidth as listed
in Table 8.4.

Audiometers usually have some means for adjusting the levels of the
masking noise. Measurement and adjustment of the masking noise are often

[10]IEC 645 (1971), *Audiometers.*
[11]ISO 389-4.

considered part of the usual calibration routine. Some audiometers (IEC type 1) allow the noise to be combined with the tone in the same earphone. This provides a quick and reliable method of checking the effective masking level of the noise — the tone should be just inaudible when the hearing and masking levels are numerically the same.

Users may sometimes fit alternative transducers for masking that differ from those originally provided by the manufacturer, and it may then be impossible to bring the level of the masking noise into correct adjustment. This will not matter provided that the masking-level scale is treated as an arbitrary decibel scale rather than one expressing the effective masking level of the noise. Users will often treat the scale as arbitrary even when it is correctly calibrated, preferring to find the correct masking level by the 'plateau-seeking' method.

Signal Switch

The switch that turns the tone on or off is usually called the *tone switch* or the *interrupter*. The act of switching a sinusoidal tone inevitably produces transient signals at frequencies above and below that of the tone itself. It is obviously important that these unwanted signals should be inaudible; accordingly, the switching should effect a smooth rise and fall of the tone.

The *rise time* of a signal is conventionally defined as the time taken for the signal voltage to rise from 10% to 90% of its final value after the signal has been turned on; the *fall time* is defined as the time taken for the signal voltage to fall from 90% to 10% of its initial value after the signal has been turned off. When the signal being switched is an alternating voltage (in this case a tone), then these definitions apply to the signal envelope — that is, to the signal amplitude — rather than to the instantaneous values. The points defining rise and fall times are therefore exactly -20 dB and approximately -1 dB, respectively, relative to the steady level attained when the tone has been on for some time (Figure 8.8). For audiometers, the rise and fall times should each be at least 20 ms. If the tone is pulsed automatically (as it is in self-recording audiometers), the rise and fall times should not exceed 50 ms. Greater times are permitted for manual switching, but the time from turning the tone on to the instant when it reaches 90% of its final value may not exceed 200 ms. The same limit applies to the falling signal after the switch is turned off. Further details are given in the standard (IEC 60645-1). Another consideration is that the tone should never exceed its steady value by an amount that would make it momentarily audible. The steady level should not be exceeded by more than 1 dB.

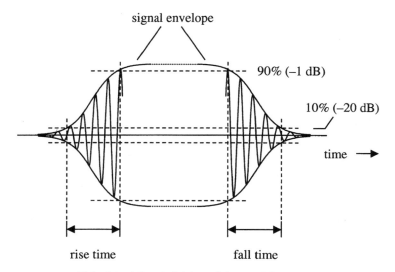

FIG. 8.8   The definition of rise and fall times.

A view of the signal on an ordinary oscilloscope may reveal gross defects in the switching circuit, such as excessive overshoot, but it is difficult to make a quantitative assessment of the rise and fall times. If these have to be measured it is best to use a storage oscilloscope. Nowadays, signal storage is accomplished digitally, which means that the signal waveform has to be sampled. Because the rise and fall times are long compared with the time for each cycle of the tone, a very large number of samples would be needed to show the entire signal during the intervals defining its rise and fall. Attempts to measure rise and fall times by sampling the signal itself are likely to lead to aliasing (see Chapter 5) and to erroneous results. The problem is avoided by rectification and low-pass filtering of the signal before digital storage. The stored quantity is then a good approximation to one-half of the signal envelope, that is, to a function showing the variation in signal amplitude with time. A suitable rectifier and filter can easily be made from standard circuit components (Figure 8.9).

Other Checks

Two-channel audiometers usually provide reference tones intended for loudness balance tests. It is worth checking to see that the reference tone and the normal test tone are produced with essentially the same sound levels

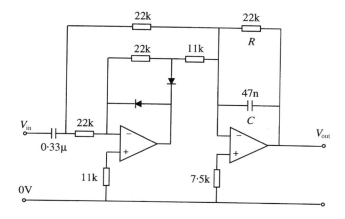

Precision full-wave rectifier
Time constant for integration = $RC = 1.03$ ms.
Low pass filter, cut-off frequency (−3dB) = 154 Hz

Example: input 4 kHz tone with 80 ms rise time; output as shown below.

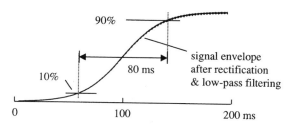

FIG. 8.9 Precision rectifier for measuring rise and fall times.

when the same hearing levels are selected. The levels should be the same within ±3 dB in the frequency range 500 to 4000 Hz, and ±5 dB at other frequencies.

Headbands can become distorted so that they become too slack or too tight. A simple spring balance can be used to check the headband tension. The test conditions for headbands with earphones are that a horizontal line joining the centres of the earphones shall lie 129 mm below the top of the headband and the earphones shall be separated horizontally by 145 mm. The force required to maintain this separation is to be 4·5 ± 0·5 N. Similar

conditions apply to the headbands for bone vibrators. The separation is to
be 145 mm for mastoid application or 190 mm for forehead application, and
the force is to be 5.4 ± 0.5 N.

Short-Duration Signals

Short-duration acoustic signals, usually tone bursts or clicks, are essential
to the production of evoked potentials and to the generation of synchro-
nized acoustic emissions from the cochlea. For evoked potentials, the source
is likely to be a supra-aural or insert earphone, but bone vibrators may
occasionally be used even though current models are ill suited to this work.
The object is usually to estimate the threshold of hearing or to examine the
waveform of the response. In either case the notional hearing level of the
stimulus is required; therefore, reference equivalent levels are needed. Although standard reference levels have not yet been agreed upon, work is in
progress with a view to publication in part 6 of ISO 389. In the meantime
users must rely on their own data.

The situation for otoacoustic emissions is different because the object
is not primarily that of finding the threshold level of the stimulus. It is
true that thresholds for the emergence of a recognizable cochlear emission
have been reported, but this is because an audiologist's first reaction to
anything is to measure its threshold. The clinical importance has yet to be
established.

The probe used to detect evoked otoacoustic emissions has both a
sound source and a microphone. Otoacoustic emissions can be measured
in a number of ways, but in the most widely practised clinical test the
sound pressure waveform in the ear canal is observed directly. The equipment is therefore capable of making its own sound pressure calibration
provided that the microphone is itself correctly calibrated. The overall
performance of the probe can usually be tested 'on site' in accordance
with the manufacturer's instructions. A simple test cavity is often supplied for this purpose. For a laboratory test, the performance of the
probe's microphone can be verified by comparison with a standard microphone exposed to the same sound pressure. The absolute calibration
of the microphone is not critical, but the linearity of its response and that
of the sound source are important, especially if the equipment is to be
operated in the nonlinear mode. In this mode it is the nonlinear component
of the ear's response that is being investigated, while the linear part is
cancelled by subtracting linearly scaled responses to different stimuli.
Nonlinearity in the measuring equipment is not likely to go unnoticed

because it will show itself as an apparent otoacoustic response where none exists (in a test cavity, for example). The absence of a response in tests on real ears is a frequent, if audiologically unwelcome, confirmation of equipment linearity.

Guidance on the measurement of short-duration signals is available from IEC 60645-3. The terms used to characterize these signals are defined and methods of relating the signal level to hearing level are suggested. The peak-to-peak equivalent sound pressure level is a particularly important characteristic that has now become a well-established expression of the signal level. Its definition as given in the IEC standard (and in Chapter 6) is, however, questionable in that it invokes a hypothetical electrical signal used to supply the sound-generating transducer. The words 'same transducer and same test conditions' are therefore necessary. But a brief *acoustic* signal has its own long-duration equivalent regardless of how it is produced. The peak equivalent is simply an arithmetical expression of a particular feature of the short-duration pressure waveform; it is the sound pressure level at the point of measurement corresponding to a sinusoidal vibration having the same peak-to-peak value as that of the signal being represented. In other words, if enough is known about the short-duration signal that its peak-to-peak sound pressure can be measured, then the continuous equivalent sinusoid is completely specified in the abstract with no requirement other than practical convenience to create such a signal in any real transducer. The significance of this is that a peak equivalent level can be measured on a sound-level meter by substituting any sinusoidal input, acoustic or electrical, that gives the same peak-to-peak response in the output.

It is manifestly more difficult to measure transient signals than steady sinusoids. A good-quality microphone is needed, and frequency response of the measuring apparatus as a whole must be uniform throughout the range of frequencies that contribute significantly to the Fourier spectrum of the signal. Phase distortion should be considered, particularly in regard to its influence on the peak levels in the amplified waveform. Possibly the simplest way of measuring the peak equivalent of the signal is to observe its waveform with a two-channel oscilloscope connected to the ac output of the sound-level meter. The second channel is supplied directly with a rectangular signal whose display is superimposed on the transient and adjusted so that both have the same peak-to-peak value. The input to the sound-level meter is then connected to an appropriate sinusoidal source whose amplitude is adjusted so that the displayed sine wave lies exactly within the limits of the rectangular wave. The indication on the sound-level meter is then the peak equivalent of the transient.

Sound-Field Audiometry

As applied to audiometry, the term *sound field* implies that the test sounds are transmitted to the listener across an open space instead of emanating from an earphone or a bone vibrator. Sound field tests have several uses, including research into basic hearing mechanisms, measurement of the attenuation of hearing protectors, testing the performance of hearing aids, and measuring the hearing thresholds of children. Audiologists will be particularly involved with the last two items in this list. In a general clinical setting it is unlikely that test rooms will be of a quality that permits absolute measurement with the accuracy possible in earphone work, but valid comparisons (for example, between different hearing aids) can be made. The uncertainties in estimating children's hearing have generally far more to do with the subject material than with the physical acoustics, so here too a somewhat imperfect test environment may be acceptable.

Sound fields for audiometric work are described in ISO 8253-2. This standard recognizes the difficulty of providing research-quality test rooms. In addition to describing the ideal free and diffuse fields, it suggests a more practical specification for what it calls a *quasi-free sound field*. The field is that of a loudspeaker, and a reference axis is specified as being perpendicular to the surface of the speaker and passing through its centre. The listener, who is to be seated facing the loudspeaker, is assigned a reference point on the reference axis and midway along a line connecting the openings of the ear canals. The reference point should be at least 1 m from the loudspeaker. The following text is taken from the standard.

To establish quasi-free sound field conditions, the following requirements shall be complied with.

a) The loudspeaker shall be arranged at the head height of a seated listener, the reference axis being directed through the reference point. The distance between the reference point and the speaker's reference point shall be at least 1 m.

b) With the test subject and the subject's chair absent and all other normal working conditions maintained, the sound pressure levels produced by the loudspeaker at positions of 0·15 m from the reference point on the left-right and up-down axis shall deviate by no more than ± 2 dB from the sound pressure level at the reference point for any of the test signals.

c) With the test subject and the subject's chair absent, the difference in sound pressure levels produced by the loudspeaker at points on the reference axis at 0·10 m in front of and 0·10 m behind the reference point shall deviate from the theoretical value given by the inverse square law by no more than ± 1 dB for any of the test signals.

The usable frequency range of the quasi-free sound field is defined by the frequency range within which these requirements are complied with.

Anyone who has attempted free-field audiometry with pure tones will appreciate the difficulties caused by standing waves. The interference patterns can be disrupted by using frequency-modulated tones or narrow bands of noise. The standard makes recommendations concerning these signals. For frequency-modulated tones, the frequency deviation should be in the range $\pm 2 \cdot 5\%$ to $\pm 12 \cdot 5\%$ with a modulation frequency of 4 to 20 Hz.

Sound field audiometers can be calibrated to indicate either the sound pressure level in the undisturbed field or the corresponding hearing level. A hearing level calibration requires reference levels, which the user may have to obtain for himself by testing a panel of audiologically normal subjects. Some standard reference thresholds[12] have been published in ISO 389-7. These are for pure tones heard binaurally in a free field and for third-octave bands of noise heard binaurally in a diffuse field. The threshold values for the two conditions are remarkably similar, differing by not more than 2 dB at any frequency below 6000 Hz. Diffuse fields are not usually used for clinical work, and, as stated above, clinical tests are usually conducted in quasi-free-field conditions for which the standard free-field reference values are probably the best available. In some behavioural tests (for example, where a head turn is an indication that the subject has heard the tone), it may be necessary to place the loudspeakers off-axis. Corrections for 45° and 90° incidence are suggested in Annex B of ISO 8253-2.

Reference sound field thresholds for pure tones and third-octave bands of noise are given in Table 8.5. The data are from ISO 389 at selected frequencies. Some audiologists use the A-weighted free-field sound pressure level as an approximate reference threshold and even report A-weighted levels as hearing levels. The table lists values of the inverse of the A-weighting function. These values are equal to the unweighted sound pressure levels (dB SPL re 20 μPa) that exist when the sound-level meter indicates 0 dB(A). They may therefore be compared directly with the tabulated reference values for pure tones.

Speech Audiometers

Speech audiometers seldom exist as commercially manufactured items in their own right. Instead, they usually comprise a number of separate components that have been combined by the user. The input is either live

---

[12]There are various sources of reference data. For a discussion, see G. Beynon and K. Munro. A discussion of current sound field calibration procedures. *B. J. Audiol.* 27: 427–435, 1993.

TABLE 8.5   Reference Thresholds of Hearing for Tones and Narrow
Bands of Noise for Sound Field Audiometry (ISO 389-7)

Frequency (Hz)	Pure tones (0° incidence) (dB SPL)	Third-octave noise, diffuse field (dB SPL)	Inverse A-weighting (dB)
250	11·0	11·0	8·6
500	4·0	3·5	3·2
750	2·0	1·0	1·1
1000	2·0	0·5	0·0
1500	0·5	−1·0	−0·9
2000	−1·5	−1·5	−1·2
3000	−6·0	−4·0	−1·2
4000	−6·5	−5·0	−1·0
6000	2·5	−0·5	0·0
8000	11·5	5·5	1·1

voice, for which a microphone and possibly a preamplifier are needed, or more probably, prerecorded speech material on tape or compact disc. A clinical or diagnostic audiometer is usually used to amplify the signal and control its level and to provide masking noise. The output is either from earphones (preferably audiometric earphones) or from loudspeakers, in which case additional power amplification is sometimes needed. Audiologists may well assemble such systems themselves, but they would be wise to seek help from acoustics specialists to ensure that the final product performs in an appropriate way. A standard specification for speech audiometers is provided by IEC 60645-2, which should be considered mandatory (but rather difficult) reading for anyone who is serious about speech audiometry. The following paragraphs are based on information in the standard. They are intended only to explain the principles underlying the calibration of speech equipment. For detailed information, particularly in regard to tolerances, the reader must consult the original.

Speech audiometers are classified into two main groups: type A and type A-E provide a wide range of facilities, while type B and type B-E have only the basics. The letter *E* specifies the type of calibration that is used when the output of the audiometer is from an earphone.[13] If the letter *E* is not included, then the calibration is in terms of the sound pressure in a coupler or artificial ear without correction. If the *E* is included, then the earphone calibration is in terms of what is called the *equivalent free field*. With this

---

[13]Bone vibrators are occasionally used for speech tests. For completeness, the word *earphone* may be understood to imply 'earphone or bone vibrator' when the context is appropriate.

type of calibration the listener wears earphones but perceives sound as though it had come from a source of plane progressive waves. The principle is that speech sounds are radiated from the human speaker into a free field, recorded, and replayed through earphones. The frequency response of the audiometer is such that the sound spectrum at the eardrum is the same for earphone listening as it would be for listening in the free field of the original source. Correction factors are therefore needed to take into account the frequency response of the earphone and the transformation of the sound pressure in the undisturbed field to the sound pressure at the eardrum when the listener is present.

In practice, it is not necessary to measure the sound pressure at the eardrum; instead, a loudness balance technique is used. To see how this works, we need the following definitions:

- *Coupler sensitivity:* The ratio of the sound pressure generated by the earphone in an acoustic coupler (or artificial ear) to the voltage at the terminals of the earphone.
- *Coupler sensitivity level:* The coupler sensitivity expressed in decibels relative to 1 Pa/V. Symbol G_C.
- *Free-field sensitivity:* At a given frequency, the ratio of the sound pressure in frontally incident plane progressive waves to the voltage that must be applied to the earphone so that the source of the plane waves and the sound heard in the earphone are equally loud. The results of loudness judgements are to be averaged for at least 10 otologically normal listeners. Listening may be binaural, but the resulting sensitivity is deemed to be that of a single earphone.
- *Free-field sensitivity level:* The free-field sensitivity expressed in decibels relative to 1 Pa/V. Symbol G_F.

Audiologists will be relieved to find that they are not required to undertake the difficult task of determining these sensitivities for themselves. Values of $G_F - G_C$ for commonly used earphones in combination with the standard coupler or artificial ear are given in Annex A of the IEC publication. Adding these values to the sound pressure levels measured in a coupler will give the corresponding sound pressure levels in a free field. The merit of calibration in terms of the free-field equivalent output is that speech tests using different types of transducer are directly comparable. This form of calibration is not mandatory, however, and despite its advantages, it is probably more usual to find that speech audiometers are calibrated directly in terms of the sound pressure in a coupler without the free-field correction. This provides what the IEC standard calls the 'uncorrected earphone output'.

The frequency response of a speech audiometer is an important characteristic that has to be established before it can be calibrated. If we assume that a clinical audiometer is used to amplify and control the speech signal (but not to record or to play back), then the frequency response of the audiometer for speech signals can be investigated by applying an electrical test signal with a constant rms voltage to the input and measuring the sound pressure level at the output. For calibrations using the uncorrected coupler response, the sound pressure in the coupler should be the same at all frequencies; for a free-field equivalent calibration, it should have a constant value plus the frequency-dependent correction described earlier; for loud-speakers, the frequency response should be uniform. The frequency response should be measured using test signals comprising third-octave bands of noise. Further details of the test signals and the tolerances on frequency response are given in the standard.

The measurement of frequency response using an electrical test signal at the input to the audiometer is valid for the speech system as a whole provided that the output from the playback system is a faithful representation of the sound pressure in the original speech signal at the recording microphone. It is therefore necessary to ensure that the appropriate equalization networks are included in the playback system.

Calibration of the output level of a speech audiometer is analogous to the calibration of a pure tone audiometer, but difficulties soon become apparent. Speech itself is a complex signal; furthermore, the test material can be recorded in several different ways — for example, as lists of single words or as words or phonemes within a carrier phrase. Various methods have been used to express the sound level in the test material. The following quotation is taken from IEC 60645.

Speech level   The sound pressure level or the vibratory force level of the speech signal measured in an appropriate coupler, ear simulator or in a sound field with a specified frequency weighting and specified time weighting.

Notes

1. For example, the speech level may be expressed as the equivalent continuous sound pressure level or vibratory force level determined by integration over the duration of the speech signals with frequency weighting C. For speech tests based on single test items separated by silent intervals, the integration should not include these intervals. For test lists based on single test items with carrier phrase, the integration should include test items only.

2. For lists of single test items the equivalent continuous sound pressure level may be estimated by subtracting 5 dB from the average of the maximum measured sound pressure levels using frequency weighting C and time weighting I. [What this note means is that the greatest C-weighted impulse response is obtained for each item in the list and then averaged for all items].

Speech level is analogous to the sound pressure level in a pure tone. In pure tone audiometry, the threshold of hearing is the level of the tone when it can just be detected; in speech audiometry, the *speech recognition threshold level* is the lowest speech level of the test material at which 50% of the material can be correctly identified by the listener. The *reference speech recognition threshold level* is then the median value of the recognition level for otologically normal listeners. In principle, then, a standard speech recognition scale can be constructed in a similar way to the hearing level scale for tones, subtracting the reference recognition level from the speech level. A level on this scale is called the *hearing level for speech.* The zero of the scale is the level at which otologically normal listeners achieve a recognition score of 50%. There is, however, no universally defined reference level and no universal hearing level for speech. The user or the manufacturer of the test material must establish a reference level by testing the material on normal listeners. The purpose of standardization is, however, to ensure so far as is possible that users of the same test material will be able to obtain comparable results, and in principle a reference recognition level should have universal validity for a given material.

Audiometer manufacturers need some guidance as to how speech equipment should be calibrated, given that it will be used with different speech materials. The IEC standard makes a simple recommendation in which the output is specified in relation to a calibration signal applied at the input. Unfortunately, it seems that it has not been possible to agree on an exact specification of the calibration signal other than its level. The signal, which should be continuous, can be a frequency-weighted random noise such as that used for masking speech, a band of noise, or a frequency-modulated tone centred on 1 kHz having a bandwidth of at least a third-octave. This signal should be recorded together with the test material at a level equal to the average speech level in the record.

The speech audiometer has a signal-level indicator (a VU meter[14]) to monitor the level of the signal at a stage in the amplifier before the output-level control. The scale of the VU meter is marked with a reference point that is normally used to set the gain of the amplifier when replaying the calibration signal prior to audiometry. A laboratory calibration is required to ensure that the audiometer produces a defined output in response to an input test signal when the gain is adjusted to give the reference VU indication. The output depends, of course, on the setting of the output-level control (the signal attenuator). On many audiometers the output indication is an arbitrary scale in decibels, but for audiometers

[14]A VU meter is a meter for measuring the level of signals such as music and speech, with a response characteristic approximating that of the human ear. The time constant is 65 ms. See IEC 60268-17.

complying with the IEC standard the scale should indicate either sound pressure level (types A-E or B-E) or hearing level for speech (types A or B). If the scale is marked in sound pressure level, the output sound pressure level in a free field, or its equivalent in a coupler or ear simulator, should be as shown on the scale when the input produces the reference VU indication. If the scale is marked in hearing level, the output sound pressure level is numerically 20 dB above the scale indication; that is, 20 dB SPL corresponds to 0 dB HL for speech. These requirements are expressed in the standard by defining a 'reference position' of the output control at 20 dB for types A-E and B-E and at 0 dB for types A and B. With the control at the reference position and the input giving the reference VU indication, the output should be 20 dB SPL (or 55 dB re 1 μN for bone vibrators). Although, as stated earlier, a standard reference threshold level for speech does not exist, it is generally the case that a sound pressure level of 20 dB corresponds approximately to a speech recognition of 50% for easily recognizable material. The standard hearing level calibration for types A and B is therefore a useful starting point that can be established by laboratory calibration. The user may add or subtract a correction according to the results of speech tests with the particular material being used.

The sound pressure level of 20 dB is given for reference purposes, but it is too low for actual calibration. In practice, a much greater output is needed to achieve an adequate signal-to-noise ratio in the test equipment. The standard recommends that the calibration should be made with the level control set to 70 dB for earphones or loudspeakers, or 40 dB for vibrators. The calibration will normally be made with an input signal consisting of a third-octave band of noise centred on 1 kHz. The input gain is set to give the reference VU indication. For a loudspeaker, the free-field sound pressure at a distance of 1 m on the reference axis should be 70 dB SPL; the sound pressure in a coupler or artificial ear should be $70 + (G_F - G_C)$ dB SPL for a type E calibration, or 90 dB SPL for an uncorrected calibration.

Masking Noise

Masking is used in pure tone audiometry to exclude the nontest ear. It is sometimes needed in speech audiometry for the same reason. Another use of masking noise in speech audiometry is to provide a competing sound, making the listener's task that of identifying speech in background noise. Often special types of noise such as 'speech babble' are used for this purpose, but broad-band noise with a spectrum approximating that of the long-term speech spectrum is a frequently used alternative. Speech-in-noise tests can be made in a free field using loudspeakers, but masking, in the sense of excluding the nontest ear, applies only to earphone tests.

The ability to generate speech-weighted noise is mandatory for all speech equipment specified in IEC 60645. The masking noise should ideally have a constant spectrum level in the frequency range 125 to 1000 Hz and fall at 12 dB per octave for frequencies between 1000 and 6000 Hz. The level for frequencies beyond 6000 Hz is not specified. If loudspeakers are used, the specification applies to measurements made acoustically in a free field; for earphones, it applies to coupler or artificial ear measurements with or without the equivalent free-field correction, depending on type.

The masking-level control should indicate either sound pressure level of the noise (types A-E and B-E) or its *effective masking level* (types A and B). The latter is defined as a level numerically equal to the increase in the speech recognition threshold due to the presence of the noise. It may be determined empirically for a group of normal listeners by combining the masking noise and the speech signal in the same earphone. For loudspeakers, the sound pressure level in the field should be the same as that indicated on the masking-level control. The same is true for a type E calibration with earphones, but allowance has to be made for the correction terms. For uncorrected coupler measurements, the masking-level control should be calibrated in terms of effective masking level, and the sound pressure in the coupler or artificial ear should be numerically 20 dB greater than the indication on the level control.

Tympanometers

The word *tympanometers* is a convenient shorthand for *instruments for the measurement of aural acoustic impedance/admittance*, which is the title of IEC 61027. These instruments have a system for measuring the acoustic admittance (or sometimes impedance) of the ear, a system for controlling the static pressure in the external ear canal, and a system to provide tones or bands of noise that evoke the acoustic reflex. There must also be a means of displaying and recording the test results.

Their maintenance and calibration is usually a matter of following the instructions in the makers' service manual. A point to bear in mind is that it is inadvisable to expose the calibrating microphone to large changes in static pressure. For this reason, the pneumatic system should be disabled prior to acoustic measurements in a coupler or artificial ear. This may involve disconnecting the pressure tube that supplies the probe, but care should be taken to see that the disconnection does not in itself change the sound pressure in the coupler by creating an acoustic leak. The usual laboratory inspection includes checking the frequency and level of the probe tone, checking the admittance or compliance indicated when the probe is connected to standard calibrating cavities, checking the static pressure, and checking the stimulus system. Stimulus levels and frequencies may be tested in the same way as they would be for audiometers. Calibration is in terms of hearing level. There are no standard reference thresholds for some types of stimuli that are often provided, but the requisite sound pressures for calibration will be specified by the manufacturer.

Calibrating the Calibrators

The equipment for calibrating audiometers has itself to be tested and calibrated from time to time. This may, on occasion, mean having microphones and mechanical couplers calibrated by national laboratories, but there are less expensive in-house methods that can reveal a significant change in performance between visits to the laboratory. A sound-level meter, excluding its microphone, can be tested by applying a known electrical input. An internal reference source is usually provided for this purpose, but it is worth making an occasional test with an external source. A suitable adapter may be needed to remove the microphone polarizing voltage. The microphone can be tested using a pistonphone. Its sensitivity (in volts per pascal) should agree with the manufacturer's specification, and there should be no significant change between measurements. The performance of the measuring system as a whole should be checked periodically by measuring the output of a reference earphone or bone vibrator kept for this purpose. National laboratories will usually be willing to calibrate the reference earphones and vibrators on their own couplers, ear simulators, and mechanical couplers. Provided that these calibrations can be replicated, there is no need to repeat the exercise.

Some medical equipment laboratories may wish to undertake their own calibrations of microphones and mechanical couplers. Methods for calibrating microphones were described in Chapter 6. The necessary equipment and advice on how to use it can be obtained from acoustic instrument manufacturers. In-house calibration of mechanical couplers is also possible. Brüel & Kjaer, currently the principal manufacturer, describe how the performance of the mechanical coupler may be measured, but there is little in the audiological literature regarding this work. The following brief account is based on the methods described by Brüel & Kjaer (B&K) with some modifications suggested by the author.

The coupler under test is driven through an impedance head by a high-quality vibrator. The impedance head (B&K type 8000) comprises an accelerometer and a force gauge.[15] The gauge is applied to the surface of the mechanical coupler with the standard coupling force (5·4 ± 0·5 N), and its output gives an accurate indication of the vibratory force at the input to the coupler. Measurement of the corresponding output of the coupler provides the desired calibration. It is, however, necessary to correct the output of the force gauge for the voltage produced by virtue of the mass of the gauge itself. This mass, described by the manufacturer as 'the mass below the gauge', is typically about 1·1gm. It is part of the vibrating system and therefore subject to the same acceleration. A correction, proportional to acceleration, can be derived from the output of the accelerometer and subtracted electronically from the force signal. When this is done correctly the force gauge should register no significant output if, having been removed from the mechanical coupler, it is made to vibrate without external load.

If money is no object, then the mechanical coupler, the force gauge, and the accelerometer can each be provided with a conditioning amplifier (preamplifier) and a sound-level meter or measuring amplifier — an arrangement favoured by the manufacturer. A cheaper alternative is to dispense with the conditioning amplifiers (direct input to the sound-level meters is usually quite acceptable) and to work with just two-sound level meters. One of these is used to measure the applied vibratory force (the force-level meter) and the other is used either to measure acceleration or to measure the output of the mechanical coupler. It is necessary to construct a simple circuit to compensate for the mass of the force gauge. The circuit, which is inserted between the external filter sockets of the force-level meter, requires two inputs, one from the ac output of the meter measuring acceleration and one from the output-to-filter socket of the force-level meter. The signal levels have to be suitably adjusted (using potentiometers) and then subtracted with a differential amplifier whose output is returned to the input-from-filter socket. This method of compensation works well but its operation is restricted to predetermined settings of the input gains of the sound-level meters.

The accelerometer does more than provide a source of correction for the force gauge. It enables the velocity of the vibration to be calculated (velocity equals acceleration divided by angular frequency). The ratio of the vibratory force to the vibratory velocity is the mechanical impedance of the coupler. The impedance can be resolved into its real and imaginary parts by measuring the phase difference between the acceleration and the force.

[15]The calibration of the accelerometer can be verified at low frequencies using the acceleration due to gravity as a reference. If the acceleration is known the calibration of the force gauge can also be checked by loading it with a known mass. (P. M. Haughton. *B. J. Audiol.* 19: 53–56, 1985.)

The driving vibrator for calibrating the mechanical coupler is likely to be quite heavy. For example, the B&K type 4810 'minishaker' weighs 1·1 kg, so that a special attachment is needed to apply it to the mechanical coupler. A good engineering solution is to suspend the vibrator from a cord that passes over pulleys to a counterweight. This weight is made equal to that of the static components of the vibrator. A loading weight equal to 5·4 N (550 gm) less the weight of the dynamic components (armature, 18 gm, and impedance head, 29 gm) is suspended by elastic rings from an attachment at the base of the impedance head. The specified coupling load will then be exerted when the force gauge is brought into contact with the top of the mechanical coupler so that it supports the impedance head and the loading weight, and the axial force between the case of the vibrator and its armature will be removed (Figure 8.10).

Because of limitations in the performance of currently available impedance heads, traceable force calibrations of mechanical couplers can only be provided at frequencies below 6 kHz.

Audiometric Zero: Problems in its Definition and Standardization

There would be little clinical interest in the threshold of hearing were it not for the fact that nearly all hearing disorders are accompanied by a loss of auditory sensitivity. Indeed, the patient with a normal audiogram who presents with hearing difficulty is likely to be diagnosed as having an 'obscure auditory dysfunction'. So, clinically at least, the threshold concept is of great practical value. The clinician would be justified in thinking that the measurement of hearing thresholds is a rather simple procedure in which the patient is presented with a series of defined sounds and asked if they are audible. As such, it is little more than a tuning fork test, refined and made quantitative. Yet the subject is bedevilled with methodological and technical difficulties that have not been fully resolved despite the many research publications that have been devoted to it. Part, if not all, of the difficulty is that audiometry has become much more than a simple clinical test. Its role has been expanded to include studies in auditory physiology, psychology, epidemiology, occupational health, and forensic medicine, with ever-increasing demands for better precision, better standardization, and more reliable calibration. Underlying this effort has been the need to refine the definition and realization of audiometric zero—that physical stimulus that corresponds to the normal threshold of hearing.

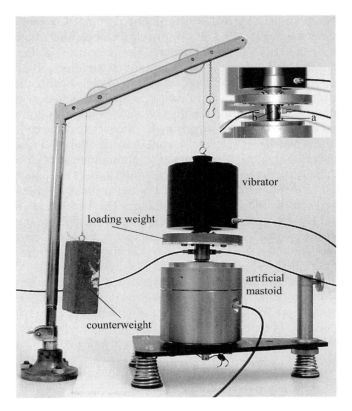

FIG. 8.10 Calibrating an artifical mastoid: a method for applying the vibrator, accelerometer, and force gauge to the mastoid with the specified static force but without static axial load on the vibrator. Inset: impedance head, (a) force gauge, (b) accelerometer.

Readers who undertake calibration work for themselves will soon find that the standards specifying audiometric zero have a long pedigree. The plethora of standards, constantly under revision, and the existence of numerous inconsistencies may seem bewildering. It may be astonishing that after so many years there are still two independent definitions of the air conduction zero: a universal one involving the artificial ear and an exclusive one applying to just two types of earphone with the 9A coupler. If a manufacturer were to produce a new model of earphone that was acousti- cally identical to the TDH-39 but which had a different name, it would have to be calibrated on the artificial ear rather than the coupler. Indeed, when the TDH-39's two-piece MX41/AR earphone cushion was replaced by a single-piece version, calibrating laboratories were uncertain whether to use the coupler or the artificial ear. Common sense prevailed and the standard now includes the single-piece model 51 cushion as an alternative in coupler measurements. In principle, the artificial ear should have made the acoustic coupler obsolete and only one set of reference values should now be needed. But the path to standardization is a difficult one, requiring compromise and pragmatism. It is therefore not surprising to discover that the calibration of the Telephonics and Beyer earphones would change slightly if it were made on the artificial ear instead of the coupler.[16] An audiometer with a TDH-39 that was perfectly calibrated on the coupler would appear to be slightly out of calibration if its earphone were transferred to the artificial ear, even allowing for the differences in the reference thresholds. The widespread use of the TDH-39 and DT-48 earphones may have provided the political justification for retaining the coupler as a standardizing device, but its days are numbered.

Bone conduction also has its anomalies. The mechanical coupler used to calibrate bone vibrators is specified in IEC 373 (now IEC 60373). In the first edition of this standard the mechanical impedance of the coupler was given in terms of its mechanical resistance and reactance, providing a complete physical specification. Because of manufacturing difficulties, the specification was later changed. The current (1990) version of the standard gives only the modulus of the mechanical impedance (which differs slightly from that in the original version) and gives the phase angle at just one frequency. We are told that the desired mechanical response will be obtained when the viscoelastic rubbers used in the coupler have the specified mix and are cured in the specified manner. Physics has been replaced by cookery 'to meet the urgent needs of users in the audiometric and hearing aid fields'.

Technical problems aside, there remains a fundamental obstacle to the definition of audiometric zero in that *threshold* itself is an elusive quantity. Hearing, in the sense of detecting the presence of sound, can be regarded as a noise-limited process. The noise is physiological, inherent in the sensory apparatus and perhaps the neural mechanisms that come after it. It is analogous to the inherent noise in physical apparatus, such as a sound-level meter. If this view is correct, then measuring the normal threshold of hearing is analogous to measuring the inherent noise in a laboratory instrument. The essential difference is that instrument noise is self-evident — we do not have to probe a sound-level meter with test sounds to find the point at which it responds; we simply observe the indication when external sounds are removed. But so far as we know, there is no alternative method for finding the limiting performance of the human ear.

[16]D. W. Robinson et al. Audiometric zero for air conduction. *Audiology* 20: 409–431, 1981.

Let us consider briefly, and in a somewhat simplified way, the steps leading to the practical definition of the normal threshold of hearing. They are as follows:

1. Select a group of individuals whose hearing is normal.
2. Measure the hearing thresholds of each individual. Each threshold value is expressed as the rms voltage across the terminals of the earphone when it is producing a tone at the threshold of hearing.
3. Undertake a statistical analysis of the results so obtained and thereby derive a value of the earphone voltage corresponding to what shall be taken as the normal threshold of hearing.
4. Couple the earphone to a microphone using a standard acoustic coupler and, applying the voltage derived in the previous step, find the sound pressure level corresponding to the normal threshold of hearing.
5. Steps 1 to 4 are also undertaken at other laboratories. Results are compared and a consensus reached as to the sound pressure that will represent the normal threshold of hearing. This sound pressure, expressed in decibels, is what we have called the reference equivalent sound pressure level (RETSPL). Reference pressure levels are required for each frequency used in audiometry.
6. To calibrate an individual audiometer, the earphone is applied to a standard coupler and the output of the audiometer is adjusted so that the sound pressure level in the coupler is equal to the indicated hearing level plus the reference equivalent pressure level for the frequency concerned.

Steps 4 and 6 involve only the direct measurement of a physical quantity (sound pressure). Any difficulties here are merely technical. On the other hand, the methods leading to the definition of a reference sound pressure are if anything closer to the biological and behavioural sciences than they are to physics. They are associated with problems that are almost, though perhaps not entirely, absent in purely physical measurement.

To illustrate this, consider an example of the kind of measurement that lies entirely in the domain of the physical sciences. Suppose a physicist is asked to measure the density of pure gold. The physicist would probably accept without question that there is such a substance as pure gold, or that samples could be obtained for which impurities were present in such small amounts as to have a negligible influence on the result, and would consider that the task was the technical one of finding the volume and mass of a sample in terms of the standard metre and the standard kilogram. Colleagues might be asked to perform similar measurements. They would, of course, obtain different results because no measurement is perfect, but they would accept the proposition that there is such a thing as a 'true' result that could be approached ever more closely as methods of measurement are refined. But is there such a thing as 'true' threshold of hearing or a 'truly normal' person? The answer has to be no. In the first place, the threshold of hearing requires a statistical definition. It is not an absolute. There is no sound level below which a tone is always inaudible and above which it is always audible. Moreover, the method of presentation (the audiometric test method) influences the result to an important degree. *Threshold* therefore has to be defined both statistically and with reference to the method of test. By contrast, the density of gold is imagined to be the same regardless of how it is measured.

We may imagine that 'normal hearing' is something that can readily be understood, and so it can be in a colloquial sense, but when a rigorous definition is needed for scientific purposes, normality seems hard to define. The audiological standards have, over the years, described the 'otologically normal person' in various ways. They amount to the same thing. The following is from ISO 389-1 (1998):

Otologically normal person — a person in a normal state of health who is free from all signs or symptoms of ear disease and from obstructing wax in the ear canal, and who has no history of undue exposure to noise, exposure to potentially ototoxic drugs, or familial hearing loss.

This definition, however well intentioned, does little but make some obvious exclusions. It is a definition that has been created retrospectively on combining data from disparate studies. The only further qualification is that the reference population should comprise 'a sufficient number of ears of otologically normal persons, of both sexes, aged between 18 and 30 years'.[17]

One of the most cogent signs of ear disease is a raised hearing threshold. Should audiometric zero be defined with respect to the hearing of persons selected for their good hearing? Exclusion criteria such as a hearing threshold that exceeds the modal value by more than 15 or 20 dB have often been applied, but the case against such criteria is that they create an arbitrary modification of the threshold statistics. On the other hand, the subject's history and the results of otological examination are legitimate and important considerations.[18] In other respects the specification for the test subjects is quite liberal. It makes no mention of anatomical size, or ethnic factors, or even the mental ability to perform an exacting hearing test under laboratory conditions. Nor does it mention factors such as socioeconomic class that

---

[17]The age range is sometimes given as 18 to 25 years (e.g. in IEC 60645-2).
[18]Robinson et al. Audiometric zero. See also ISO/TC43/WG1. Preferred test conditions for determining hearing thresholds for standardization. *Scand Audiol.* 25: 45–52, 1996.

TABLE 8.6 International Standards Relating to the Calibration of Audiometric Equipment

Standard	Title
IEC 60126 (1973)	IEC reference coupler for the measurement of hearing aids using earphones coupled to the ear by means of ear inserts
IEC 60225 (1966)	Specification for octave and one-third octave band-pass filters
IEC 60268-3 (1992)	Sound system equipment—Part 3: Methods for specifying and measuring the characteristics of sound system amplifiers
IEC 60268-17 (1990)	Sound system equipment—Part 17: Standard volume indicators
IEC 60318-1 (1998)	Electroacoustics—Simulators of human head and ear—Part 1: Ear simulator for the calibration of supra-aural earphones
IEC 60318-2 (1998)	Electroacoustics—Simulators of human head and ear—Part 2: An interim acoustic coupler for the calibration of audiometric earphones in the extended high-frequency range
IEC 60318-3 (1998)	Electroacoustics—Simulators of human head and ear—Part 3: Acoustic coupler for the calibration of supra-aural earphones used in audiometry
IEC 60373 (1990)	Mechanical coupler for measurements on bone vibrators
IEC 60601-1 (1989)	Medical electrical equipment—Part 1: General requirements for safety
IEC 60645-1 (1992) (revised 2001)	Audiometers—Part 1: Pure tone audiometers
IEC 60645-2 (1993)	Audiometers—Part 2: Equipment for speech audiometry
IEC 60645-3 (1994)	Audiometers—Part 3: Auditory test signals of short duration for audiometric and neuro-otological purposes
IEC 60645-4 (1994)	Audiometers—Part 4: Equipment for extended high-frequency audiometry
IEC 60651 (1979)	Sound level meters [Under review: Amendment 2 October 2000]
IEC 60711 (1981)	Occluded-ear simulator for the measurement of earphones coupled to the ear by means of ear inserts
IEC 61027 (1991)	Instruments for the measurement of aural acoustic impedance/admittance
ISO 389-1 (1998)	Acoustics—Reference zero for the calibration of audiometric equipment—Part 1: Equivalent threshold sound pressure levels for pure tones and supra-aural earphones
ISO 389-2 (1994)	Acoustics—Reference zero for the calibration of audiometric equipment—Part 2: Equivalent threshold sound pressure levels for pure tones and insert earphones

TABLE 8.6   Continued

Standard	Title
ISO 389-3 (1994)	Acoustics — Reference zero for the calibration of audiometric equipment — Part 3: Equivalent threshold force levels for pure tones and bone vibrators
ISO 389-4 (1994)	Acoustics — Reference zero for the calibration of audiometric equipment — Part 4: Reference levels for narrow-band masking noise
ISO 389-5 (1998)	Acoustics — Reference zero for the calibration of audiometric equipment — Part 5: Reference equivalent sound pressure levels for pure tones in the frequency range 8 kHz to 16 kHz
ISO 389-6	[Not yet published (2001)]
ISO 389-7 (1996)	Acoustics — Reference zero for the calibration of audiometric equipment — Part 7: Reference threshold of hearing under free-field and diffuse-field listening conditions
ISO 8253-1 (1989)	Acoustics — Audiometric test methods — Part 1: Basic pure tone air and bone conduction threshold audiometry
ISO 8253-2 (1992)	Acoustics — Audiometric test methods — Part 2: Sound field audiometry with pure tone and narrow-band test signals
ISO 8253-3 (1996)	Acoustics — Audiometric test methods — Part 3: Speech audiometry

may be considered to have a bearing on audiometric performance.[19] It would probably be pointless to insist on a more restrictive definition. The general sense is only that the test population should consist of young, healthy people.

Having chosen the test subjects it is then necessary to measure their hearing thresholds. There are obviously several ways of doing this, but ISO 389 is perversely silent on the matter, although its bibliography cites ISO 8253, where a detailed account of audiometric test methods can be found. Even here there is little guidance as to how results should be interpreted in relation to audiometric zero. It is generally held that the bracketing or ascending methods of manual audiometry lead to results in line with the current expression of audiometric zero, whereas self-recording (Békésy) audiometry is said to yield thresholds that are lower by approximately 3 dB.

---

[19]A. Davis. *Hearing in Adults.* Whurr Publishers, London, 1995. This work reports a survey of hearing in the adult population of England, Scotland, and Wales. Data are analysed by age, sex, and socioeconomic class (manual worker versus nonmanual worker).

A further consideration is that the signal level for manual audiometry is usually varied in 5-dB steps. This feature is not mentioned in the definition of audiometric zero, and it is tempting to assume that reference value is therefore independent of step size. The ascending and bracketing methods introduce an average 2·5-dB rounding error when a 5-dB step is used, and this is sometimes subtracted in the belief that audiometric zero should be the unbiased estimate of the normal threshold.

Although the distribution of hearing thresholds within the populace at large is highly skewed in the direction of impaired hearing, a far more symmetrical distribution is found within the selected test populations used to establish the normal values. Accordingly, it may make little difference whether the mean, the median, or the modal value of threshold is taken as representative, but the lack of consistency in the standards is irksome. In fact all three measures are quoted:

Modal value, ISO 389-1, RETSPL for air conduction
Mean value, ISO 389-3, RETFL for bone conduction
Median value, IEC 60645-3, clause 6 (hearing level for short-duration signals)

The difficulties and inconsistencies just described raise questions that are difficult or even impossible to answer. In one sense it is important that reference thresholds should be an accurate representation of 'normality', at least as it is generally understood. This ensures that the normal audiogram is a more or less horizontal line from which departures have an obvious clinical interpretation. But in another sense it does not really matter if the reference values are 'wrong', provided that everyone uses the same values and that all calibrations lead to the same stimulus regardless of the particular earphone or bone vibrator that generates it.

Questions and Exercises

8.1   Explain what is meant by a *hearing level* for pure tones. What is the essential difference between the hearing level scale and the sound pressure level scale? What is meant by a *reference equivalent sound pressure level*?

8.2   An audiometric earphone is fitted to an artificial ear (type 60318-1). The audiometer is connected to the earphone, and the signal-level control is set to 60 dB HL for an 8-kHz tone. The sound-level meter indicates a sound pressure level in the artificial ear of 72·2 dB SPL. The meter has been calibrated using a pistonphone so that its indication is correct to the nearest 0·1 dB at 250 Hz. The microphone sensitivity is known to be 0·9 dB lower at 8 kHz than at 250 Hz. Using the appropriate reference sound pressure given in Table 8.1, find the true hearing level of the tone.

8.3 An audiometer, normally used with supra-aural earphones, is to be used occasionally with ER-3A insert earphones. The insert earphone is found to produce the following sound pressure levels in an occluded-ear simulator (IEC 60711) when the signal-level control on the audiometer is set to 60 dB HL. Refer to Table 8.2 to find what corrections should be *added* to the values indicated on the signal-level control to obtain the correct hearing levels. Round the results to the nearest 1 dB. *Note:* The sound pressure levels given below are entirely fictitious.

Frequency (Hz)	500	1000	2000	4000
Sound pressure level in simulator (dB re 20 μPa)	75·3	70·1	68·4	82·8

8.4 An audiometer is set to deliver a 250-Hz bone-conducted signal at 35 dB HL. The force level, measured on an artificial mastoid, is 102 dB re 1 μN. Is the audiometer correctly calibrated? When the signal is filtered to remove the 250-Hz component, the force level falls to 78 dB. Calculate the total harmonic distortion. Is it acceptable? Refer to Tables 8.1 and 8.3.

8.5 Calculate the lower and upper cut-off frequencies for third-octave masking noise centered on 1000 Hz. If the noise is intended to mask a 1-kHz tone at 50 dB HL, what sound pressure level should be observed if the masking earphone (type TDH-39) is applied to a 6-cc coupler (IEC 60318-3)? See Tables 8.1 and 8.4.

8.6 In response to a rectangular pulse, an earphone produces a transient in which the sound pressure rises to 0·32 Pa It then falls to −0·20 Pa before returning to the baseline. Calculate the peak-to-peak equivalent sound pressure level of the signal. What is the peak sound pressure level?

8.7 For sound field audiometry, an audiometer is et to deliver a 250-Hz, frontally incident tone at 60 dB HL. In the absence of the test subject, the A-weighted sound pressure level at the reference point is 62·0 dB(A). What is the corresponding sound pressure level without weighting? If the ISO reference applies, what is the hearing level of the tone? (Refer to Tables 8.5 and 6.1.)

8.8 In the context of speech audiometry using earphones, what is meant by an *equivalent free field*?

8.9 What is meant by *hearing level for speech*? A speech audiometer complying with IEC 60645, type A, is set so that the output-level control indicates a hearing level of 50 dB. The test signal is a narrow band of noise centred on 1 kHz. If the amplifier is adjusted so that this signal produces an indication at the reference point on the VU meter, what sound pressure would you expect to observe when the earphone is applied to an appropriate acoustic coupler?

8.10 Sketch a graph to show the shape of the sound spectrum for standard speech-weighted noise used for masking in speech audiometry. What is meant by the *effective masking level* of the noise?

9

Audiometric Test Rooms

The accurate measurement of hearing requires an extraordinarily quiet test environment, particularly when the subject has normal hearing and the test is being made by bone conduction. The reason is, of course, that environmental sounds mask the threshold that is being measured. To understand the problem, we need to know the relationship between the acoustics of these sounds and the increase in hearing threshold that they effect. We shall refer to the environmental sound as *noise* and assume that the test signal is a pure tone.

Hearing Pure Tones in Noise

It is well established that the auditory system behaves as though it possesses a series of band-pass filters. The bandwidth of each is called a *critical band*, and together the critical bands form a contiguous set that covers the entire range of audible frequencies (Table 9.1). It should be understood that the bands listed in the table constitute an arbitrary set insofar as the frequency of any one member of the set can be chosen without restriction. We can make an analogy with the tuning circuit on a radio — one that has a variable capacitor connected to the tuning dial to allow any frequency to be selected within the displayed range. Similarly, we could choose any frequency and discover a psychoacoustic critical band centred on it. Having chosen one band, the others can then be selected so that the upper cut-off frequency of each coincides with the lower cut-off frequency of its neighbour. In psychoacoustic terms, the width of a critical band is approximately equal to 60 difference limens of frequency. Physiologically, the position of the maximum vibration on the basilar membrane moves basally by approximately 1·3 mm as the frequency is increased from the lower to the upper limit of the band. The psychoacoustic filters differ in some respects from the filters that we are familiar with in electrical engineering. They are asymmetric, having a steeper cut-off on the high-frequency side than on the low-frequency side, and their bandwidth (so far as it can be defined) is probably not entirely independent of the level of the signals that are being passed. A consequence of the asymmetry (which is

333

TABLE 9.1   Examples of Critical Bands

Centre frequency (Hz)	Bandwidth (Hz)	Lower cut-off (Hz)	Upper cut-off (Hz)	Width as fraction of octave
50	—	—	100	—
150	100	100	200	1.00
250	100	200	300	0.58
350	100	300	400	0.42
450	110	400	510	0.35
570	120	510	630	0.30
700	140	630	770	0.29
840	150	770	920	0.26
1000	160	920	1080	0.23
1170	190	1080	1270	0.23
1370	210	1270	1480	0.22
1600	240	1480	1720	0.22
1850	280	1720	2000	0.22
2150	320	2000	2320	0.21
2500	380	2320	2700	0.22
2900	450	2700	3150	0.22
3400	550	3150	3700	0.23
4000	700	3700	4400	0.25
4800	900	4400	5300	0.27
5800	1100	5300	6400	0.27
7000	1300	6400	7700	0.27
8500	1800	7700	9500	0.30
10500	2500	9500	12000	0.34
13500	3500	12000	15500	0.37

The data in this table are from B. Scharf. Critical bands. In *Foundations of Modern Auditory Theory*, Vol. 1., ed. J. V. Tobias. Academic Press, New York, 1970.

mirrored in Békésy's travelling wave) is that signals whose frequencies are above the band are rejected more effectively than signals whose frequencies are below it.

Suppose a pure tone is heard against a background of noise that has a more or less uniform spectrum over a wide range of frequencies (white noise). The threshold of hearing for the tone is of course greater than it would be if the noise were absent, and we can describe the noise as a *masker*. To a first approximation the tone is masked only by those frequencies of the noise that fall within a critical band centred on the frequency of tone. This is, however, only an approximation because the psychoacoustic filter does not have an infinitely rapid cut-off. Noise components whose frequencies are outside the critical band do contribute to masking. Their effect is called *remote masking*, as distinct from *local masking*, which is masking by components within the critical band containing the tone.

The performance of the psychoacoustic filter is demonstrated in a well-known diagram by Zwicker, reproduced here in Figure 9.1. Each curve in the diagram shows, as a function of frequency, the sound pressure level in a pure tone at the threshold of hearing. The tone is heard against a background consisting of a 200-Hz band of noise, centred on 1200 Hz. A family of threshold curves is shown, each for a constant noise level between 20 and 110 dB SPL. The curves converge at their base to the nonmasked threshold (the *absolute threshold*). It is immediately evident from the diagram that the masking effect of a band of noise is greatest for tones at the centre frequency of the band. The peaks of the masking curves are separated by 10-dB intervals on the vertical axis, showing that the threshold for a tone in the centre of a critical band increases by 10 dB for each 10-dB increase in the level of the masker. This is also nearly true for tones of lower frequency

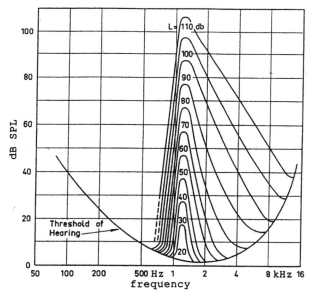

FIG. 9.1 The sound pressure level in a tone at threshold of hearing in the presence of a narrow band of noise centre on 1200 Hz. The parameter *L* is the sound pressure level of the noise. (Reproduced from E. Zwicker, Über psychologische und methodische grundlagen der lautheit. *Acustica* 8: 237–258, 1958.)

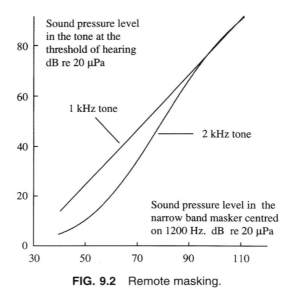

FIG. 9.2   Remote masking.

(Figure 9.2). However, if the frequency of the tone is above the upper limit of the band, 2 kHz for example, the increase in threshold level with masking level is nonlinear. Above the critical band the downward slope of the curves in Figure 9.1 becomes progressively less steep as the noise level is increased, and intercepts on a vertical line at 2 kHz are not uniformly spaced. The increase in masking effect with noise level is generally more rapid than it would be for tones within or below the band.

We have yet to consider the masking effect of a noise whose level is close to the threshold of hearing. Again, the signal being masked is a pure tone. We shall suppose that the noise comes from a wide-band source whose masking is attributed to components within a critical band centred on the tone. If the noise is present at a very low level, it produces no discernible masking. As its level is raised the threshold of the tone begins to rise, slowly at first, but eventually matching the increase in the noise as previously described. The masking function is shown in Figure 9.3. An explanation of the nonlinear form of this function is that the threshold of the tone is determined by the total noise level at some (unspecified) place within the auditory system. The noise comprises internal physiological noise together with noise due to the masker. Let T_0 dB be the threshold in the absence of masking, and let T dB be the threshold when sound pressure level in the band of masking noise is L_N dB. We suppose that the internal noise is equivalent to an external noise whose level is L_0 dB. Then, working in

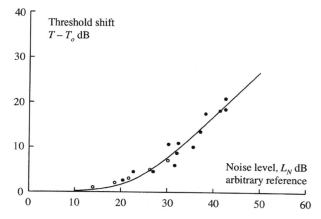

FIG. 9.3 Elevation of the hearing threshold for a pure tone as a function of the level of masking noise. The curve is derived from Equation 9.1. Open circles: data from *NPL Acoustics Report Ac 69*, 1975. Filled circles: data from Hawkins and Stevens, *J. Acoust. Soc. Am.* 22: 6–13, 1950.

absolute units rather than decibels, we have

$$10^{T_0/10} = k\,10^{L_0/10}$$

and

$$10^{T/10} = k(10^{L_0/10} + 10^{L_N/10})$$

where the signal-to-noise ratio k is assumed to be constant.

Dividing the second equation by the first and converting the result to decibels gives

$$T - T_0 = 10\log[1 + 10^{(L_N - L_0)/10}] \qquad 9.1$$

This equation gives the increase in threshold $T - T_0$ of a pure tone as a result of masking by a noise whose sound pressure level is L_N dB above L_0 in the critical band. It is based on the assumption that the listener's ability to detect the presence of the tone depends on the sound energy in the masker added to the energy associated with internal physiological noise that is assumed to determine the absolute limit of hearing. The derivation implies that the signal-to-noise ratio, namely, the energy in the tone to that in the noise, is constant at the threshold of detectability. Not everyone would agree, and it has been argued that characteristics other than the acoustic energy of the signal are implicated. This is true insofar as we recognize a tone for what it is — something that has a tonal quality that distinguishes it from the noise. If the energy of the signal is not itself the determining factor, then signal-to-noise ratio is not necessarily constant. Whatever the truth of the matter, the masking function near the absolute threshold is at least approximately as described by Equation 9.1. The masking it describes is shown by the curve in Figure 9.3. The points in the figure are experimental values. The levelling at the foot of the curve shows that external noise has a masking effect even when the sound pressure in the noise (within a critical bandwidth) is less than that in a tone at the absolute threshold of hearing. Moreover, it is known that pure tone signals can be detected when the signal-to-noise ratio in the critical band is as little as −4 dB. Taken together these observations suggest that environmental noise levels have to be exceedingly low if audiometric thresholds are to be unaffected. A further difficulty is that the normal threshold of hearing is an average, and many normal individuals have hearing thresholds that are 5 or even 10 dB below audiometric zero at some frequencies.

Measurement of masking by ambient noise close to the threshold of hearing requires an exceptionally quiet test environment in which the noise can be carefully controlled, and very precise audiometry capable of detecting small changes in threshold. Despite the difficulties, such measurements have been made, notably in experiments at the National Physical Laboratory. They show that unless the noise spectrum in the environment falls rapidly with increasing frequency, only local masking need be considered at frequencies above the lowest test frequency to be used for audiometry. Below this frequency, remote masking does have to be considered, but the restrictions on permissible noise are relatively light.

It will be evident from the foregoing discussion that limiting ambient noise levels for audiometry cannot be defined by a single number such as an overall sound pressure even when this is A-weighted. Instead, it is necessary to specify the spectrum of the noise over a wide range of frequencies. This is best done with reference to measurements in third-octave bands because, as already stated, they correspond approximately to the critical bands. Various recommendations have been published specifying the maximum level of noise that can be tolerated. The most stringent are probably those given in ISO 8253-1, yet even this standard allows the possibility of a 2-dB error when measuring down to 0 dB HL. Correspondingly lower levels would be needed for measurements below audiometric zero (which by definition should include approximately half those on young individuals with normal hearing). The ISO recommendations are reproduced, in part, in Table 9.2.

TABLE 9.2 Maximum Permissible Noise Levels re 20 μPa in One-Third Octave Bands

Centre frequency (Hz)	Air (dB)	Bone (dB)	Attenuation (dB)
31·5	66	63	0
40	62	56	0
50	57	49	0
63	52	44	1
80	48	39	1
100	43	35	2
125	39	28	3
160	30	21	4
200	20	15	5
250	19	13	5
315	18	11	5
400	18	9	6
500	18	8	7
630	18	8	9
800	20	7	11
1000	23	7	15
1250	25	7	18
1600	27	8	21
2000	30	8	26
2500	32	6	28
3150	34	4	31
4000	36	2	32
5000	35	4	29
6300	34	9	26
8000	33	15	24

Note: These levels allow pure tone thresholds to be measured by air or bone conduction as indicated down to 0 dB HL with a maximum error due to masking by ambient noise of 2 dB provided that the lowest test frequency for audiometry is 250 Hz. The numbers in the fourth column show the estimated attenuation for typical earphones (TDH-39 and Beyer DT-48). Data in the table are taken from ISO 8253-1 (1989). The standard should be consulted for more complete information.

As can be seen, very low background levels are necessary for bone conduction audiometry (and therefore for sound field audiometry). Fortunately for audiologists, most measurements are made with earphones. The earphones themselves provide useful attenuation of environmental sounds in the mid and high frequencies, but they are quite useless in this regard at low frequencies, where, paradoxically, they are likely to raise the threshold

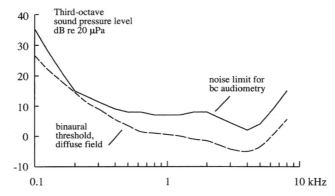

FIG. 9.4 Ambient noise limits for bone conduction (bc) audiometry compared with sound levels at the binaural threshold of hearing for tones in a diffuse sound field.

of hearing by occluding the external ear.[1] It will be noticed that the air conduction limits are not, according to the authors of the ISO standard, quite the same as the bone conduction limits plus the earphone attenuation.

When the ears are not covered (in bone conduction tests, for example), we might expect the ambient noise limits to have approximately the same frequency dependence as the minimum audible free field. The comparison is made in Figure 9.4 for the frequency range 250 to 8000 Hz. This Figure shows the limiting background noise levels given in Table 9.2 for bone conduction audiometry and the threshold of hearing for third-octave bands of noise heard binaurally in a diffuse sound field (ISO 389-7). The curves do have approximately the same shape, but it may seem surprising that the permitted ambient noise levels are greater than the hearing thresholds. The reason is perhaps that the comparison should really be with a diffuse sound field that is heard monaurally. Monaural thresholds are up to 3 dB less sensitive than binaural thresholds. Also, as previously stated, tones can be detected when the signal is 4 dB below the noise level in the critical band. It can be seen that as the frequency is reduced below the lowest audiometric test frequency (250 Hz), the noise limit rises more rapidly than the hearing threshold because in this region only remote masking is involved.

[1] The increase in ear canal noise caused by an earphone is believed to account for a discrepancy between the sound pressure in the ear canal when measurements are made in a free field and the corresponding sound pressure under the cushion of an earphone. The effect, which is present at low frequencies, remained unexplained for many years and was called 'the missing 6 dB' before Anderson and Whittle found physiological noise to be the likely cause. C. M. B. Anderson and L. S. Whittle. Physiological noise and the missing 6 dB. *Acustica* 24: 261–272, 1971.

Measuring Noise in Audiometric Rooms

The audiologist has to contend with two kinds of noise. One consists of transients or at least relative short-lived disturbances, such as the sound of footsteps of people walking near the test room, or the sound of their conversations. Although such noise is undesirable, it is generally unlikely to make audiometry impossible, provided that there are frequent quiet intervals in which the test signals can be presented. It is the other kind of noise — the steady background — that is the most detrimental. Background noise has many possible sources, including noise from lighting and ventilation equipment, from electrical machines and office equipment, and from general activity in the building. Environmental sources such as distant road traffic may also be significant, especially at low frequencies.

The measurement of low sound levels is by no means easy, and special microphones and amplifiers are generally needed. Such equipment is not cheap; audiologists will probably find it better to hire equipment rather than purchase it, or to have noise surveys of their test rooms made by acoustic specialists. The principal technical difficulty is that microphones and amplifiers are not themselves entirely noise free. It can be shown that the inherent noise in a condenser microphone is given by

$$p^2 = 4kTR_a\Delta f$$

where p is the root mean square sound pressure in a noise band of width Δf Hz, k is a fundamental physical constant called Boltzmann's constant $(1.38 \times 10^{-23}$ J/K), T is the absolute temperature, and R_a is an acoustic resistance. This resistance is dominated by the internal damping applied to the microphone diaphragm. A significant reduction in microphone noise can therefore be achieved if this damping is made smaller than normal. The drawback is that diaphragm resonance then leads to a peak in the frequency response. With careful design, however, this peak can be cancelled electrically by an equalization circuit within the preamplifier. The best one-inch microphones and preamplifiers achieve inherent noise levels below -10 dB SPL in third-octave bands throughout the auditory range. The remainder of the measuring equipment, including of course the band-pass filters, must also be of a high quality so that it does not contribute significantly to the overall instrument noise.

Few audiologists will have access to equipment needed to measure ambient noise to the levels required for bone conduction tests. Fortunately, diagnosis seldom depends on the ability to measure thresholds by bone conduction down to audiometric zero. One can usually accept that at some frequencies the limit may have to be somewhat higher, 10 dB HL perhaps.

A practical solution to the difficulty of making sound-level measurements is to measure the hearing of otologically normal individuals. If thresholds below 10 dB HL can be readily obtained in bone conduction audiometry, then it is likely that the test environment is adequate for most clinical work. Air conduction poses far less of a problem because the noise limits are within a range that is accessible to many general-purpose sound-level meters. It may be necessary to correct for inherent noise as described in Chapter 6. Measurements are best made using the slow response.

The limits on ambient noise are most restrictive when audiometry is required for reasons other than routine otological diagnosis. Research makes obvious demands, but less academic work such as establishing local norms for special tests will also require measurement below 0 dB HL. The noise limits listed in Table 9.2 define a test environment that is only just acceptable; the ideal is a test environment where threshold measurements can be made down to -10 dB HL without reliance on attenuation by earphone cushions. It is an objective difficult to achieve and difficult to confirm.

Room Acoustics: Reverberation

Sound is present in any room. It comes from sources within the room and from external sources by transmission through the walls and other boundaries. The sound waves are reflected endlessly by the boundaries and by objects inside the room, but some sound energy is absorbed on each reflection, so that if, hypothetically, all sources were removed, the sound vibrations would decay to become indistinguishable from the random thermal movements of the air molecules. The rate at which reflections occur must be proportional to the speed of sound. We can therefore expect the rate at which sound decays after sources are turned off to depend on the speed of sound and the absorption of the reflecting surfaces. The persistence of sound due to multiple echoes is called *reverberation*. The decay of sound after sources are turned off is expressed in terms of a quantity called the *reverberation time*.

If all sound sources are maintained constant, the sound pressure in the room will reach an equilibrium value such that all sound energy entering the room and emitted by sources within it is exactly balanced by the energy absorbed by the internal reflections. The boundaries of the room present a barrier to external sounds, but the sound pressure due to transmitted sound and to internal sources is, on average, greater than it would be if the radiation were unobstructed. This increase may be beneficial in auditoria because less vocal effort is required of speakers for their voices to be heard, but it is obviously undesirable in rooms intended for audiometry. The practical solution is to line the internal surfaces with sound-absorbing materials. If the surfaces are not absorbent, reflections will tend to produce a sound field that is diffuse, that is, a field in which the sound energy density is uniform and in which sound waves travel in random directions. High absorption is therefore a requirement in audiometric rooms intended for free-field work.

The problems of describing and predicting the acoustic properties of enclosures are generally complex, but some useful approximations can be made. If the absorption is relatively low, the sound field in the room will be approximately diffuse and the reverberation time will be long. Rooms with these characteristics are called *live rooms*. At the other extreme, where absorption is high, the distribution of sound energy in the room is generally not uniform and the reverberation time is relatively short. Such rooms are described as *dead*.

The first systematic investigation of the acoustics of enclosed spaces is attributed to W. C. Sabine (1868–1919). He defined the reverberation time of a room as the time required for the sound level to fall from a sensation level of 60 dB to the threshold of hearing. This was a practical definition in an age when many acoustic measurements relied on the listening skills of the observer. Nowadays we would use microphones and electronic instruments. The modern definition is as follows:

Reverberation time: of an enclosure, for a sound of a given frequency or frequency band, the time that would be required for the sound pressure level in the enclosure to decrease by 60 decibels, after the source has been stopped. (IEC 60050).

In order to relate reverberation time to the sound absorption in the room, suppose that the room has surfaces of areas S_1, S_2, S_3, ..., each having an absorption coefficient α_1, α_2, α_3, and so on. The *absorption coefficient* is defined as the fraction of the sound power in the incident sound that is not reflected. Unless otherwise stated, it is assumed that the coefficient is specified for a diffuse sound field, that is, for sound incident from random directions. Examples for various materials are given in Table 9.3.

Notice that in the definition, the sound that is 'not reflected' includes sound transmitted as well as that physically absorbed in the surface. In Sabine's work an open window was regarded as a perfect absorber. The *room absorption* can be expressed as the area A of a perfectly absorbing surface that provides the same sound absorption as the room itself; that is,

$$A = [\alpha_1 S_1 + \alpha_2 S_2 + \alpha_3 S_3 + \cdots]$$

TABLE 9.3 Effective Sound Absorption Coefficients

Material	125 Hz	250 Hz	500 Hz	1000 Hz	2000 Hz	4000 Hz
Acoustic panel	0·05	0·14	0·38	0·76	0·98	0·90
Acoustic panel with air gap and mineral wool packing	0·36	0·79	0·97	0·97	0·99	1·10
Mineral wool alone (50 mm thick)	0·14	0·36	0·65	0·80	0·76	0·70
Acoustic plaster	0·30	—	0·50	—	0·55	—
Ordinary plaster	0·04	—	0·05	—	0·05	—
Floor, carpeted	0·11	—	0·37	—	0·27	—
Floor, wood	0·06	—	0·06	—	0·06	—
Wood panels, pine	0·10	—	0.10	—	0·08	—

Note: This table gives the *effective* absorption coefficient. This is based on the effect of the given material on reverberation in a diffuse sound field. It is therefore not exactly equivalent to the acoustic energy absorbed. For strongly absorbing surfaces, the effective coefficient is greater than the true absorption coefficient, and in some cases it is greater than 1. The data are from manufacturers' publications and from Table 14.1 in L. E. Kinsler and A. R. Frey. *Fundamentals of Acoustics*, 2nd ed. John Wiley & Sons, New York, 1962.

The room absorption is sometimes expressed in *sabins* (imperial units of area, ft^2) or in *metric sabins* (m^2). These units are helpful because they make it clear that the area referred to is that of the equivalent perfect absorber rather than the actual area of the absorbing surfaces in the room.

Suppose that the sound created in the room (and entering from external sources) amounts to an acoustic power W. It can be shown that if all sources are turned on at time $t = 0$, the growth of sound intensity in the diffuse field inside a live room is given by

$$I = \frac{W}{A}\left[1 - \exp\left(-\frac{Ac}{4V}t\right)\right]$$

where c is the velocity of sound and V is the volume of the room. It will be seen from this, that as time passes the intensity approaches a final steady value equal to W/A. Correspondingly, if the sound sources are turned off at time $t = 0$, the subsequent fall in sound intensity from an initial value I_0 is given by

$$\frac{I}{I_0} = \exp\left(-\frac{Ac}{4V}t\right)$$

If we put $t = T$, the reverberation time, then by definition

$$-60 = 10 \log_{10} \frac{I}{I_o} = 10 \log_{10}\left[\exp\left(-\frac{AcT}{4V}\right)\right]$$

Therefore,

$$-60 = \frac{-AcT \times 10 \log_{10}e}{4V}$$

Substituting numerical values with $c = 343$ ms^{-1} and rearranging gives

$$T = \frac{0{\cdot}161V}{A} \text{ seconds} \qquad\qquad 9.2$$

This is Sabine's equation for reverberation time when metric units are used.

Example

As an illustration,[2] suppose that a rectangular room has the dimensions $3 \times 5 \times 9$ metres. The volume of the room is 135 m^3 and its surface area is 174 m^2. If the average sound absorption coefficient is 0·1, the room absorption A is 17·4 metric sabins. The reverberation time is therefore

$$T = \frac{0{\cdot}161 \times 135}{17{\cdot}4} = 1{\cdot}25 \text{ seconds}$$

After a few seconds, the sound intensity level due to a constant 10-μW source would become

$$10 \log_{10}\left[\frac{10 \times 10^{-6}}{17{\cdot}4 \times 10^{-6}}\right] = -2{\cdot}4 \text{ dB re } 1\,\mu\text{W m}^{-2}$$

This corresponds to a sound pressure level of 58 dB re 20 μPa. The average sound power produced in conversational speech is about 20 μW, so this example suggests that a person's voice would be clearly audible throughout the room. But the reverberation time is such that some speech sounds would be masked by preceding speech sounds that had not decayed sufficiently. A reverberation time of about 0·6 s would be acceptable.

Dead Rooms

Equation 9.2 is not valid for acoustically dead rooms. In an extreme case where the absorption coefficient approaches unity, there should be no

---

[2]This example is based on one given by Kinsler and Frey, *Fundamentals of Acoustics*, p. 424.

reverberation at all (just as there is no reverberation outdoors). The equation, on the other hand, gives a reverberation time of 0·161 V/S for a surface area S with $\alpha = 1$. The reason is that its derivation assumes that the sound field is diffuse, which is not the case if the absorption is high. In these circumstances alternative theories due to Eyring (and also to Norris) apply. They lead to results that can be expressed in terms of an average absorption coefficient, known as the Eyring absorption coefficient, which is calculated as

$$\bar{\alpha} = \frac{1}{S}[\alpha_1 S_1 + \alpha_2 S_2 + \alpha_3 S_3 + \cdots]$$

where $S = S_1 + S_2 + S_3 + \cdots$, the total surface area in the room. The equations previously given for a live room now apply if A is replaced by

$$-S \log_e(1 - \bar{\alpha})$$

The Eyring reverberation time is therefore given by

$$T = \frac{0\cdot161 V}{-S \log_e(1 - \bar{\alpha})} \qquad\qquad 9.3$$

Example

An audiometric test room has a square floor measuring 3×3 metres, and its walls are 2·5 metres high. The carpeted floor has an absorption coefficient of 0·3 at 500 Hz; the ceiling and walls are lined with acoustic panels for which the absorption coefficients are respectively 0.55 and 0.70 at this frequency. What reverberation time is expected for sounds whose frequencies are in the neighbourhood of 500 Hz?

The working is as follows:

Floor	area 9 m^2	$\alpha = 0\cdot30$	$\alpha S = 2\cdot70$ m^2
Ceiling	area 9 m^2	$\alpha = 0\cdot55$	$\alpha S = 4\cdot95$ m^2
Walls	area 30 m^2	$\alpha = 0\cdot70$	$\alpha S = 21\cdot0$ m^2

Total surface area = 48 m^2
Room volume = 22·5 m^3
$\Sigma \alpha S = 2\cdot7 + 4\cdot95 + 21\cdot0 = 28\cdot65$ m^2
Mean absorption coefficient = $28\cdot65/48 = 0\cdot597$
Reverberation time = $0\cdot161 \times 22\cdot5/[-48 \times \ln(1-0\cdot597)] = 0\cdot083$ seconds

According to design criteria recommended some years ago in the United Kingdom,[3] the reverberation time in audiometric rooms should be less than 250 ms for frequencies below 1 kHz, and less than 200 ms at higher frequencies. The preceding calculation suggests that these criteria are not too difficult to achieve with modern sound-absorbing materials. In a good-quality audiometric room the reverberation time is likely to be less than 100 ms at frequencies from 1 kHz upwards, and typically 240, 190, and 100 ms at 125, 250, and 500 Hz, respectively.

Measuring Reverberation Time

We are offered the following caution:

> Reverberation time T is defined as the length of time in seconds it takes for the energy in the steady-state sound field in a room to decay by 60 dB after the source of the sound excitation is suddenly turned off. The simplicity of this definition vanishes when one begins the process of physically measuring T.[4]

The problem is that a great many modes of vibration are possible within an enclosure, and although all modes decay exponentially when the source is turned off, they do so at different rates. At any instant the distribution of sound in the room is a complex pattern created by the interference of numerous vibrations of different frequency and phase. As the sound decays, the pattern changes in ways that are generally unpredictable, so that at any point in the field the time course of the decay is irregular. Nevertheless, it is usually sufficiently regular to permit meaningful measurement, especially if filters are used to restrict observations to a small band of frequencies.

Within a specified band, suppose that the sound decays exponentially, that is,

$$p = p_0 e^{-nt}$$

where p is the sound pressure at time t, p_0 is the sound pressure at the start of observations at time $t = 0$, and n is a constant. It is assumed that the decay is slow compared with the period of the sound vibrations within the band. It is then meaningful to represent sound pressure by a root mean square value where the average is taken over a time that is short compared with $1/n$ but long enough to smooth out statistical fluctuations in sound

[3] *Hospital ENT Services: A Design Guide*, DS 32/74. DHSS, London, 1974. Also *A Standard Procedure for Testing and Qualification of Audiology Test Rooms*, CE(76)14. DHSS, London, 1976.

[4] L. Beranek. *Acoustical Measurements*. American Institute of Physics, New York, 1988, p. 782.

pressure. From the exponential relationship, it follows that

$$20 \log_{10} p/p_r = 20 \log_{10} p_0/p_r - (20n \log_{10} e)t$$

where p_r is the reference pressure, 20 μPa. If, therefore, the decay of the reverberation is a smooth exponential, a graph of sound pressure level against time will be a straight line whose negative slope in decibels per second is equal to the term in the parentheses. Reverberation time T is then given by

$$T = 60/20n \log_{10} e = 60 \div \text{slope of the decay curve}$$

An example of a smooth exponential decay is shown in Figure 9.5. An irregular decay is shown in Figure 9.6.

The traditional method for measuring reverberation time was to plot the decay curve using a high-speed sound-level recorder in conjunction with a sound-level meter and filters. Mechanical recorders are now obsolete because modern equipment provides real-time digital analysis and data storage, allowing the decay curve to be held in memory and displayed or printed as required. The initial sound field is generated either by a random

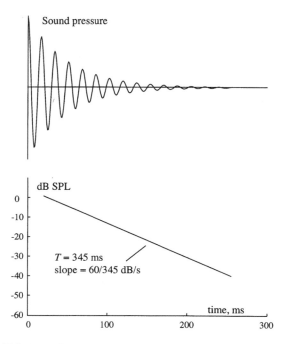

FIG. 9.5   Reverberation: simple exponential decay.

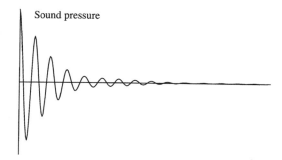

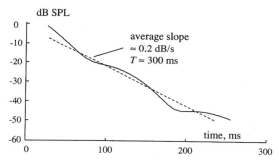

FIG. 9.6 Reverberation: irregular decay.

noise source or by an impulsive source such as a starting pistol. The
response time of the measuring equipment has, of course, to be considerably
shorter than the reverberation time being measured. With simple exponential averaging (see Chapter 6), the meter indication falls at a rate determined
by the discharge of the capacitor after the signal has been turned off.
Expressed in decibels per second, this rate is

$$10 \log(e^{-1/RC}) = -4.3/RC \text{ dB/s}$$

In contrast to this, the sound field decays at $60/T$ dB/s. The averaging time
has to be less than a twentieth of reverberation time for a satisfactory
performance.

Audiometric Rooms

Audiologists, particularly if they work in hospitals, usually have to accept
the test environment provided for them, but on rare occasions, perhaps
when planning a new building, their views may be sought. When designing

test rooms, it is generally advisable to seek the advice of acoustic consultants or companies that specialize in the construction of acoustic rooms. Architects undoubtedly have some knowledge of acoustics, but they are unlikely to be experts. It is difficult — perhaps impossible — to remedy acoustic faults once the building has been constructed, so appropriate consultation and careful planning are worth the effort and expense.

Some general considerations may influence decisions at the planning stage. The first is that structure-borne vibration is often the worst enemy, whose defeat requires the test room to be located (so far as one has a choice) well away from obvious sources of vibration such as lift shafts and plant rooms. Isolation from structural vibration can be accomplished by a room-within-a-room construction in which the test room itself is contained within a brick or concrete enclosure, separated from it by an air gap and, if possible, supported on antivibration mountings. Airborne sound is transmitted by passing vibrations through the walls, floor, and ceiling of the test room. It is therefore imperative that these structures have high mass and rigidity in order to impede their vibration. Good test rooms are made of heavy-gauge steel sheets that form a double skin. The gap may be filled with mineral wool, which is both dense and acoustically absorbent. Windows, if they have to be present at all, should be double or triple glazed. Doors must be heavy and must seal firmly. Ventilation, a necessity for the occupants, is potentially fatal to the acoustics, but fortunately, well-designed silencers are very effective. Interior surfaces must absorb sound effectively. Apart from reducing the general sound level in the room, high absorption will inhibit the formation of standing waves. This will be important if the room is intended for sound field audiometry. Nowadays there are many companies offering acoustic lining materials that are both excellent sound absorbers and pleasing to the eye. Rooms made to order by specialist firms are probably best, but if responsibility has to be given to general builders and architects, they would be well advised to attend to the foregoing principles.

Questions and Exercises

9.1 In psychoacoustics, what is a *critical band*? What is the importance of critical bands in the theory of masking?

9.2 Distinguish between *local masking* and *remote masking*.

9.3 Hearing thresholds of a 1000-Hz tone are to be measured in the presence of narrow-band noise. The noise is centred on one of the following frequencies: (a) 1000 Hz, (b) 700 Hz, and (c) 1400 Hz, with the same sound pressure level in each case. The corresponding thresholds are T_a, T_b, and T_c. Rank the thresholds in ascending order.

9.4 Explain what is meant by *reverberation* and define *reverberation time*. In a particular room, the sound pressure level is found to decay at a constant rate after a steady source has been turned off. If the sound pressure level is initially 95 dB, what sound pressure level is expected after 90 ms if the reverberation time is 200 ms?

9.5 Explain the meaning of the terms *live room* and *dead room* in architectural acoustics.

9.6 What is meant by *room absorption* (in sabins)? Why should audiometric test rooms have a high absorption?

9.7 A room has the dimensions length × width × height = 5 × 4 × 2·5 metres. It has windows whose combined area is 2 m². If the walls are covered with bare plaster (sound absorption coefficient 0·05) and the floor is made of wood (sound absorption coefficient 0·06), calculate the room absorption in sabins when the windows are (a) closed and (b) open. Assume that glass and plaster have the same absorption coefficient. In each case, use Sabine's equation (Equation 9.2) to calculate the reverberation time.

9.8 An audiometric test room has a volume of 22·5 m³ and a surface area of 48 m². The surfaces of the room have an average sound absorption coefficient of 0·3. Use Eyring's formula (Equation 9.3) to calculate the reverberation time of the room, assuming it to be empty. What reverberation time is expected if the average absorption coefficient is doubled?

10

Hearing Aids: Basic Electroacoustic Characteristics

The measurement of the acoustic characteristics of a hearing aid is central to hearing aid technology. Audiologists must be able to understand manufacturers' data sheets and be able to use test equipment effectively. This chapter is intended as a brief introduction to what is involved. An extensive literature describing all aspects of hearing aid design and performance is available; although many of the books and research papers on the subject are written for experts, there are many accounts that will be readily accessible to the more general reader. Some suggestions are given at the end of this chapter.

In an attempt to alleviate the burden of deafness, a hearing aid has to perform two functions: it has to raise the level of sound reaching the ear so that the threshold of hearing is exceeded and audibility is restored, and it has to modify the acoustic signal in ways that compensate as far as possible for defects in the ear's analytical capacity. Amplification is relatively easy to accomplish, but compensation for impaired auditory processing is very difficult indeed — in some cases even impossible. The problem is deep rooted. It is, perhaps, that we learn to understand the world of sound at a very early age (some even suggest that the process begins before birth), but once the 'code' has been learnt, the learning cannot easily be repeated. Familiar signals become difficult to interpret when degraded by sensory impairment, and often the new representation of the auditory environment that a hearing aid can provide is just as difficult to understand. With digital processing, the technology now exists to modify the incoming acoustic signal in almost any way we wish. The difficulty is that no one knows quite what modifications should be made. New strategies continue to be tried, and although at times they offer great promise, more often than not they achieve only marginal results. It may be an act of faith to believe that a way will be found to deliver clear speech to a damaged ear, but so far we have seen only stumbling progress in this regard. On the other hand, there have been great technical advances in the construction and electroacoustic performance of hearing aids. The technology has culminated in the development and mass production of truly wearable devices that operate for long periods using only a miniature cell as the power source, producing, if required, sound outputs at levels on the limit of what is physiologically acceptable.

It may be fashionable to deride 'analogue' types, and they may one day give way entirely to the new digital instruments, but even analogue circuits are capable of important forms of signal processing, particularly in shaping the frequency response of the aid and in regulating its output. But whatever the technology, all hearing aids produce an amplified version of the sound at their input. The characterization and measurement of this process has a well-established importance in hearing aid technology and rehabilitative audiology. In the brief account that follows, most descriptions have a certain general validity but apply particularly to analogue hearing aids because universal test methods for digital aids have yet to be developed. Some characteristics, such as gain, are meaningful regardless of the technology, but others will probably become obsolete in the digital era.

Standard Tests of Gain, Output, and Frequency Response

Test methods are specified in the various parts of the international standard IEC 60118, but readers, particularly those in the United States and Canada, should be aware that parts of the international standard are not technically compatible with the American standard, ANSI S3.22 (1987). The American hearing aid industry has made a major contribution to hearing aid technology, and the American National Standards Institute (ANSI), under the auspices of the Acoustical Society of America, plays an important part in acoustic metrology and standardization. It is therefore regrettable that despite efforts to harmonize the ANSI and IEC standards, technical differences remain.[1] In what follows, however, we will make direct reference only to the IEC standard.

The introduction to part 0 of the IEC standard reads as follows:

> The object of this standard is to describe methods of measurement for the evaluation of the electroacoustical characteristics of hearing aids.

> The methods are chosen first of all to be practical and reproducible, and consequently they are based on fixed parameters chosen, to a certain extent, arbitrarily. This should be taken into consideration when comparisons are

---

[1]For a discussion, see D. A. Preves. Standardizing hearing aid measurement parameters and electroacoustic performance tests. In *Hearing Aids: Standards, Options and Limitations*, ed. M. Valente. Thieme Medical Publishers, New York, 1996. Revisions of IEC 60118-7 bring this standard very close to ANSI S3.22, and work is in progress to further harmonize the international and American standards.

being made between test results for hearing aids of different models and manufacturer, and in each case it is advisable to examine to what extent the arbitrarily chosen parameters will influence the comparison of such test results.

In general, the test methods relate the sound pressure in a free field in which the aid is placed and the sound pressure developed by the aid in an occluded-ear simulator (IEC 60711). The use of the simulator is important because it facilitates comparison between different hearing aids. The same methods can be applied when a 2-cc coupler (IEC 60126) is used, and manufacturers usually supply performance data for both devices. Comparisons based on coupler data are less reliable and the data give a less satisfactory indication of the likely performance when the aid is worn by a real subject. The coupler is, however, a more robust and far less expensive instrument than the ear simulator and is considered adequate for quality inspection as described in part 7 of IEC 60118. Further information about the ear simulator and acoustic coupler can be obtained from the relevant standards and from the descriptions in Chapters 7 and 8.

One of the most important characteristics of a hearing aid is its gain, so we need a clear understanding at the outset exactly what this term means. In its most general sense, *gain* is the ratio of some physical quantity at the output of a device to the same quantity at the input. We can therefore speak of voltage gain, current gain, power gain, pressure gain, and so on. When describing the performance of an electronic amplifier, gain usually means voltage gain; in hearing aid technology, it usually means the gain in sound pressure. It is standard practice to express the voltage gain of an amplifier and the sound pressure gain of a hearing aid in decibels. The gain G in decibels is then given by

$$G = 20 \log(V_o/V_i) \quad \text{or} \quad G = 20 \log(p_o/p_i) \quad \text{decibels}$$

where V_o and V_i are the output and input voltages of an amplifier and p_o and p_i are the output and input sound pressures of a hearing aid. These formulae apply only to sinusoidal signals, so that values of gain are necessarily frequency specific. Moreover, a sinusoidal input will give rise to a sinusoidal output only if the relationship between output and input is linear. It is not meaningful to talk about the gain of a hearing aid if its operation is nonlinear. By *nonlinear* we mean here that a sinusoidal input leads to a nonsinusoidal output. This is the kind of nonlinearity associated with peak clipping. On the other hand, automatic gain control and compression (to be described later) do preserve the waveform and do allow gain to be expressed according to the preceding definitions.

The voltages and pressures in the gain formulae will normally be expressed as root mean square values, but because ratios are involved, other corresponding values such as peak values (amplitudes) are equally valid. It may be noted that the logarithm is multiplied by 20 on the assumption that the gain is equivalent to a power gain proportional to the square of the voltage or pressure ratio. In general, the electrical impedance at the output of an amplifier is not the same as that at its input, and the acoustic impedance at the output to a hearing aid is not the same as the acoustic impedance at its microphone. Accordingly, the squared voltage or pressure ratios are not identical to the corresponding ratios of electrical or acoustic power. But the gain formulae are retained because they provide logarithmic conversions that are highly convenient and widely used even if they are not true to the ideal definition of the decibel.

The gain of a hearing aid has to be specified for its manufacture and testing, and it has to be known in order to predict the acoustic performance that the aid will have when it is worn. When gain is measured according to the standard test methods, the input to the hearing aid is derived from what is essentially a free field. The field is usually generated by a loudspeaker within a small enclosure lined with sound-absorbing material. The sound pressure is measured at a specified *test point* in the field. A *reference point* on the hearing aid is nominated, and the gain of the aid is determined when the aid is placed in the sound field with the reference point at the test point. The orientation of the aid has to be specified; the standard *reference orientation* corresponds to the orientation that the aid would have when worn by a person facing the sound source. The gain of the aid is defined as follows:

Acoustic gain — at a specified frequency and under specified operating conditions, the difference between the sound pressure level developed in the ear simulator by the hearing aid and the sound pressure level measured at the test point.

Four methods of specifying the input sound level are provided in the IEC standard:

- *Substitution method:* The hearing aid and the microphone used to measure the free-field sound pressure are alternately placed at the same point in the field
- *Comparison method:* The hearing aid and the measuring microphone are placed at acoustically equivalent points in the field
- *Pressure method:* The sound pressure is controlled at a point close to the sound entry to the hearing aid by a controlling microphone

- *Simulated in situ method:* The hearing aid is mounted on an artificial head and torso to simulate the diffraction that occurs when the aid is worn

The pressure method (as used in a hearing aid test box) is probably the one best known to audiologists. Because the sound pressure is controlled at a point close to the microphone of the aid, the effect of diffraction by the aid itself is eliminated. At the other extreme, the in situ method provides as complete a simulation as possible of the performance of the aid when it is worn by an average adult user. In this situation diffraction does matter and is an inherent part of the test procedure. Provisional specifications for a head-and-torso simulator are given in IEC 60959.[2]

Step-by-step instructions on how to measure the gain and output of a hearing aid are given in Section 7 of IEC 60118-0. Acoustic gain will normally be measured over a range of frequencies. When the measurements are made as a function of frequency, the input sound pressure is held constant at a level, usually 50 or 60 dB SPL, consistent with a linear response. The volume control should be turned full on and the settings of other controls should be stated. Gain measured in this way is called *full-on gain,* and the curve describing its frequency dependence is best called the *full-on gain response curve.*

If we want to measure the output of a hearing aid, it is usually necessary to set the gain in some standard way. The standard gain is established at what is called the *reference test frequency.* The reference frequency is normally 1600 Hz, but for some high-tone hearing aids 2500 Hz is preferred. The standard gain at the reference frequency is called the *reference test gain.* The first step in setting this gain is to measure the output of the aid in response to a high-level input. The input specified for this task is a tone having a level of 90 dB SPL at the reference frequency. The measurement is made with the volume control full on and other controls set to give the greatest available output. The output so obtained is called the $OSPL_{90}$. It is the limiting output that occurs when the aid is operating under nonlinear conditions beyond its normal working range. To find the reference test gain, the input is now reduced to 60 dB SPL (at the reference frequency). If the output has decreased by less than 8 dB, the volume control should be adjusted to bring the output to 15 dB below $OSPL_{90}$; otherwise, the adjustment should set the output to 7 dB below the full-on value. This procedure for establishing the reference test gain may seem slightly complicated, but it has the virtue of putting the

[2]A widely used simulator is known as KEMAR, an acronym for Knowles Electronics Manikin for Acoustic Research. The head, which is hollow, contains the ear simulator.

hearing aid in a condition that is comfortably within its intended operational range.[3]

Having set the volume control as described, and with other controls set for maximum gain, the frequency can be varied while holding the input level constant at 60 dB SPL. With this input, the curve showing the way the output of the aid depends on frequency is called the *basic frequency-response curve*. Similar curves can be obtained for different input levels to give a family of response curves that illustrate the performance of the aid over its entire range of operation. In addition to the output curves obtained at the reference test gain, manufacturers usually provide a curve showing the OSPL$_{90}$ as a function of frequency. This curve therefore shows the output of the aid in response to a 90-dB input when the volume control is set full on and when other controls are set for maximum gain. It usually represents the maximum output that can be obtained from the aid.

Manufacturers generally provide data consistent with either the recommendations of the IEC or ANSI, but there are minor variations in the way results are presented. For example, some manufacturers label as 'frequency response' curves that show the variation of gain with frequency, whereas others apply this name to the variation of output with frequency. It is usually obvious which measure is being used because the difference, which is equal to the input sound pressure level, is quite a large number of decibels (typically 60 dB). Note that gain curves are valid only for linear operation, but this restriction does not apply to output measures. With automatic gain control circuits disabled, essentially linear input-output conditions are said to exist if a 10-dB change in the input level leads to the same change in the output level (± 1 dB) at any frequency in the range 200 to 8000 Hz.

Examples of typical gain and output characteristics are shown in Figure 10.1. The frequency response is designed with the following in mind: it should bring the output to a level that is comfortably between the threshold of hearing and the level that the user will find too loud, it should reduce the masking of medium-frequency and high-frequency components by stronger components at lower frequencies, and it should emphasize those components that contribute most to speech intelligibility. A typical response establishes an output that rises at about 12 dB per octave from 200 to 1000 Hz. Above 1000 Hz the response is likely to be approximately uniform except for peaks and valleys associated with the acoustics of the ear and its acoustic coupling to the hearing aid. At the high-frequency end of the range, usually from 3500 Hz, the response falls rapidly. Many users have little serviceable hearing at these frequencies, and in any event the high-frequency

---

[3]The procedure described here follows Amendment A1 (1994) of IEC 60118-0. The previous version required the volume control to be placed in the full-on position if an output 15 dB below OSPL$_{90}$ could not be achieved. The 7-dB excess was not required.

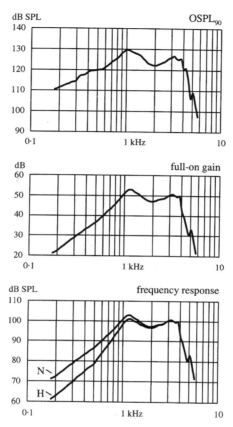

FIG. 10.1   The output, gain and frequency response typical of a simple post-aural hearing aid. In the bottom panel, 'N' is the normal response and 'H' is a high tone response (low tones are reduced).

components above 3500 Hz contribute little to speech intelligibility. Some authorities, however, are in favour of extending the working range to higher frequencies.

Output-Limiting Methods

The great majority of hearing aid users have sensory hearing impairments that are almost always accompanied by loudness recruitment. This means that although the threshold of hearing is abnormally high, loudness increases rapidly with sound level once threshold is exceeded. If the output of the hearing aid is not carefully controlled, strong signals are likely to produce an output that is uncomfortably loud.

The output of any amplifier is ultimately limited by the supply voltage and the range over which the circuit components are able to function normally. If the input signal is increased, there comes a point where no further increase in output is possible. The amplifier is then said to be *saturated*. In a simple amplifier without automatic gain control, *peak clipping* occurs as saturation is approached. Peak clipping is often introduced deliberately as a way of controlling the output of a hearing aid. In its simplest form, this type of control produces an output waveform that is an amplified version of the input waveform below a limiting level. Peaks in the waveform that would exceed this level are removed so that the output is held constant at the limit for the duration of the peak (see Chapter 5, Figure 5.15). Peak-clipped signals are obviously distorted and they sound 'fuzzy' to the normal ear, but despite this, the intelligibility of speech signals may not be severely degraded. The advantage of peak clipping over other methods of output control is that its action is instantaneous. The user is therefore well protected against the occurrence of unexpectedly intense sounds in the environment. The disadvantage is that information contained in clipped portions of the waveform is lost.

The alternative to peak clipping is *automatic gain control* (AGC).[4] As the name suggests, this method can be used to regulate the output of an amplifier by varying its gain according to the level of the signal at some point in the system. Automatic gain control has had a long history. It was probably first used in radiotelephony to compensate for unintended changes in the strength of the carrier signal arriving at the receiver. With amplitude modulation, the amplitude of the audio signal is directly proportional to the amplitude of the carrier, so that control of the carrier level is advantageous. Many varieties of automatic gain control have been tried in hearing aids. In all cases it is necessary to sample the signal at some point in the amplifier to obtain a voltage that represents the average signal level. This voltage is then passed back to a control circuit that regulates the gain in some way.

The simplest form of control is one that attempts to maintain a constant average level at the sampling point. This may be ideal in a radio receiver but would be inappropriate in a hearing aid because it would deprive the user of the natural variation in environmental sound level that should be experienced. In a quiet environment, the gain would become excessive and

---

[4]Another term for AGC is *compression*. Some authorities and some manufacturers make a distinction between these terms and reserve AGC for the description of compression that is characterized by a low compression threshold and a high compression ratio. Others use the terms synonymously.

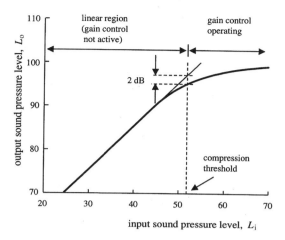

FIG. 10.2 Generalized input-output function for a hearing aid with automatic gain control (compression).

the output would be heavily contaminated with noise generated by the amplifier itself. Hearing aid gain control is far more sophisticated. It generally allows unrestricted linear amplification to a specified level (the so-called *knee point* or *compression threshold*[5]). Above this level the gain is progressively reduced as a limiting output is approached. It should be understood that it is the average signal that is limited. Gain control, unlike peak clipping, allows the audio-frequency variation in the magnitude of the output signal to continue under all conditions. Unless the averaging time of the control circuit is so short as to be commensurate with the period of audio-frequency components, the output of the amplifier will remain undistorted.

The input-output characteristics of a hearing aid can be shown graphically, plotting the sound pressure level at the input on the horizontal scale and the corresponding sound pressure level at the output on the vertical scale (Figure 10.2). Let L_i and L_o be the input and output sound levels. The gain of the hearing aid is $L_o - L_i$ decibels at any point on the input-output curve. This difference is constant when the gain is constant. For this condition, a change in the number of decibels at the input is matched by the same decibel change at the output. The corresponding input-output function

---

[5]ANSI S3.22 defines *compression threshold* as the input level at which the input-output function falls 2 dB below the value it would have in the absence of compression, that is, 2 dB below the projection of the linear portion of the function.

is therefore a straight line whose slope is unity. Elsewhere the function is a straight line of lesser slope or not a straight line at all. If it is a curve, its slope diminishes as the input level is raised and the difference between L_o and L_i becomes progressively smaller. The reciprocal of the slope of the input-output function is called the *compression ratio*. Accordingly, the change in input level for a given change in output level defines the compression ratio in a given region of the input-output function as

$$\text{compression ratio} = \frac{\text{change in input level}}{\text{change in output level}}$$

it being understood that the changes are small such that the slope of the input-output function is essentially constant over the range of sound levels for which the value of the compression ratio is required.

The range over which output signals are usable by the listener defines the *dynamic range* in which the hearing aid should operate. The hearing aid response needed to bring a given range of input signals within the dynamic range of the user depends on both the compression threshold and the compression ratio. The choice of these parameters is likely to be determined by the personal preferences of the audiologist and the user.

Automatic gain control (compression) requires a control voltage that is derived from an average value of the signal at some point in the amplifier. The average can be obtained by rectifying the signal and smoothing it with a low-pass filter to obtain a representative dc level. The averaging time is important because it determines the speed at which the amplifier will react to changes in signal level. A rapid response is needed to protect the user from the discomfort that would be caused by a sudden increase in input, but a long response time gives a stable output that is free from intrusive changes in level. The conflict between these requirements is resolved by allowing a slow response under normal conditions while imposing a rapid response whenever a sudden large increase in the signal is encountered. The shorter time constant is called the *attack time*. It can be determined by abruptly increasing the level of a previously steady input from 55 to 80 dB SPL and measuring the time needed for the output to stabilize within 2 dB of its final value. The longer time constant is called the *release time*. It is measured as the time needed for the output to return within 2 dB of its steady state following a sudden fall in input from 80 to 55 dB SPL. Attack and release times are typically in the range 10 to 50 ms and 150 to 2000 ms, respectively, but some hearing aids incorporate systems that respond more rapidly to give what is called *syllabic compression* (attack less than 5 ms, release 10 to 100 ms).

It should be noted that the compression characteristics of a hearing aid amplifier are likely to depend on frequency because the underlying frequency response is deliberately nonuniform. Moreover, some hearing aids are designed for multichannel operation. They filter the incoming signal to separate its components into two or more channels according to frequency. The channels work independently to apply different gains and different output characteristics before their outputs are combined in the final stage of the amplifier. The aim is to improve the compensation for sensory hearing loss. This type of hearing loss is strongly dependent on frequency and, in theory at least, is best aided by applying different amplification characteristics in different parts of the spectrum.

The place in the amplifier where gain control is applied is important. There are two principal options. In the first of these the sampling and the gain control are established before the volume control to give what is called *input compression*. The action of the volume control is then to displace the input-output function vertically. The knee point therefore occurs at the same input level for all settings of the control (Figure 10.3). The other commonly used method is called *output compression*. In this configuration, the signal is sampled in the output stage of the amplifier, and the feedback is

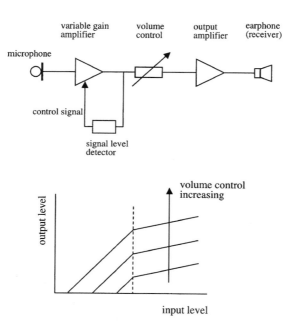

FIG. 10.3 Input-controlled compression.

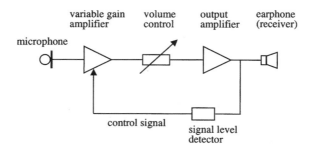

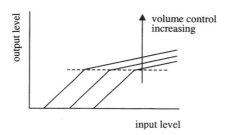

FIG. 10.4   Output-controlled compression.

taken to one of the first stages, where the gain control is effected. Changing the volume control displaces the input-output function parallel to the horizontal axis, but the output level at the knee-point remains the same (Figure 10.4).

Distortion

The IEC recommendation is that distortion should be measured with the volume control set to produce the reference test gain. The position of other controls should be stated. They should preferably be set to give the widest bandwidth.

Harmonic distortion is measured with a test tone at 70 dB SPL in the frequency range 200 to 5000 Hz. For each frequency of the test tone, the levels of components at the harmonic frequencies can be measured using filters or a frequency analyser. The harmonic distortion at each frequency and the total harmonic distortion can then be calculated as described in Chapters 5 and 8.

Portable test equipment (hearing aid test boxes) usually provides tests of harmonic distortion. The results are often misinterpreted. A quiet test environment is essential because environmental noise can produce significant sound levels in the output of the aid in frequency bands containing the harmonics. Self-generated noise in the hearing aid may be similarly misinterpreted.

Intermodulation distortion may also be measured. It is recommended that two test tones, each at 64 dB SPL, should be used as the input. The tones should have frequencies f_1 and f_2 such that the difference $f_2 - f_1 = 125$ Hz, though other differences are acceptable. The principal distortion products to be measured are those with frequencies $f_2 - f_1$ and $2f_2 - f_2$, where f_2 is in the range 475 to 5000 Hz.

Harmonic and intermodulation distortion are related, but it generally is difficult to derive one from the other.

Internal Noise

Internal noise is unlikely to be a problem for the hearing aid user. It will probably be inaudible unless the gain is set to a high level and the aid is worn in a quiet environment. Internal noise should be measured at the reference test frequency or in third-octave bands with centre frequencies in the range 200 to 5000 Hz. Internal noise is expressed as an equivalent external noise at the input to the hearing aid. This is done by measuring, in the ear simulator, the output caused by internal noise and then subtracting the gain of the hearing aid. The procedure is as follows. The volume control is set at approximately the reference gain position, but the setting is not critical. The aid is then presented with an input signal, which should be a tone at 60 dB SPL, either at the reference test frequency (1600 Hz) or at the centre frequency of the band being examined. It is important to have the aid operating in a region where the input-output characteristic is linear. This means that the input must be below the level at which automatic gain control (if present) becomes active. If necessary a test level below 60 dB should be used. It is also important, as with distortion measurements, that ambient noise does not contribute significantly to the sound levels measured at the output.

Using the notation in IEC 60118, let L_S be the sound pressure level in the ear simulator when the test tone is on and let L_2 be the level when it is off. Let L_1 be the level of the test tone. It may be assumed that internal noise makes a negligible contribution to L_S so that the gain of the aid is simply $L_S - L_1$. The internal noise referred to the input of the aid is then L_N, where

$$L_N = L_2 - (L_S - L_1) \text{ dB}$$

Direct Electrical Input and Induction Pick-Up Coil

Hearing aids used by children in schools for the deaf often take their input directly from a radio receiver or by magnetic induction from a neck loop or a loop system in the classroom. Advice on appropriate test methods should be sought from the hearing aid manufacturers and from the relevant standards (IEC 60118 parts 2 and 6).

TABLE 10.1   Standards for Hearing Aids

Standards	Title
IEC 60118-0 (1983)	Hearing aids — Part 0: Measurement of electroacoustical characteristics
IEC 60118-1 (1999)	Hearing aids — Part 1: Hearing aids with induction pick-up coil input
IEC 60118-2 (1983)	Hearing aids — Part 2: Hearing aids with automatic gain control circuits
IEC 60118-3 (1983)	Hearing aids — Part 3: Hearing aid equipment not entirely worn on the listener
IEC 60118-4 (1981)	Hearing aids — Part 4: Magnetic field strength in audio-frequency induction loops for hearing aid purposes
IEC 60118-5 (1983)	Hearing aids — Part 5: Nipples for insert earphones
IEC 60118-6 (1999)	Hearing aids — Part 6: Characteristics of electrical input circuits for hearing aids
IEC 60118-7 (1983)	Hearing aids — Part 7: Measurement of performance characteristics of hearing aids for quality inspection for delivery purposes
IEC/TR 60118-8 (1983)	Hearing aids — Part 8: Methods of measurement of performance characteristics of hearing aids under simulated in-situ working conditions
IEC 60118-9 (1985)	Hearing aids — Part 9: Methods of measurement of characteristics of hearing aids with bone vibrator output
IEC/TR 60118-10 (1986)	Hearing aids — Part 10: Guide to hearing aid standards
IEC 60118-11 (1983)	Hearing aids — Part 11: Symbols and other markings on hearing aids and related equipment
IEC 60118-12 (1996)	Hearing aids — Part 12: Dimensions of electrical connector systems
IEC 60118-13 (1997)	Hearing aids — Part 13: Electromagnetic compatibility (EMC)
IEC 60118-14 (1998)	Hearing aids — Part 14: Specification of a digital interface device

TABLE 10.1   Continued

Standards	Title
IEC 60959 (1990)	Provisional head and torso simulator for acoustic measurements on air conduction hearing aids
IEC 61669 (2001)	Electroacoustics — Equipment for the measurement of real-ear acoustical characteristics of hearing aids
ISO 12124 (2001)	Acoustics — Procedures for the measurement of real-ear acoustical characteristics of hearing aids
ANSI S3.22 (1996)	Specification of hearing aid characteristics
ANSI S3.35 (1985) (R1997)	Methods of measurement of performance characteristics of hearing aids under simulated in-situ working conditions
ANSI S3.37 (1987) (R1997)	Preferred earhook nozzle thread for postauricular hearing aids
ANSI S3.42 (1992) (R1997)	Testing hearing aids with a broad-band noise signal
ANSI S3.46 (1997)	Methods for measurement of real-ear performance characteristics of hearing aids
ETSI ETS 300 381 (1994)	Telephony for hearing impaired people — inductive coupling of telephone earphones to hearing aids
ETSI ETS 300 679 (1996)	Terminal equipment (TE) — telephony for the hearing impaired: electrical coupling of telephone sets to hearing aids

Further Reading

1. Staab, W. J., and Lybarger, S. F. Characteristics and use of hearing aids. In *Handbook of Clinical Audiology*, 4th ed. J. Katz, ed. Williams & Wilkins, Baltimore, 1994, pp. 657–722. This chapter provides an excellent introduction. It is written in nontechnical language and includes most topics that beginners will need to know about.
2. Valente, M. (ed.).*Hearing Aids: Standards, Options, and Limitations.* Thieme Medical Publishers, New York, 1996. This book provides a very detailed account of several important topics in hearing aid technology. The language is technical but not mathematical. The coverage goes beyond what could be considered introductory, but beginners will find much of the text understandable and relevant to their needs.
3. Dillon, H. *Hearing Aids.* Australia and New Zealand, Boomerang Press, 2001. This is a remarkable book written by an acknowledged expert. It will appeal to beginners and to seasoned practitioners.

Appendix A
Supplementary Mathematics

This appendix provides help with those elements of mathematics that occur frequently in the text or that are needed for special purposes. A knowledge of mathematical notation and simple algebra is assumed. Particular effort should be given to understanding the sine function because it is an almost indispensable part of acoustics. An understanding of exponentials and logarithms is also important, particularly as applied to decibel units. Help in working with decibels is given in Appendix B. It is anticipated that readers who consult the appendices will have different requirements. The technical difficulty is not uniform. Some parts deal with simple matters such as scientific notation, while others are concerned with more difficult issues such as trigonometric functions and complex numbers. There is no intention to develop a mathematical theme; topics are for the most part presented arbitrarily in alphabetical order. The result is something of a ready reference. Where a more thorough treatment is needed, the reader is encouraged to consult mathematics textbooks.

Contents

Angles

An angle expresses the inclination of one straight line to another straight line in the same plane. An angle is generated when a line is rotated about a fixed point. The usual convention is to give a positive sign to an angle produced by an anticlockwise rotation and a negative sign to one produced by a clockwise rotation. For everyday purposes we usually measure angles in degrees, 360° being equal to the anticlockwise rotation that returns the line to its initial direction. The *degree* is a valuable unit, not only because we are so used to it, but because it conveniently expresses commonly used fractions of a circle in whole numbers. For example, a quarter of a circle is 90°; 5 minutes on a clock face is 30°, and the angular speed of the earth's rotation is 15° per hour.

For all its virtues, a degree is a quite arbitrary unit,[1] which is its downfall so far as mathematics is concerned. To see the problem, consider the following statement,[2] which is true for all values of x:

$$\sin x = x - \frac{x^3}{3!} + \frac{x^5}{5!} - \frac{x^7}{7!} + \frac{x^9}{9!}\cdots$$

Because the exponents of x differ from one term to the next, this expansion of $\sin x$ is only possible if x is dimensionless, that is, just a number, independent of the units of measurement. We therefore have to define a unit angle as the ratio of like quantities, and the obvious choice is the ratio of two lengths. The scientific unit is the *radian*, which is defined as the ratio of the circumference of a circle to its radius. Because the circumference is 2π times the radius, a complete rotation generates an angle of 2π radians. One

[1]Another arbitrary unit is the *grad*, which divides the circle into 400 parts. The distance over the earth's surface measured along the arc of a great circle is 60 nautical miles for an angle of 1° and 100 km for an angle of 1 grad.
[2]The exclamation marks denote factorials; for example, $3! = 3 \times 2 \times 1$, $5! = 5 \times 4 \times 3 \times 2 \times 1$, and so forth.

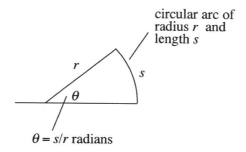

$\theta = s/r$ radians

FIG. A.1   Measuring angles.

radian is equivalent to $360/2\pi = 57{\cdot}3°$. If r is the radius of a circle and s the length of an arc subtending the angle θ radians, then

$$s = r\theta$$

as shown in Figure A.1.

Angular Frequency

If a point P moves with constant speed v around the circumference of a circle, the angle between the radius to P and a fixed line OX will increase at the constant rate v/r radians per second (Figure A.2). The rate of change of angle is called *angular velocity*. It is usually represented by the letter ω. In acoustics, as distinct from the physics of rotating bodies, angular velocity is associated with the frequency of harmonic oscillation, and the term *angular frequency* is used instead. The angular frequency is the rate of change of angle needed to achieve one complete revolution in the time of one cycle of oscillation.

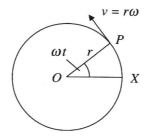

FIG. A.2   Angular velocity.

Exponents

Writing a million as 10^6 is a convenient shorthand, as is writing a^3 for $a \times a \times a$. In the term a^n, the number n (that is, the power to which a has been raised) is called an *exponent*.

The rules governing the operation of exponents are as follows:

$a^n \times a^m = a^{(n+m)}$ Exponents are added when terms involving powers of the same variable are multiplied.

$a^n/a^m = a^{(n-m)}$ Exponents are subtracted when terms involving powers of the same variable are divided.

$a^0 = 1$ Any number raised to the power zero is equal to unity $[1 = a^n/a^n = a^{(n-n)} = a^0]$.

$a^{-n} = 1/a^n$ A negative exponent indicates that the reciprocal of the number is required.

$a^{1/n} = \sqrt[n]{a}$ The exponent is a fraction. The denominator is n, showing that the nth root should be taken.

$a^{m/n} = [\sqrt[n]{a}]^m$ The numerator is m, showing that the nth root must be raised to the power m.

Exponential Functions

If an equation has the form $y = a^x$ where x is a variable and a is a constant, we say that y is an *exponential function* of x. An important special case exists for the function e^x, where the mathematical constant e is a number which, like π, is known in the limit but which can never be evaluated exactly (i.e., a *transcendental* number). The function is called *the exponential function*. It is often written as $\exp(x)$. The number e is given by

$$e = L_{t \to 0}(1 + t)^{1/t}$$

$$= 1 + \frac{1}{1!} + \frac{1}{2!} + \frac{1}{3!} + \cdots$$

$$= 2.71828\ldots$$

and

$$e^x = 1 + \frac{x}{1!} + \frac{x^2}{2!} + \frac{x}{3!} + \cdots$$

The functions e^x and e^{-x} are shown in Figure A.3. A very important property of the exponential function is that the rate of increase of e^x with x is itself equal to e^x. If we draw a straight line $y = mx + c$ through a point

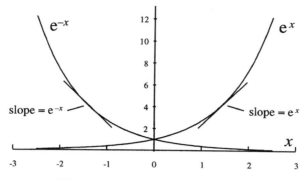

FIG. A.3 The exponential function.

(x_1, y_1) on the curve $y = e^x$, we find that the slope m of the line is e^{x_1}. The same is true for the function e^{-x}. This function decreases as x increases; the rate of increase is $-e^{-x}$. Exponential functions occur frequently in physics and biology because it often happens that the rate of change of 'X' is proportional to 'X'. For example, the rate at which an organism grows is, at least initially, proportional to how big it is. In radioactive processes, the number of nuclear disintegrations per second must be proportional to the number of nondisintegrated nuclei remaining, so that if the disintegrations are counted, the change in the count rate with time is proportional to the prevailing count rate.

Figure A.4 shows the functions $1 - e^{-t/\tau}$ and $e^{-t/\tau}$, where t represents time and τ is a constant known as the *time constant*. Examples are abundant: the rise or decay of the charge on a capacitor through a series resistance following a step change in the applied voltage, and the increase or decrease in amplitude of a forced harmonic vibration following a step change in the level of the driving force are described by these functions.

Functions

When we first learn to use algebra, we are told that constants and variables can be represented by letters of the alphabet. The value of this representation is that general relationships can be explored without the need to work with specific numerical values. Here is a simple example. Let h and w be the height and width of a rectangle whose perimeter and area are respectively P and S. Then $P = 2(h + w)$ and $S = hw$. Therefore,

$$S/P = hw/2(h + w)$$

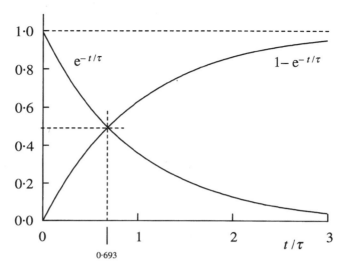

FIG. A.4   Exponential growth and decay functions.

What is important here is that the ratio of the area to the perimeter of *any* rectangle is given by this formula. A valuable benefit of the algebraic form is that it becomes easy to manipulate the formulae to arrive at new ways of expressing the same ideas. If, for example, we replace w by S/h, we find that $P = 2(h + S/h)$. Now if, instead of a rectangle, we had chosen an isosceles triangle of height h, we would have arrived at the formula

$$P = 2\frac{S}{h} + 2\sqrt{h^2 + \frac{S^2}{h^2}}$$

The area within a regular hexagon can be divided into six isosceles triangles. If we stand the hexagon on one of its sides, its height h is twice that of each of the triangles. The formula for the perimeter in terms of the area is $P = 4S/h$.

The three formulae just given are all different, which is not surprising because they apply to the three differently shaped figures that we have described, but they do have one thing in common. They each give the perimeter of the figure in terms of its area and one specified dimension. It can therefore be said of rectangles, symmetrical triangles, and regular hexagons that the perimeter can be calculated if the height and area are known. This truth can be expressed symbolically by saying that the perimeter is a *function* of area and height and writing this as

$$P = f(S, h)$$

This simply states that P depends on S and h without telling us exactly what the dependence is. Unless we know more about the function f, we cannot calculate P from S and h. We have used the letter f as a name for the function (because it reminds us of the word *function*), but any letter would do as well, including the letter P itself. We could represent the perimeter of one of these figures by $P(S, h)$ and give the function a name such as 'perimeter function'.

Functions are usually printed as a name or a letter followed by parentheses. The terms within the parentheses constitute the *argument* of the function. An argument can be a single variable, a list of variables separated by commas, or an algebraic expression. For example, the particle displacement ξ in a progressive acoustic wave is given variously as

$$\xi = f(c, t, x)$$

and

$$\xi = f(ct - x)$$

where c is the speed of sound, x the distance from the source, and t the time. Notice that the second expression supplies more information than the first by telling us that particle displacement really depends on the value of $ct - x$ rather than the individual values of c, t, and x. We could replace this argument with a single variable (call it d, where $d = ct - x$) and write $\xi = f(d)$.

It is important to understand that when a function is evaluated, each component of the argument is treated in its entirety as though it were a single variable. For example, suppose we have a function $g(x)$ such that

$$g(x) = a + bx^2$$

Then

$$g(2x) = a + b(2x)^2$$

and

$$g(y + z) = a + b(y + z)^2$$

Simplified Notation for Functions

When using computers to evaluate standard or user-defined functions, it is nearly always necessary to include the opening and closing parentheses around the arguments. In printed text, however, the algebra often looks less cluttered if the parentheses are omitted. Their omission is perfectly acceptable provided that the result is correct and unambiguous. For example, $\sin(x)$ can be written as $\sin x$, but $\sin(x + \phi)$ cannot be written as $\sin x + \phi$ because this would be confused with $\sin(x) + \phi$. On the other hand, multiplication or division can bind terms to the function name when parentheses are omitted. For example, $\log(p_1/p_2)$ can be written as $\log p_1/p_2$. Similarly

$$2a \sin \left[\frac{x + y}{2} \right] \cos \left[\frac{x - y}{2} \right]$$

and

$$2a \sin \frac{x + y}{2} \cos \frac{x - y}{2}$$

mean the same thing.

Inverse Functions

Suppose that y is a function of x. Then we can find y given x. The inverse problem is to find x when we are given y. The function of y that gives us x is said to be the *inverse* of the function of x that gives us y. Suppose, for example, that there is a function called $P(x)$ whose inverse is $Q(y)$ such that

$$P(x) = ax^2 + bx + c$$

Then if $y = P(x)$,

$$x = Q(y) = \frac{-b \pm \sqrt{b^2 - 4a(c - y)}}{2a}$$

In this example, as often happens, the inverse function does not give a unique value. Inverse functions are important in trigonometry, where they sometimes have special names. For example, $\arctan x$ is the inverse of $\tan x$. Because $\tan \pi/4 = 1$, $\arctan(1) = \pi/4$. It also has other values (see the section Trigonometric Functions later in this appendix.)

Imaginary and Complex Numbers

The numbers that we use in normal arithmetic and in the description of physical events are called *real* numbers. In ordinary arithmetic and ordinary algebra a positive number multipled by itself is positive, and a negative number multiplied by itself is also positive. There is therefore no real number whose square is negative; the square roots of negative real numbers do not exist. But suppose that the roots of negative numbers can exist — they are called *imaginary*. Imaginary numbers may have started life as a mathematical curiosity, but they are now an indispensible part of mathematical analysis and of great importance in all branches of physcial science.

It is customary[3] to use the letter i or j to represent the positive square root of -1, that is,

$$j = +\sqrt{-1}$$

and

$$j^2 = -1$$

A number formed by adding a real number and an imaginary number is called a *complex number*. A complex number Z can be written as

$$Z = a + jb$$

where a and b are real numbers.[4]

It is found that complex numbers obey the standard rules of arithmetic and algebra. This means that the following rules which normally apply to ordinary numbers x, y, and z also apply to complex numbers:

Commutative
$$x + y = y + x$$
$$xy = yx$$

Associative
$$(x + y) + z = (x + z) + y$$
$$x(yz) = (xy)z$$

Distributive
$$x(y + z) = xy + xz$$

An important property of complex numbers is that if they are equal then their real and imaginary parts are equal independently. To see why this should be, let

$$Z_1 = a + jb \quad \text{and} \quad Z_2 = c + jd$$

[3]There are differences in some conventions involving the choice of i or j, but the distinction is seldom recognized. The use of j in the theory of electrical circuits allows i to be retained as the conventional symbol for electric current. See L. Beranek. *Acoustical Measurements.* Acoustical Society of America, New York, 1988, p. 298.

[4]As with vectors, we shall show complex numbers in ordinary rather than bold type and use modulus signs to identifiy their magnitudes. For example Z, $|Z|$.

then if $Z_1 = Z_2$,

$$a + jb = c + jd$$

Therefore,

$$(a - c) = -j(b - d)$$

So that, on squaring both sides,

$$(a - c)^2 = -1 \times (b - d)^2$$

Now $(a - c)^2$ and $(b - d)^2$ are each the squares of real numbers and therefore positive or zero, but they cannot be greater than zero because a positive number cannot equal -1 times another positive number. Therefore $a = c$ and $b = d$. It follows that whenever complex numbers are equal, we also know that their real parts are equal and their imaginary parts are equal. For example, if

$$a + jb = 2 - 5j$$

then $a = 2$ and $b = -5$.

It follows further that when complex numbers are added, their real and imaginary parts are added separately:

$$(a + jb) + (c + jd) = (a + c) + j(b + d)$$

However, if two complex numbers are multiplied, the real part of the product contains elements from both the real and imaginary parts of the factors, and the same is true for the imaginary part of the product:

$$(a + jb)(c + jd) = (ac - bd) + j(ad + bc) \qquad \text{A.1}$$

If the sign of the imaginary part of a complex number is reversed, we obtain a new complex number called the *complex conjugate* of the original number. The numbers $(a + jb)$ and $(a - jb)$ are conjugates. The asterisk symbol is often used to identify the complex conjugate. So,

$$Z = a + jb \qquad \text{and} \qquad Z^* = a - jb$$

We see therefore that

$$ZZ^* = (a + jb)(a - jb) = a^2 + b^2$$

The square root of ZZ^* is a real number called the *modulus* of the complex number Z. That is,

$$|Z| = |Z^*| = \sqrt{ZZ^*} = \sqrt{a^2 + b^2}$$

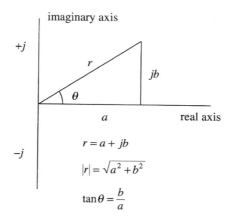

FIG. A.5 The Argand diagram.

The properties of a complex number, particularly the separate addition of its real and imaginary parts and the expression of its modulus, are neatly reproduced if it is thought of as a vector. The vector representation can be shown graphically by means of an *Argand diagram*[5] (Figure A.5). In this diagram the imaginary part is displayed on the vertical axis, and the real part is displayed on the horizontal axis. The complex number $a + jb$ is shown as a line of length r drawn from the origin and making an angle θ with the real axis. It can be written in polar form as (r, θ), where

$$|r| = \sqrt{a^2 + b^2} \qquad \text{and} \qquad \tan \theta = \frac{b}{a}$$

The length $|r|$ represents the *modulus* or *magnitude* of the complex number. The angle θ is sometimes called the a*rgument* of the number (not to be confused with the argument of a function).

The vector addition of the complex numbers $Z_1 = a + jb$ and $Z_2 = c + jd$ is shown in Figure A.6. By extension, an indefinite number of complex vectors can be added in the same way. The process would be represented in the Argand diagram by constructing a polygon, and the sum of the vectors would be the line from the origin of the first vector to the end of the last one. It is important to notice that the same result is obtained regardless of the order in which the addition is performed. (The addition of complex numbers is associative and commutative.)

[5]After the mathematician J. R. Argand (1768–1822).

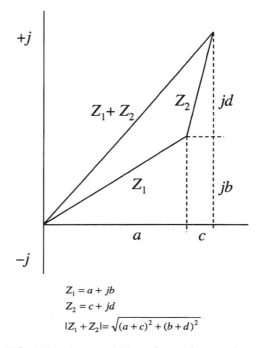

$Z_1 = a + jb$

$Z_2 = c + jd$

$|Z_1 + Z_2| = \sqrt{(a+c)^2 + (b+d)^2}$

FIG. A.6   Vector addition of complex numbers.

Multiplication of complex numbers does not lead to the standard scalar or vector products (see the section on vectors). The product of two complex numbers is a new complex number having real and imaginary parts as defined in Equation A.1. Multiplication by the unit vector $+j$ or $-j$ is an important special case. If the number $Z = a + jb$ is multiplied by $+j$, the result is $jZ = -b + ja$. As Figure A.7 shows, the effect of multiplying by $+j$ is to rotate the vector Z anticlockwise by $\pi/2$. Multiplying by $-j$ produces a clockwise rotation of $\pi/2$. This way of looking at multiplication by j is particularly important when complex numbers are used to express impedance, because the rotation of the vector through $\pm\pi/2$ implies a corresponding change in phase.

Multiplication of a complex number by its conjugate leads to a useful method for rationalizing fractions that have imaginary parts in the denominator. Suppose, for example, that we require the complex number Y where $Y = 1/Z$ and where $Z = a + jb$. Then

$$Y = \frac{1}{Z} = \frac{1}{a+jb} \qquad \text{and} \qquad Y^* = \frac{1}{Z^*} = \frac{1}{a-jb}$$

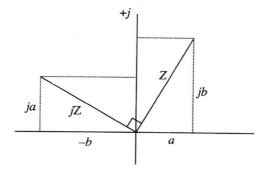

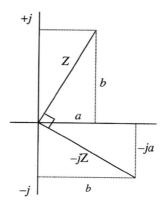

FIG. A.7   Multiplying a complex number *Z* by *j* and by −*j*.

We are likely to feel uncomfortable with the imaginary term jb in the denominator because it prevents us from seeing Y as the sum of a real part and an imaginary one. Fortunately the difficulty is easy to deal with. If the top and bottom of the fraction are each multiplied by the complex conjugate, the bottom becomes a real number:

$$Y = \frac{1}{a+jb} = \frac{1}{a+jb} \times \frac{a-jb}{a-jb} = \frac{a-jb}{a^2+b^2}$$

The real and imaginary parts of Y are therefore

$$a/(a^2+b^2)$$

and

$$-jb/(a^2+b^2)$$

Notice that the modulus of Y is

$$|Y| = \sqrt{YY^*} = \sqrt{\frac{1}{a+jb} \cdot \frac{1}{a-jb}} = \sqrt{\frac{1}{a^2+b^2}} = \frac{1}{|Z|}$$

The arguments θ_Y and θ_Z of Y and Z are given respectively by $\tan \theta_Y = -b/a$ and $\tan \theta_Z = b/a$, and therefore $\theta_Y = -\theta_Z$. We see therefore that if complex numbers are related such that one is the reciprocal of the other, then their moduli are similarly related and their arguments have opposite signs. It is important to be aware of these relationships when working with complex impedance and admittance.

Multiplication and Division

The reciprocal relationship of the moduli of a complex number and its inverse is a particular case of a more general relationship. Suppose that

$$Z = Z_1 Z_2$$

where $Z_1 = a + jb$ and $Z_2 = c + jd$. Then

$$Z = (a + jb)(c + jd) = ac - bd + j(ad + bc)$$

Therefore,

$$\begin{aligned}
|Z|^2 &= (ac - bd)^2 + (ad + bd)^2 \\
&= a^2c^2 + b^2d^2 - 2acbd + a^2d^2 + b^2c^2 + 2adbc \\
&= a^2c^2 + b^2d^2 + a^2d^2 + b^2c^2 \\
&= (a^2 + b^2)(c^2 + d^2) \\
&= |Z_1|^2 |Z_2|^2
\end{aligned}$$

so that

$$|Z| = |Z_1||Z_2|$$

This means that the modulus of the product of complex numbers is equal to the product of their moduli.

A similar rule applies to division. Let

$$Z = \frac{Z_1}{Z_2}$$

Then

$$Z = \frac{a + jb}{c + jd} = \frac{(a + jb)(c - jd)}{c^2 + d^2} = \frac{ac + bd - j(ad - bc)}{c^2 + d^2}$$

Therefore,

$$|Z|^2 = \frac{(ac + bd)^2 + (ad - bc)^2}{(c^2 + d^2)^2} = \frac{(a^2 + b^2)(c^2 + d^2)}{(c^2 + d^2)^2} = \frac{(a^2 + b^2)}{(c^2 + d^2)} = \frac{|Z_1|^2}{|Z_2|^2}$$

so that

$$|Z| = \frac{|Z_1|}{|Z_2|}$$

Therefore the modulus of the quotient of complex numbers is equal to the quotient of their moduli.

Components of Complex Impedance and Admittance

The real and imaginary components of impedance Z are resistance R and reactance X. The corresponding components of admittance Y are conductance G and susceptance B. Therefore

$$Z = R + jX \qquad \text{and} \qquad Y = G - jB$$

The admittance Y is equal to $1/Z$ so that

$$Y = \frac{1}{R + jX} = \frac{R - jX}{(R + jX)(R - jX)} = \frac{R - jX}{R^2 + X^2}$$

But

$$Y = G - jB$$

If complex numbers are equal, the real part of one is equal to the real part of the other and the imaginary part of one is equal to the imaginary part of the other. Therefore

$$G = \frac{R}{R^2 + X^2} = \frac{R}{|Z|^2}$$

and

$$B = \frac{X}{R^2 + X^2} = \frac{X}{|Z|^2}$$

Similarly,

$$R = \frac{G}{G^2 + B^2} = \frac{G}{|Y|^2}$$

and

$$X = \frac{B}{G^2 + B^2} = \frac{B}{|Y|^2}$$

Logarithms

The logarithm is an important mathematical function in its own right, but before the invention of electronic calculation it had a valuable practical application as an almost indispensable aid to multiplication and division. Slide rules[6] were to engineers what stethoscopes are to doctors, and every schoolboy would carry a book of log tables in his satchel. The logarithm has survived the electronic revolution and is alive and well in the decibel and in all places where it is a practical convenience to compress the range of values needed to express the magnitude of the object of our attention.

Logarithms work by expressing a variable x as the exponent of a base number b. The logarithm is therefore defined by the statement that if

$$\log_b(x) = n$$

then

$$b^n = x$$

For example, $\log_2 8 = 3$ because $2^3 = 8$.

In ordinary circumstances we use numbers in base 10. Logarithms for which b is equal to 10 are called *common logarithms*. The exponential e is the base of *natural logarithms*. Unless it is necessary to be explicit, the base can be omitted in written form to simplify the notation. Unless otherwise defined, we normally write $\log_{10} x$ as just $\log x$. Also, $\log_e x$ is often written as $\ln x$. The names *log* and *ln* often appear on the keyboards of calculators and as function names in computer software (but beware that *log* implies base e in some software). The compression achieved with logarithmic scaling is evident in Table A.1.

To use logarithms, we also need a function to provide the inverse of a logarithm; in other words, we need the function inverse-log(x) that gives us the number whose logarithm is x. It will be evident from the definition at

[6]Slide rules achieve multiplication and division by adding or subtracting sections of printed scales whose lengths are proportional to the logarithms of the numbers they represent.

TABLE A.1 Logarithms in Base 10

Number	Logarithm	Number	Logarithm
1,000,000	6	2	0·3010
100,000	5	3	0·4771
10,000	4	5	0·6990
1,000	3	7	0·8451
100	2	0·5	−0·3010
10	1	π	0·4971
1	0	e	0·4343
0·1	−1		
0·01	−2		
0·001	−3		
0·000,1	−4		
0·000,01	−5		
0·000,000,1	−6		

the start of this section that the inverse function is simply $b^{\log_b(x)}$. For example, if the common logarithm of a number is 1·7, then the number must be $10^{1·7}$, which is 50·11. The name *antilogarithm* is often used in tables to mean the inverse of a logarithm.

Multiplication and Division

Numbers may be multiplied by adding their logarithms and then taking the inverse of this sum. Division involves the same thing except that the logarithm of the divisor has to be subtracted. If the basis of these operations is not self-evident, consider the following. Let

$$m = \log x \quad \text{and} \quad n = \log y$$

Then

$$x = 10^m \quad \text{and} \quad y = 10^n$$

so that

$$xy = 10^m \times 10^n = 10^{(m+n)}$$

By definition, $10^{(m+n)}$ is the inverse logarithm of $(m + n)$.
  Notice that

$$\log xy = \log [10^{(m+n)}] = m + n = \log x + \log y$$

In general, the logarithm of a product is the sum of the logarithms of the factors, that is,

$$\log xyz \ldots = \log x + \log y + \log z + \cdots$$

By the same reasoning,

$$\log x/y = \log x - \log y$$

and

$$\log(x^n) = n \log x$$

This last observation should be remembered when working with decibels, where the argument of the log function is often squared. For example, $\log(p^2) = 2 \log p$.

It may also be noticed that $\log(1/x) = -\log x$. This is because

$$\log(1/x) = \log 1 - \log x$$

$$= 0 - \log x$$

As shown in Table A.1, $\log 2 = 0 \cdot 301$ and $\log (1/2) = \log 0 \cdot 5 = -0 \cdot 301$.

Change of Base

Having found the logarithm of a number, we may occasionally need to find its logarithm in a different base. This happens, for example, in conversions between common and natural logarithms.

Let

$$m = \log_a x$$

Then

$$x = a^m$$

Therefore,

$$\log_b x = \log_b(a^m)$$

$$= m \log_b a$$

so that

$$\log_b x = (\log_a x)(\log_b a)$$

Also, letting $x = b$ and noting that $\log_b b = 1$, we see that

$$1 = (\log_a b)(\log_b a)$$

That is,

$$\log_a b = 1/(\log_b a)$$

To give examples,

$\log_{10} 2 = 0.3010$
$\log_e 2 = \log_{10} 2 \times \log_e 10 = 0 \cdot 3010 \times 2 \cdot 3026 = 0 \cdot 6931$
$\log_e 10 = 2 \cdot 3026$
$\log_{10} e = 1/\log_e 10 = 1/2 \cdot 3026 = 0 \cdot 4343$

Mantissa and Characteristic

The decimal part of a common logarithm is called the *mantissa*;[7] the integer part before the decimal point is the *characteristic*. These terms are not much used these days but were well known when logarithms were needed for calculation. It is still helpful to be aware of the role of these two parts of the logarithm, particularly when working with numbers written in scientific notation. Consider the following.

$$\log(186,000) = \log(1 \cdot 86 \times 10^5)$$
$$= \log(1 \cdot 86) + \log(10^5)$$
$$= 0 \cdot 2695 + 5$$
$$= 5 \cdot 2695$$

If the mantissa is positive, its contribution to the inverse logarithm is always a number greater than 1, with only one digit before the decimal point. Its antilog corresponds to the first part of a number written in scientific form. We can always arrange things so that the mantissa is positive by changing the characteristic. For example, suppose that a number has the negative logarithm $-2 \cdot 3$. This can be written as $(-3 + 0 \cdot 7)$, and the inverse logarithm is then $10^{-3} \times 10^{0 \cdot 7} = 5 \cdot 01 \times 10^{-3}$.

Logarithmic Scales

Nonlinear scales are often used in graphical work. There are two reasons for this: the first is to turn curves into straight lines (Figure A.8), and the second is to accommodate data encompassing a wide range of values. As an example in the first category, suppose we want to show values of x and y

---

[7]*Mantissa* = 'makeweight' (Latin).

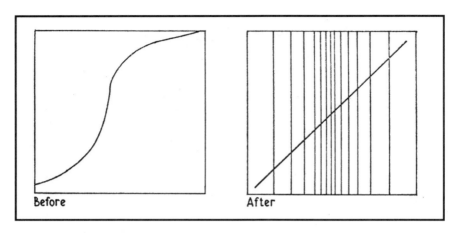

FIG. A.8 The art of finding the right graph paper to get a straight line. 'For the statistician, there is always probability paper, which will turn a normal ogive into a straight line or a normal curve into a tent. It is especially popular with staticians because it makes their work look precise' (S. A. Rudin. The art of finding the right graph paper to get a straight line. *Journal of Irreproducible Results* 12, 1964). (Figure copied from Rudin as reproduced in R. L. Weber and E. Mendoza, eds. *A Random Walk in Science*. Institute of Physics, London, 1973, p. 99.)

when these are related as

$$y = ae^{bx}$$

A graph of x and y on linear scales is the curve shown in Figure A.9a. If, however, we find the logarithm of each side of the equation, we obtain

$$\log y = \log a + bx \log e$$

Because $\log a$ and $\log e$ are constants, a graph of $\log y$ plotted against x is a straight line whose slope is $b \log e$ and whose intercept on the $\log y$ axis is $\log a$ (Figure A.9b). Instead of labelling the vertical axis in log units, we could keep the same scaling but mark the axis in units of y (Figure A.9c). Chart papers with one or both axes marked out in proportion to the logarithms of the numbers they represent were widely used in science and engineering, but their use as a technical aid has declined now that computer processing is universally available.

Logarithmic scaling is particularly helpful when a wide range of values is to be presented graphically. It is nearly always necessary in acoustics and in audiological work when showing variation with frequency; the audiogram is an obvious example (Figure A.10). To understand this scaling better, consider first what happens on an ordinary linear scale. Here equal intervals

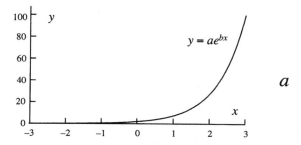

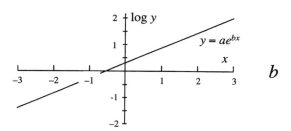

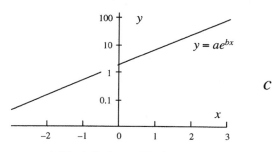

FIG. A.9 Logarithmic scales.

represent equal additions to the variable, so starting with x_0, the scale is marked at intervals

$$x_0, x_0 + d, x_0 + 2d, x_0 + 3d, \ldots$$

On frequency scales, we want equal intervals to represent equal frequency ratios. Starting from f_0, the scale is marked at intervals

$$f_0, nf_0, n^2f_0, n^3f_0, \ldots$$

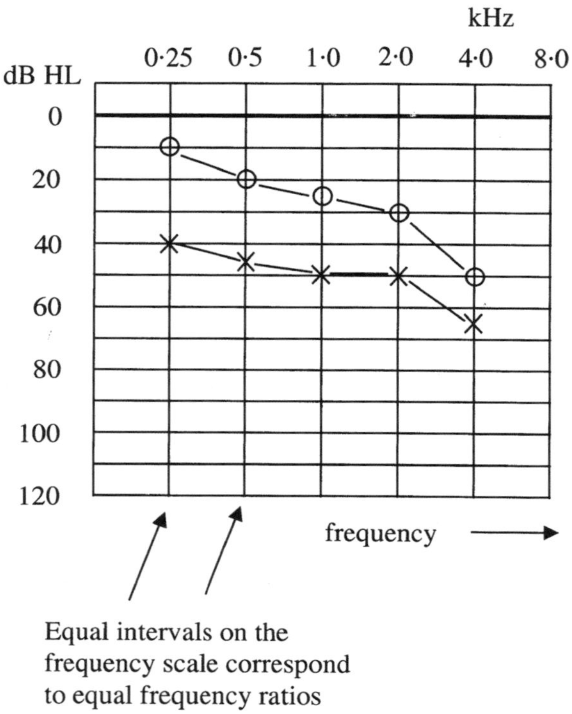

FIG. A.10 Logarithmic scales: the audiogram.

For example, if $f_0 = 125$ Hz and $n = 2$, we have

$$125, \ 2 \times 125, \ 2^2 \times 125, \ 2^3 \times 125 \ldots = 125, \ 250, \ 500, \ 1000, \ldots \text{Hz}$$

Taking the logarithms of these numbers gives

$$\log \ f_0, \ \log n + \log f_0, \ 2\log n + \log f_0, \ 3\log n + \log f_0, \ldots$$

which is a linear scale in intervals of $\log n$. This illustrates the important fact that equal intervals in a logarithmic scale are equivalent to equal ratios of the variable represented. In a scale of decibels, for instance, equal decibel differences correspond to equal ratios of acoustic power.

On scales that represent frequency of vibration, it is sometimes necessary to know the frequency f_c that is located graphically at the centre of the interval between frequencies f_1 and f_2. In accordance with this requirement,

$$f_c/f_1 = f_2/f_c$$

so that

$$f_c = \sqrt{f_1 f_2}$$

That is, f_c is the geometric mean of f_1 and f_2.

Trigonometric Functions

The sine function and its relatives (cosine, tangent, and so on) are given by the ratios of the lengths of sides of a right triangle. Accordingly, the sine and cosine functions can be generated from the vertical or horizontal projection of the radius of a circle. Figure A.11 shows a circle whose centre is at O on the line $X'X$. P is a point on the circumference of the circle, and PQ is the perpendicular from P to $X'X$. We may regard PQ as the vertical projection of the radius OP, and OQ as the horizontal projection. Let θ be the angle POQ. The sine and cosine functions are defined as follows:

$$\sin \theta = \frac{QP}{OP}$$

and

$$\cos \theta = \frac{OQ}{OP}$$

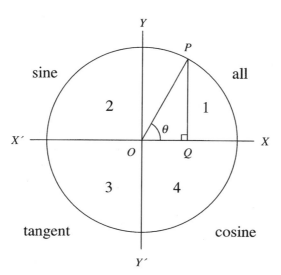

FIG. A.11 The trigonometric functions.

The ratio of these functions is the tangent function:

$$\tan \theta = \frac{\sin \theta}{\cos \theta} = \frac{QP}{OQ}$$

The reciprocals of these functions have the names *secant*, *cosecant*, and *cotangent*:

$\sec \theta = 1/\cos \theta$
$\operatorname{cosec} \theta = 1/\sin \theta$
$\cot \theta = 1/\tan \theta$

The circle in Figure A.11 can be divided into quadrants by drawing the line $Y'Y$ perpendicular to $X'X$ through the centre. The quadrants are numbered 1 to 4 as shown. We use the usual convention that distances measured upwards from $X'X$ are positive while distances measured downwards are negative. The same applies respectively to distances measured to the right and left of $Y'Y$. The radius OP is positive in all quadrants. The diagram shows P in the first quadrant, but it could be anywhere on the circle. From the definitions given earlier we see that the value of the sine function is positive when OP is in the first and second quadrants and negative when OP is in the third and fourth quadrants. We also see that the sine of a negative angle created by a clockwise rotation of OP from the horizontal is minus the sine of the corresponding positive angle; that is, $\sin(\theta) = -\sin(-\theta)$. Similarly, we see that the cosine function is positive in the first and fourth quadrants and negative in the others, and that $\cos(\theta) = \cos(-\theta)$. The rule for remembering which of the three principal trigonometric functions are positive is '*all, sin, tan, cos*' applied in quadrants 1 to 4, respectively.

Suppose that the point P moves around the circumference to generate an ever-increasing angle, either positive if the rotation is anticlockwise, or negative if it is clockwise. Projecting the radius on the vertical gives the continuous sine function shown in Figure A.12. Points to notice are the following:

$\sin \theta = 0$ for $\theta = \pm n\pi$ where $n = 0, 1, 2, 3, \ldots$
$\sin \theta = \sin(\theta \pm 2n\pi)$ where $n = 0, 1, 2, 3, \ldots$
$\sin(\theta + \pi) = -\sin \theta$
$\sin(\theta - \pi) = -\sin \theta$
$\sin(\theta) = -\sin(-\theta)$

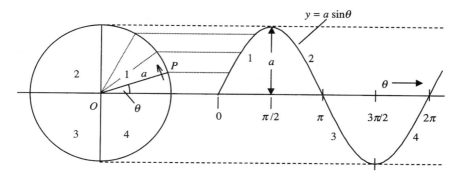

FIG. A.12 The sine function.

Sin θ itself varies between $+1$ and -1 in each cycle. The maxima and minima occur once in each cycle. The least positive angle for which sin θ is equal to 1 is $\pi/2$; the least positive angle for which it is equal to -1 is $3\pi/2$. The greatest absolute value of the sine wave that has been constructed, is the length of the radius OP. If the length of the radius is a, its vertical projection y is the function $a\sin\theta$. The length of the radius is the *amplitude* of the wave.

The cosine function is similar to the sine function except that it is displaced towards the origin by a quarter of a cycle. We can generate a cosine by projection of OP onto the horizontal. The result is shown in Figure A.13. Points to notice are the following:

$$\cos\theta = 0 \text{ for } \theta = \pm n\pi/2 \text{ where } n = 1, 3, 5, \ldots$$
$$\cos\theta = \cos(\theta \pm 2n\pi) \text{ where } n = 0, 1, 2, 3, \ldots$$
$$\cos(\theta + \pi) = -\cos\theta$$
$$\cos(\theta - \pi) = -\cos\theta$$
$$\cos(\theta) = \cos(-\theta)$$

The cosine function is equal to 1 if the angle is zero or an even multiple of $\pm\pi$, and -1 if it is an odd multiple of $\pm\pi$.

Note also that:

$$\cos(\theta + \pi/2) = -\sin\theta$$
$$\cos(\theta - \pi/2) = \sin\theta$$
$$\sin(\theta + \pi/2) = \cos\theta$$
$$\sin(\theta - \pi/2) = -\cos\theta$$

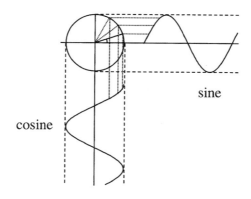

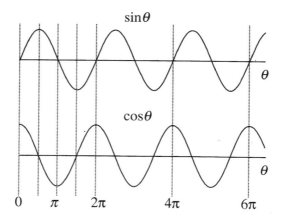

FIG. A.13   Comparing sine and cosine functions.

Phasor Diagram

In the theory of vibrations, we often require the sine or cosine of an angle ωt that increases progressively with time t at ω radians per second. Moreover, we often wish to compare two vibrations that have the same frequency but differ in phase by an angle ϕ. For simplicity let us suppose that these vibrations are

$$u = a \sin \omega t$$

and

$$v = b \sin(\omega t + \phi)$$

In the previous discussion we used the rotating line OP to generate the sine functions. We should regard this line as a vector whose length is proportional to the amplitude of a vibration and whose direction represents the angle for which the sine is required. The first of the vibrations just given can be represented in circle diagram with a radius vector of length a rotating anticlockwise at the rate ω. The second can be similarly represented by a radius vector of length b. If we take the horizontal to represent the angle at time zero, then at time t these vectors point in the directions ωt and $\omega t + \phi$, respectively. We can draw both vectors as lines from a common centre as in Figure A.14. Because they have the same angular frequencies, they rotate

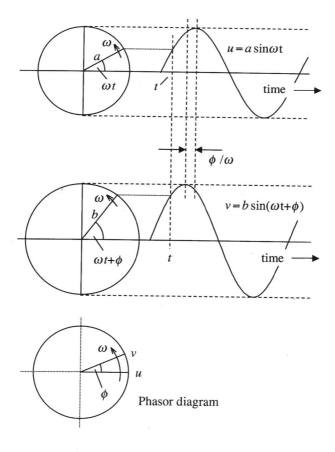

FIG. A.14 Phasor diagram.

together at the rate ω. The important aspect of the diagram is that the angular separation of the vectors is constant. It is equal to the phase difference ϕ of the two vibrations. The diagram is called a *phasor diagram*, and its purpose is to depict phase relationships. We can see immediately that vibration v leads vibration u because ϕ is positive. Amplitudes are not usually represented fully, if only because they often refer to different physical quantities and so cannot be drawn to scale in the same diagram. We can think of the rotating vectors in a phasor diagram as having unit length, so that in the example just given they represent $\sin \omega t$ rather than $a \sin \omega t$, and $\sin(\omega t + \phi)$ rather than $b \sin(\omega t + \phi)$.

Relationships Between the Trigonometric Functions

Figure A.15 shows a right-angled triangle. The length of the hypotenuse is c, and the lengths of the sides opposite the interior angles A and B are a and b, respectively. By definition, $\sin A = a/c$, $\cos A = b/c$, and $\tan A = b/c$. Therefore,

$$\tan A = \frac{\sin A}{\cos A} \qquad \text{A.2}$$

But

$$a^2 + b^2 = c^2$$

so that

$$(a/c)^2 + (b/c)^2 = 1$$

and therefore

$$[\sin A]^2 + [\cos A]^2 = 1$$

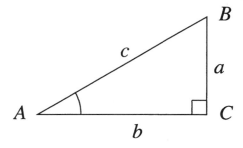

FIG. A.15   Right triangle to define the trigonometric functions of the angle A.

This important relationship is true for all values of the angle A. A neater way of writing it is

$$\sin^2 A + \cos^2 A = 1 \qquad\qquad \text{A.3}$$

By similar reasoning we find that

$$\sec^2 A = 1 + \tan^2 A$$

and

$$\operatorname{cosec}^2 A = 1 + \cot^2 A$$

Inverse Trigonometric Functions

The function that gives the value of the angle whose sine is x is called the *inverse sine* of x. There are similar inverse functions corresponding to cosine and tangent and the other trigonometric functions (see also the section Inverse Functions). There are several ways of naming these functions. Taking the inverse of $\sin A$ as an example, we have

$x = \sin A$
$A = \text{inv } \sin x$
$A = \sin^{-1} x$
$A = \arcsin x$

The first of these is not often seen in print, but it is found on pocket calculators to show the keystrokes needed to obtain the inverse function. The second form is quite common but has the disadvantage of possible confusion with $(\sin x)^{-1}$. The third form is frequently used in print. The shortened version ASIN( ) is often a named function in computer software.

The trigonometric functions are periodic. The corresponding inverse functions therefore have multiple values. For instance:

$\cos(2\pi/3) = \cos(-2\pi/3) = -1/2$
$\cos(4\pi/3) = \cos(-4\pi/3) = -1/2$
$\arccos(-1/2) = \pm n\pi/3$ where $n = 2, 8, 14, \ldots$ or $n = 4, 10, 16, \ldots$

Because the inverse functions provide multiple answers, care is needed to see that the appropriate answers are returned when they are evaluated using calculators or computers. In most cases the smallest result in the range $-\pi$ to $+\pi$ will be offered. In the preceding example, a calculator or computer would give the value of $\arccos(-1/2)$ as $2\pi/3$ ($120°$).

The inverse function that we are most likely to need in elementary acoustics is the arctangent. This is required when finding the phase angle whose tangent is the ratio of the imaginary and real parts of the acoustic impedance. In this context the required angle must lie in the range $-\pi$ to $+\pi$. Although this narrows the choice, there is still room for ambiguity if a simple inverse tangent is calculated electronically. Some calculators and some computer software provide arctan functions of fractions for which the numerator and denominator have to be entered separately.[8] There is then no ambiguity and the calculated result is necessarily correct.

The Compound Angle Formulae

If A and B are any two angles, their sum or difference is called a *compound angle* when it appears as the argument of one of the trigonometric functions. We often need to express the result in terms of functions of these angles taken singly. The basic formulae are as follows:

$$\sin(A + B) = \sin A \cos B + \cos A \sin B \qquad \text{A.4}$$

$$\sin(A - B) = \sin A \cos B - \cos A \sin B \qquad \text{A.5}$$

$$\cos(A + B) = \cos A \cos B - \sin A \sin B \qquad \text{A.6}$$

$$\cos(A - B) = \cos A \cos B + \sin A \sin B \qquad \text{A.7}$$

These formulae are the bedrock of analysis. Here are some examples of their use.

Example 1

Suppose that $A = \omega t$ and $B = \phi$. Equation A.4 then tells us that

$$R \sin(\omega t + \phi) = R \sin \omega t \cos \phi + R \cos \omega t \sin \phi$$

where R is a constant.

If we now let $P = R \cos \phi$ and $Q = R \sin \phi$, we have, after some rearrangement,

$$R \sin(\omega t + \phi) = P \sin \omega t + Q \cos \omega t = P \sin \omega t + Q \sin(\omega t + \pi/2)$$

This means that any harmonic oscillation is equivalent to the sum of two oscillations of the same frequency that differ in phase by $\pi/2$. From Equations A.2 and A.3 and our definitions of P and Q, we see that

$$\tan \phi = Q/P \qquad \text{and} \qquad R^2 = P^2 + Q^2$$

---

[8]Calculators often provide transforms from rectangular to polar coordinates. After entering x and y values, the results $(x^2 + y^2)^{1/2}$ and $\arctan(y/x)$ can be displayed.

Example 2

By putting $A = B$ so that $A + B = 2A$, we obtain useful formulae for the sine and cosine of the double angle. From Equations A.4 and A.6 we see that

$$\sin 2A = 2 \sin A \cos A$$

and

$$\cos 2A = \cos^2 A - \sin^2 A$$

If we now make use of the relationship between sine and cosine functions given in Equation A.3 we obtain

$$\cos 2A = 1 - 2 \sin^2 A$$

and

$$\cos 2A = 2 \cos^2 A - 1$$

If we replace A by $A/2$ in this last equation and rearrange it, we obtain a result that we shall need in Example 4:

$$\cos \frac{A}{2} = \sqrt{\tfrac{1}{2} + \tfrac{1}{2} \cos A} \qquad \text{A.8}$$

Example 3

Suppose two trigonometric functions are multiplied. As an example, consider the product $\sin A \cos B$. There are times when it is useful to express this and similar products in terms of the sum of trigonometric functions. The method is to add or subtract the appropriate compound angle equations. So, for the example just given, adding Equations A.4 and A.5 gives

$$\sin(A + B) + \sin(A - B) = 2 \sin A \cos B$$

This method is used again in the next example. It is also required to explain the creation of side bands in an amplitude-modulated signal (Chapter 5).

Example 4

If we want to combine sounds from different sources, we may wish to add sinusoids having different frequencies, different phases, and different amplitudes. To do this we need a formula for the addition $a \sin x + b \sin y$, where

x is the frequency and phase of one vibration (e.g., $x = \omega t + \phi$) and y expresses the corresponding frequency and phase of the other vibration.

Let

$$A + B = x$$

$$A - B = y$$

$$\alpha + \beta = a$$

$$\alpha - \beta = b$$

Then

$$A = \tfrac{1}{2}(x + y)$$

$$B = \tfrac{1}{2}(x - y)$$

$$\alpha = \tfrac{1}{2}(a + b)$$

$$\beta = \tfrac{1}{2}(a - b)$$

If we add Equations A.4 and A.5 and multiply throughout by α, we obtain

$$\alpha \sin(A + B) + \alpha \sin(A - B) = 2\alpha \sin A \cos B$$

Similarly, subtracting and multiplying by β gives

$$\beta \sin(A + B) - \beta \sin(A - B) = 2\beta \cos A \sin B$$

The latter two equations add to give

$$(\alpha + \beta) \sin(A + B) + (\alpha - \beta) \sin(A - B) = 2\alpha \sin A \cos B + 2\beta \cos A \sin B$$

So that substituting for α, β, A, and B gives the desired result, namely,

$$a \sin x + b \sin y = (a + b) \sin \frac{x + y}{2} \cos \frac{x - y}{2} + (a - b) \cos \frac{x + y}{2} \sin \frac{x - y}{2}$$

If the vibrations have the same amplitudes, then $(a - b)$ is zero, so that this expression simplifies to

$$a \sin x + b \sin y = 2a \sin \frac{x + y}{2} \cos \frac{x - y}{2}$$

The cosine term can be replaced by the cosine of $(x - y)$ using Equation A.8 to give

$$a \sin x + b \sin y = 2a \sqrt{\tfrac{1}{2} + \tfrac{1}{2} \cos(x - y)} \sin \frac{x + y}{2}$$

This now has the same form as the equation used to describe acoustic beats in Chapter 3.

Average Values of the Trigonometric Functions

The average value of a sine or cosine function taken over a whole number of cycles is zero because the average during the negative parts of each cycle exactly cancels the average during the positive parts. However, the absolute value of a sinusoid (a rectified voltage, for example) has a waveform that lies entirely above zero (Figure A.16). It can be shown that the average absolute value during a whole number of cycles is $2a/\pi$, where a is the amplitude. If a sine function is squared, its value is again always one-sided and its average is nonzero. It can be shown that this average over a complete number of cycles is $a^2/2$, where a is the amplitude. The square root of this average is the *root mean square* (rms) value. The rms value of a sinusoid is therefore the peak value (the amplitude) divided by root 2, that is, $a/\sqrt{2}$. The square root of 2 is 1.414.

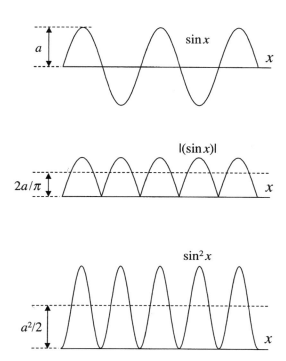

FIG. A.16 Average values of rectified and squared sine functions.

Scientific Notation

Technical writing and calculation frequently involve working with numbers whose values range from the very small to the very large. The expression and arithmetical processing of such numbers is greatly helped by writing them in *scientific notation*. In this form only one digit precedes the decimal point, and the correct magnitude is obtained by multiplication by a power of 10. For example, the speed of light is 299,790,000 ms^{-1}. This can be written more conveniently as 2.9979×10^8 ms^{-1}. A 2-mm diameter raindrop has a mass of about 0.0000042 kg, which could be written better as 4.2×10^{-6} kg. In so-called engineering notation, the powers of 10 are restricted to multiples of 3 and there are then up to three digits before the decimal point. In this form, for example, the speed of light is 299.79×10^6 ms^{-1}. The scientific and engineering formats are available on many pocket calculators. In technical work it is good manners (and good sense) to limit the number of digits in the final result to accord with the precision that is meaningful. The speed of light could legitimately be written as 299.79×10^6 ms^{-1}, but whether this was significantly different from 299.8×10^6 ms^{-1} in a physical rather than an arithmetic sense would depend on the accuracy of the measurement.

Units and Dimensions

Let us start with an example. The kinetic energy of a moving mass is found from the formula

$$E = \tfrac{1}{2}mv^2$$

where m is mass and v is velocity. In order for this, or any other formula, to be true, the quantity on the left-hand side must be of the same physical kind as the quantity on the right-hand side. In the example, the quantity on the left is energy and the quantity on the right, the product mv^2, must also be an energy. We use the word *dimension* to mean 'physical kind' and say that E and mv^2 have *dimensions* of energy. In order to confirm that the formula is dimensionally consistent, it is helpful to express the dimensions of each component in terms of the dimensions of more primitive quantities such as mass, length, and time. The dimensional interpretation is shown by the use of square brackets: $[M]$, $[L]$, and $[T]$.

Energy is defined as force multiplied by distance, and force is mass multiplied by acceleration. Accordingly,

$$[E] = [\text{force}][\text{distance}]$$
$$= [\text{mass}][\text{acceleration}][\text{distance}]$$
$$= [M][LT^{-2}][L]$$
$$= [ML^2T^{-2}]$$

On the right side of the formula the dimensions are

$$[\text{mass} \times \text{velocity}^2] = [M][L^2T^{-2}]$$
$$= [ML^2T^{-2}]$$

The dimensions of quantities on the left side of the formula are therefore the same as those on the right. We can use any units we wish to quantify E, with the proviso that they must be consistent from term to term. If E is in joules, then mass must be in kg and velocity in metres per second; if E is in foot-poundals, then mass must be in pounds and velocity must be in feet per second because the poundal is the force needed to give a mass of 1 pound an acceleration of 1 foot per second.

Vigilance in observing the correctness of dimensions and units helps to avoid mistakes in algebraic working and numerical calculation. Here is an example. Suppose we are deriving an expression for the volume V of an ellipsoid whose principal radii are a and b. We arrive at the expression

$$V = \pi b^2(2a - 2a^2/3)$$

We should notice that this cannot be true because the dimensions of the terms in the parentheses are mixed, $[L]$ and $[L^2]$, whereas only $[L]$ is required, so that the multiplication with b^2 gives a volume $[L^3]$. The last term should be $2a/3$.

Mathematical expressions often contain terms that have no associated physical dimension. Such terms are simply numbers and are said to be *dimensionless*. The constants e and π are dimensionless, as are functions such as e^x, $\sin x$, and $\log x$. Their arguments must also be dimensionless. Consider the formula for the decay of the voltage V across a capacitor C when its charge is allowed to leak through a resistance R for a time t:

$$V = V_o e^{-t/CR}$$

The exponent t/CR must be dimensionless. It is composed of two elements, a time in the numerator and CR in the denominator. From the formula $Q = CV$ where Q is electric charge, we see that capacitance has dimensions $[QV^{-1}]$. But electric current is the rate of flow of charge, so that charge is current $\times$ time and $[Q]$ can be replaced by $[IT]$, giving capacitance the dimensions $[C] = [ITV^{-1}]$. Because $V = IR$, we see that the dimensions of resistance are $[R] = [VI^{-1}]$. The dimensions of the exponent are therefore

$$[t/CR] = \frac{[T]}{[ITV^{-1}][VI^{-1}]} = \frac{[T]}{[T]}$$

The time dimension cancels leaving the result that the exponent is dimensionless and therefore independent of the units of measurement of t, C, and R provided that they are self-consistent.

Vectors

Many quantities have a direction associated with them. Velocity in the full sense of the word includes the direction of travel, a force is said to act in a specified direction, and so on. Such quantities are called *vectors* to distinguish them from *scalar* quantities that have no directional connotation. Ordinary space is three-dimensional and we would need three coordinates to specify the direction of a vector, but to keep things simple we will limit this short account to vectors lying in the same plane.

A vector can be represented diagramatically by a straight line. The length of the line is proportional to the magnitude of the vector, and the direction from the origin of the line to its termination shows the direction of the vector. A vector representing the motion of a ship travelling at 15 knots northeast would be a line whose length is scaled to represent the 15 knots and whose direction from the origin shows a 45° heading. We can take this picture further by imagining that the ship's speed has been measured relative to the water and that the northeasterly heading is the compass bearing. Suppose now that there is an ocean current such that the water is moving due west at 5 knots. What then is the course and speed of the ship relative to the ocean floor? To solve this problem we note that the ocean current can be represented by a vector of length 5 knots in the direction 270°. Adding the two vectors gives the answer to the problem. The addition, represented in Figure A.17, can be performed by scale drawing or trigonometry.

Addition and Subtraction

Vectors A and B can be added as just explained.[9] They can be subtracted by reversing the direction of one of them and then adding. This is illustrated

---

[9] As with complex numbers, we show vectors in ordinary rather than bold type and use modulus signs to identify their magnitudes. For example, A, $|A|$.

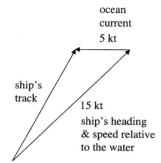

FIG. A.17   Vector addition: adding velocities.

in Figure A.18. The sum and difference of two vectors can be represented by the two diagonals of the parallelogram. To understand this, we must realise that when a line is drawn to represent a vector, the position of the line on the page is immaterial. What matters is its length and direction. A vector can be represented by any one of a number of parallel lines of the same length.

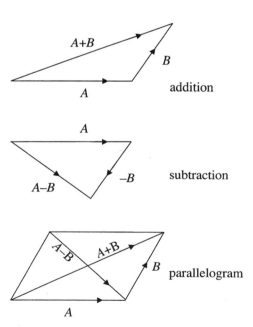

FIG. A.18   Vector addition and subtraction.

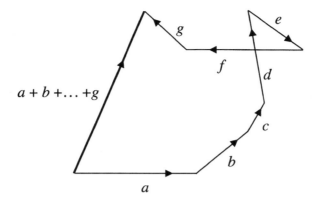

FIG. A.19   A polygon formed by adding vectors.

Figure A.19 shows the addition of several vectors. The diagram is a series of connected lines, and the vector representing the final sum is the line joining the origin of the first vector to the end of the last. The order in which the lines are drawn, that is, the order in which the vectors are added, does not affect the result.

Components

We often require the component of a vector in a specified direction. In the example given earlier, we might want to know the rate at which the ship is progressing in a specified direction — due north, for instance. If vector A is inclined at an angle θ to the x-direction, then its component in the x-direction is a vector in that direction having the magnitude $|A| \cos \theta$.

A vector is equivalent to the sum of any two of its components acting in mutually perpendicular directions, for instance, parallel to the x- and y-axes. If the angle to the x-axis is θ, then the components are $|A| \cos \theta$ and $|A| \sin \theta$, respectively (Figure A.20). The critical reader might object that the components in the form just written are not vectors at all because $|A|$ is scalar and the cosine or sine of an angle likewise. In a more formal mathematical treatment this difficulty is overcome by defining unit vectors (vectors having unit magnitude) directed along the x- and y-axes. These vectors when multiplied by the scalar quantities just given are the true component vectors. Multiplying a vector by a scalar factor simply gives a new vector whose length is increased by the factor and whose direction is unchanged.

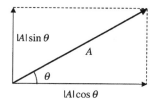

FIG. A.20 Components of a vector.

Multiplication

Can vectors be multiplied? The answer is yes, but the method and the interpretation of the result depend on which convention is applied. There are two important forms of product. The *scalar* or 'dot' product of two vectors is a scalar (nondirectional) quantity found by multiplying the magnitudes of the vectors and the cosine of the angle between them. So for the vectors A and B,

$$\text{scalar product} = A{\cdot}B = |A|\,|B|\cos\theta$$

To give an example of this form of multiplication, suppose that a force acts on an object that can move only in a direction inclined to that of the force. In other words, the force acts obliquely to the direction of movement (Figure A.21). The force and the amount of movement can be represented

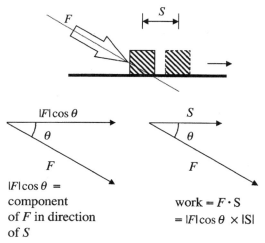

FIG. A.21 Multiplying vectors: the scalar product of the vectors *F* and *S*.

by vectors *F* and *S*. The work done by the force is scalar and equal to the product of the magnitudes of *F* and *S* multiplied by the cosine of the angle between them. A way of looking at this is to see $|F|\cos\theta$ as the magnitude of the component force in the same direction as *S*. This component moves its point of application through a distance $|S|$ in the same direction, doing the work $|F|\,|S|\cos\theta$.

We are not likely to meet the vector, or 'cross', product of vectors in audiological acoustics, and it is mentioned here only for the sake of completeness. If vector *C* is the *vector* product of *A* and *B*, then

$$C = A \wedge B$$

Vector *C* has magnitude $|A|\,|B|\sin\theta$ and lies in a direction perpendicular to the plane containing *A* and *B*.

Appendix B
Working with Decibels

The following worked examples and questions are intended to give the reader practice and confidence in working numerical problems involving decibels. Correct answers are given at the end of the appendix. It is inadvisable to attempt all the questions at one sitting or to work through them from the first to the last. It is better to do a few from each section at regular intervals. A few examples are given here in laborious detail, but this is just to illustrate the basic principles. Try to answer some of the questions in the same way before taking short cuts. Readers who are not entirely at ease with the underlying mathematics should turn to the sections on exponents and logarithms in Appendix A. A calculator or a book of log tables is essential.

To begin, it is important to become completely familiar with the calculator or tables that will be used. Notice that calculators (and computers) often provide two kinds of logarithms, either *log*, which usually (but unfortunately not always) means 'logarithm to base 10', or *ln*, which means 'logarithm to base e'. Base 10 is needed for decibels. Keys that provide the inverse function — the *antilogarithm* — are generally labelled 'inv log' or '10^x'. In computer languages and spreadsheets, the caret ($^\wedge$) is often used to denote an exponent so, for example, the formula ' $= 10^\wedge 2$' in a spreadsheet would be evaluated as 10^2, giving 100 as the result.

Tables

It is anticipated that most readers will use calculators for arithmetical work. If, however, tables are to be used, one has to remember that logarithms are tabulated only for numbers less than 10, and antilogarithms are tabulated only for the decimal part of the logarithm (the mantissa). Here are some examples.

To find $\log(256)$, write this as $\log(10^2 \times 2\cdot56)$, which is the same as $\log(10)^2 + \log(2\cdot56)$. By definition, $\log(10)^2$ is equal to 2, and $\log(2\cdot56)$ is found from tables to be $0\cdot4082$. Therefore, $\log(256) = 2\cdot4082$.

To find $\log(0\cdot00256)$, think of it as $\log(10^{-3} \times 2\cdot56)$, which is $\log(10^{-3}) + \log(2\cdot56)$. Therefore, $\log(0\cdot00256) = -3 + 0\cdot4082 = -2\cdot5918$.

To find antilog(3.16), we note that this is the same as antilog(3 + 0·16), which is equal to antilog(3) × antilog(0·16). The first factor, which is not tabulated, is 10^3; the second factor is found in the tables to be 1·445, so that the result in scientific form is $1·445 × 10^3$.

To find antilog(−3·16), we first have to replace the number in the parentheses by a negative integer added to a positive decimal. Accordingly, we write antilog(−3·16) = antilog(−4 + 0·84), which is antilog(−4) × antilog(0·84). The tables show antilog (0·84) to be 6·9183, to give the result $6·9183 × 10^{-4}$.

Preliminary Exercise

Use a calculator or tables or common sense to evaluate the following. The answers are shown in the square brackets.

(a) log(1000)   [3]
(b) $\log(10^{15})$ [15]
(c) log(343.5)   [2·54]
(d) log(1)   [0]
(e) $\log(\pi)$ [0·4971]
(f) $\log(e)$ [0·4343]
(g) log(17)   [1·23]
(h) log(1/17)   [−1·23]
(i) $\log(1·85 × 10^{-6})$ [−5·73]
(j) log(0·00211)   [−2·68]
(k) $\log(10^{7.23})$ [7·23]

(l) log(−3)   [complex]
(m) log(0)   [−∞]
(n) log(0·000155)   [−3·81]
(o) $10^{1·356}$ [22·7]
(p) $10^{13·56}$ [$3·63 × 10^{13}$]
(q) 10^0 [1]
(r) $10^{\log(2·33)}$ [2·33]
(s) $10^{-2·03}$ [$9·33 × 10^{-3}$]
(t) $10^{-1/1·23}$ [0·154]
(u) $1/10^{1/1·23}$ [0·154]
(v) $10^{18·55}$ [$3·55 × 10^{18}$]

Converting Absolute Values to Decibels

We should be aware that the quantity in decibels is always acoustic power or one of its cousins, such as sound intensity, so the basic formula is $n = 10 \log W/W_0$, where W is the power of the sound source or a quantity proportional to it and W_0 is a reference power. When, as often happens, we have to work with sound pressure or its relatives, such as voltage or particle velocity, the values of the physical quantities have to be squared to obtain the corresponding power ratios. Because $\log x^2 = 2 \log x$, the squaring is conveniently done by multiplying by 2 in the basic formula; that is, $n = 20 \log p/p_0$.

Exercise 1

Express the following in decibels relative to the reference value given. If no reference value is given, assume it to be a sound pressure of 20 μPa. Give the results to the nearest 0·1 dB.

Example

Express in decibels, 40 mPa re 1 Pa.

$$n = 20 \log p/p_0 = 20 \log(40 \times 10^{-3}/1)$$

$$= 20 \log(4 \times 10^{-2}) = 20(\log 4 + \log 10^{-2})$$

$$= 20 \times (0·602 - 2)$$

$$= -28·0 \text{ dB}$$

(a) 500 Pa re 1 Pa
(b) 2000 μPa
(c) 100 μPa re 1 mPa
(d) 78·6 μPa
(e) 16 mW re 1 μW
(f) 5 μV re 1 μV
(g) 0·8 kPa
(h) 1 μV re 1 V
(i) 1 μPa
(j) 10 μV re 1 nV
(k) 18 mW re 1 W
(l) 7·52 mA re 1 μA
(m) 0·00022 ms^{-1} re 1 m/s
(n) 4·3 $\times$ 10^4 μW re 1 μW

(o) 200 mW/m^2 re 1 μW/m^2
(p) 1 μW/cm^2 re 1 W/m^2
(q) 2 V re 1 mV
(r) 220 μPa
(s) 0·007 Pa
(t) 0·3 W re 1 μW
(u) 0·6 mV re 5 μV
(v) 8 μPa
(w) 0·325 mPa re 1 microbar
  (1 bar = 10^5 Nm^{-2})
(x) 16 μm re 100 nm
(y) 1·7 $\times$ 10^{-3} J/m^3 re 10^{-6} J/m^3
(z) 0·17 dynes/cm^2
  (1 dyne = 10^{-5} N)

Converting Quantities Expressed in Decibels to Absolute Values

This operation is the converse of that in the preceding section. If the reference quantity has units of power or its equivalent (watts or watts per m^2, for example), then the first step is to divide the number of decibels by 10 before finding the antilog. If the square of the reference quantity is proportional to power, we divide by 20 to avoid having to find a square root. This means that we divide by 20 if the answer is required in units of sound pressure, voltage, acceleration, and so on.

Exercise 2

Convert decibels to absolute units. The required unit is shown in the square brackets. Where no reference value is given, assume it to be 20 μPa. Where appropriate, give the answer to three significant figures.

Example 1

Find the acoustic power corresponding to a power level of 31·5 dB re 1 μW.

$$10 \log W/W_0 = 31 \cdot 5 \quad \text{where } W \text{ is power in watts}$$

$$10 \log(W/10^{-6}) = 31 \cdot 5$$

$$\log(W/10^{-6}) = 3 \cdot 15$$

$$W/10^{-6} = 10^{3 \cdot 15}$$

$$W = 10^{3 \cdot 15} \times 10^{-6}$$

$$= 10^3 \times 10^{0 \cdot 15} \times 10^{-6}$$

$$= 1 \cdot 41 \times 10^{-3} \text{ watts}$$

Example 2

Find the acoustic pressure corresponding to a sound pressure level of 52 dB.

$$20 \log p/p_0 = 52 \quad \text{where } p \text{ is the sound pressure in pascals}$$

$$20 \log[p/(20 \times 10^{-6})] = 52$$

$$\log [p/(20 \times 10^{-6})] = 2 \cdot 6$$

$$p/(20 \times 10^{-6}) = 10^{2 \cdot 6}$$

$$p = 10^{2 \cdot 6} \times 20 \times 10^{-6}$$

$$= 10^2 \times 10^{0 \cdot 6} \times 20 \times 10^{-6}$$

$$= 10^2 \times 3 \cdot 98 \times 20 \times 10^{-6}$$

$$= 7 \cdot 96 \times 10^{-3} \text{ Pa}$$

(a) 60 dB re 1 μW [μW]
(b) 35 dB [μPa]
(c) 120 dB re 1 μV [V]
(d) −67 dB re 1 Pa [μPa]
(e) 9·4 dB re 1 mV [V]
(f) 3 dB re 1 mPa [mPa]

(g) -41 dB re 1 W/m^2 [W/m^2]
(h) -10 dB [μPa]
(i) 94 dB re 1 μV [mV]
(j) 33·2 dB [Pa]
(k) -60 dB re 1 W [μW]
(l) -6 dB re 100 mV [mV]
(m) 12 dB re 0·01 Pa [Pa]
(n) -40·3 dB 1 Pa/V [Pa/V]
(o) 23 dB re 1 W/m^2 [W/cm^2]
(p) 1·1 dB re 1 V/Pa [mV/Pa]
(q) 87·2 dB [μPa]
(r) -3 dB re 1 W [W]
(s) 110 dB re 1 μPa [μPa]
(t) 72 dB re 100 nPa [μPa]
(u) -0·4 dB re 1 μW [μW]
(v) 33·7 dB re 1 m/s^2 [m/s^2]
(w) 33·7 dB re 1 m/s [m/s]
(x) -41 dB re 1 W/m^3 [μW/m^3]
(y) 0·1 dB re 1 mW [mW]
(z) 83·5 dB [mPa]

Changing the Reference Value

The zero point of any decibel scale is determined by the reference value. For sound pressure level, for example, the reference value is usually 20 μPa. Occasionally we may wish to use other reference values. Converting from one scale to another is usually quite straightforward. Increasing the reference value raises the zero point of the decibel scale and decreases the number of decibels that correspond to a given physical magnitude. The converse is also true. For example, a voltage level of 70 dB re 1 μV would be the same as 10 dB re 1 mV because the reference value has been increased 1000-fold, and accordingly the scale values must be reduced by 20 log(1000), that is, by 60 dB. Similarly, a sound intensity level of 17 dB re 1 W/m^2 would be 77 dB above 1 μW/m^2 because the reference has been reduced by a factor of 10^6, raising scale values by 10 log(10^6), that is, by 60 dB. The foregoing may seem obvious, but we sometimes encounter conversions that need careful thought if mistakes are to be avoided. Here is an example.

Suppose that the sensitivity M of a microphone is given by the manufacturer as -20·3 dB re 1 mV/0·01 Pa. This is a rather unfriendly unit, so we would like to change it to V/Pa. We may argue as follows. Let M_0 be the reference sensitivity. Then

$$M_0 = 1 \text{ mV}/0.01 \text{ Pa}$$
$$= 10^{-3} \text{ V}/10^{-2} \text{ Pa}$$
$$= 0.1 \text{ V/Pa}$$

A change to 1 V/Pa will therefore increase the reference sensitivity by a factor of 10 and so decrease the number of decibels by 20 log 10, that is, by 20. The sensitivity of the microphone is therefore $(-20.3 - 20) - = 40.3$ re 1 V/Pa.

Exercise 3

Refer the levels in the following examples to the new reference values shown in the square brackets.

(a) 35 dB re 20 μPa [1 μPa]
(b) -42 dB re 1 μW [1 mW]
(c) 63 dB re 1 cm/s² [1 m/s²]
(d) -22 dB re 1 N [1 μN]
(e) 6.3 dB re 1 V [1 mV]
(f) 51 dB re 1 μbar [20 μPa]
    (1 bar = 10^5 Pa)
(g) -16 dB re 1 mN [1 μN]

(h) 105 dB(A) re 1 mPa [20 μPa]
(i) 27 dB re 1 W/m² [1 μW/cm²]
(j) 87 dB re 1 Pa²h [1 pa²s]
(k) 19 dB re 1 dyne/cm² [20 μPa]
    (1 dyne = 10^{-5} N)
(l) 73 dB re 1 erg/cm³ [1 J/m³]
    (1 erg = 10^{-7} J)

Adding and Subtracting Quantities Expressed in Decibels

It may seem that questions on 'adding decibels' are contrived to torment students, but the need to add or subtract energy-related quantities often arises in practice. In Chapter 6 we saw that sound-level readings could be corrected for inherent noise in the measuring instruments or for background noise by subtracting the noise from the signal plus the noise. The arithmetical difficulty in this process is that the measured quantities are expressed in decibels. A similar problem occurs when addition is required. For example, we may wish to combine sound levels measured in third-octave bands to find the level in a one-octave band. Another example is the addition of sound exposures occurring in different periods of work. The method is as follows:

1. Convert decibels to absolute values.
2. Add or subtract absolute values as required.
3. Convert absolute values to decibels.

Steps 1 and 3 have already been practised in Exercises 1 and 2, so the only further requirement is the addition or subtraction in step 2. It should be noted, however, that the quantities to be added or subtracted must be proportional to power (or energy).

We start with a simple example. Two independent sources, A and B, produce noise levels of 12 dB and 14 dB respectively when acting alone. What level is expected when both sources are on at the same time?

$$L_A = 10 \log \frac{p_A^2}{p_0^2} \quad \text{and} \quad L_B = 10 \log \frac{p_B^2}{p_0^2}$$

where $L_A = 12$ dB and $L_B = 14$ dB. Therefore,

$$\frac{p_A^2}{p_0^2} = 10^{L_A/10} \quad \text{and} \quad \frac{p_B^2}{p_0^2} = 10^{L_B/10}$$

where p_A^2 and p_B^2 are the mean squared sound pressures corresponding to the action of sources A and B, respectively. The squares of the sound pressures are proportional to the acoustic power of the sources, and they may be added:

$$p_{AB}^2 = p_A^2 + p_B^2$$

Therefore,

$$\frac{p_{AB}^2}{p_0^2} = \frac{p_A^2}{p_0^2} + \frac{p_B^2}{p_0^2}$$
$$= 10^{L_A/10} + 10^{L_B/10}$$

Therefore,

$$10 \log \frac{p_{AB}^2}{p_0^2} = 10 \log[10^{L_A/10} + 10^{L_B/10}]$$

so that the sound level when both sources are on at the same time is

$$L_{AB} = 10 \log[10^{L_A/10} + 10^{L_B/10}]$$
$$= 10 \log[10^{12/10} + 10^{14/10}]$$
$$= 10 \log[10^{1\cdot 2} + 10^{1\cdot 4}]$$
$$= 10 \log[15\cdot 85 + 25\cdot 12]$$
$$= 16\cdot 1 \text{ dB}$$

This arithmetic is very easily accomplished using an electronic calculator. Models differ, but the sequence of keystrokes shown in Table B.1, or something like it, will give the required result.

TABLE B.1   Adding Decibels on a Calculator

Keystroke	Display	Comment
1·2	1·2	12 ÷ 10 done mentally
INV LOG	15·848932	$10^{1·2}$
+		
1.4	1·4	
INV LOG	25.118864	$10^{1·4}$
=	40·967796	$10^{1·2} + 10^{1·4}$
LOG	1·6124426	$\log(10^{1·2} + 10^{1·4})$
×		
10	10	
=	16·124426	$10 \times \log(10^{1·2} + 10^{1·4})$ *Result:* 16·1 dB

Exercise 4

(a) Two independent sources of noise produce sound pressures of 85·0 dB and 89·5 dB respectively when acting alone. What sound pressure will they produce together?

(b) With the earphone disconnected from its electrical supply, the sound pressure level in an acoustic coupler is 10·3 dB. When the earphone is reconnected the sound pressure level is 11·6 dB. Estimate the true sound pressure level due to the earphone alone.

(c) Noise immission levels during three periods of employment are estimated as 87, 85, and 91 dB(A), respectively. What is the noise immission level for the total exposure?

(d) Measurements of a sound spectrum give the following third-octave levels in the range 400 to 1250 Hz.

Centre frequency (Hz)	400	500	630	800	1000	1250
dB SPL	83	81	80	80	79	73

Calculate (i) the total sound level for the range 400 to 1250 Hz and (ii) the levels in the octave bands centred on 500 and 1000 Hz.

(e) The sound pressure level of a broad-band noise is 93 dB. When the noise source is filtered to remove all components between 1000 and 1200 Hz, the sound pressure level falls to 89 dB. What was the sound pressure level of the filtered components?

(f) It can be shown that the energy in a periodic signal is equal to the sum of the energies in each of its harmonic (Fourier) components. A sinusoidal signal is distorted by the presence of harmonics. The voltage levels of the second and third harmonics are respectively $-2·8$ and $-9·7$

dB relative to the level of the fundamental (the first harmonic). Other harmonics are negligible. The voltage level of the fundamental is 72·2 dB re 1 mV. What voltage level can be expected in the signal as a whole?

(g) An employee in a noisy factory is subject to a personal daily exposure of 81 dB(A) during the first four days of the week and 85 dB(A) on the fifth day. What is his average exposure over the five days?

(h) The inherent noise in a sound-measuring system is 11·3 dB referred to its input. What is the least sound pressure level that can be measured such that the error due to noise in the equipment will not exceed 1 dB?

Answers

Exercise 1

(a) 54·0 dB	(j) 80·0 dB	(s) 50·9 dB
(b) 40·0 dB	(k) −17·4 dB	(t) 54·8 dB
(c) −20·0 dB	(l) 77·5 dB	(u) 41·6 dB
(d) 11·9 dB	(m) −73·2 dB	(v) −8·0 dB
(e) 42·0 dB	(n) 46·3 dB	(w) −49·8 dB
(f) 14·0 dB	(o) 53·0 dB	(x) 44·1 dB
(g) 152·0 dB	(p) −20·0 dB	(y) 32·3 dB
(h) −120·0 dB	(q) 66·0 dB	(z) 58·6 dB
(i) −26·0 dB	(r) 20·8 dB	

Exercise 2

(a) 10^6 μW	(j) $9·14 \times 10^{-4}$ Pa	(s) $3·16 \times 10^5$ Pa
(b) $1·12 \times 10^3$ μPa	(k) 1 μW	(t) 398 μPa
(c) 1 V	(l) 50·1 mV	(u) 0·912 μW
(d) 447 μPa	(m) $3·98 \times 10^{-2}$ Pa	(v) 48·4 m/s^2
(e) $2·95 \times 10^{-3}$ V	(n) $9·66 \times 10^{-3}$ Pa/V	(w) 48·4 m/s
(f) 1·41 mPa	(o) 0·0200 W/cm^2	(x) 79·4 μW/m^3
(g) $7·94 \times 10^{-5}$ W/m^2	(p) $1·14 \times 10^3$ mV/Pa	(y) 1·02 mW
(h) 6·32 μPa	(q) $4·58 \times 10^5$ μPa	(z) 0·300 mPa
(i) 50·1 mV	(r) 0·501 W	

Exercise 3

(a) 61·0 dB re 1 μPa	(g) 44 dB re 1 μN
(b) −72 dB re 1 mW	(h) 139·0 dB(A) re 20 μPa
(c) 23 dB re 1 m/s^2	(i) 47 dB re 1 μW/cm^2
(d) 98 dB re 1 μN	(j) 122·6 dB re 1 pa^{2}s

(e) 66·3 dB re 1 mV   (k) 93·0 dB re 20 μPa
(f) 125·0 dB re 20 μPa   (l)   63 dB re 1 J/m^3

Exercise 4

(a) 90·8 dB   (e) 90·8 dB
(b) 5·7 dB   (f) 74·3 dB re 1 mV
(c) 93·2 dB(A)   (g) 82·1 dB(A)
(d) (i) 88·0 dB, (ii) 86·3 and 83·0 dB   (h) 17·2 dB

Answers to Numerical Problems

Chapter 2

(2.1) 2.73×10^3 radian s^{-1}
(2.2) 204, 3.96×10^3, 3.14×10^3, 1.26×10^4, 5.03×10^4 radian s^{-1}
(2.3) 2.55, 25.0, 239, 1000, 5250 Hz
(2.4) $\pi/2$
(2.5) $\pi/4$
(2.9) 4
(2.12) 0.66 N, 1.49 Hz
(2.14) 5.3
(2.16) rms value $= 1/\sqrt{2} = 0.707$
(2.17) 4.44×10^{-3} ms^{-1}, 79.0 ms^{-2}
(2.18) 4.67×10^{-2} ms^{-1}, 1.26 W

Chapter 3

(3.2) $7.4 \times 10^{-7} \times$ atmospheric pressure
(3.4) 2.12 μm, 0.0133 ms^{-1}
(3.8) 340.3 ms^{-1}, 761 mph
(3.10) 7.46 minutes
(3.13) 68, 17, 8.5 cms
(3.14) 1.771 kHz
(3.15) 31.4 m^{-1}
(3.16) 29.6 m^{-1}, 0.503 radian (28.8°)
(3.17) 0.217 μm
(3.20) (a) 0.6 W, (b) 60 W, (c) 0.54 μWm^{-2}, (d) 0.060 Wm^{-2}
(3.26) 4500 Hz
(3.27) 651 mm (It is in fact slightly shorter.)
(3.28) 221 ms^{-1}
(3.29) 0.29%
(3.33) (a) 0.01 Pa, (b) 0.02 Pa
(3.37) 28.6°
(3.38) 353 ms^{-1}

Chapter 4

(4.3)   (a) $3 \cdot 40 \times 10^{-3} \ m^3 s^{-1}$, (b) $1 \cdot 06 \times 10^{-4} \ m^3 s^{-1}$
(4.4) $0 \cdot 0202$ Pa
(4.7) $0 \cdot 259$

Chapter 5

(5.3)   890.9 Hz, 1122.5 Hz, 695 μW
(5.7) $4/\pi$, $4/3\pi$
(5.9)   4 kHz
(5.10) 92 to 100 kHz; 100 to 108 kHz

Chapter 6

(6.2) $2 \cdot 67 \times 10^{-4}$ V/Pa, $3 \cdot 08 \times 10^{-4}$ V/Pa, $1 \cdot 2$ dB
(6.3)   increase, negative
(6.10) (a) $0 \cdot 0085$ V/Pa, $-6 \cdot 0$ dB, (b) $0 \cdot 0030$ V/Pa, $-15 \cdot 0$ dB
(6.14) $70 \cdot 6$ dB SPL
(6.15) $-35 \cdot 1$ dB
(6.16) $-0 \cdot 28$ dB
(6.21) $44 \cdot 1$ dB(A)
(6.23) $0 \cdot 110$ mV, 288 ms
(6.24) $e^{-2 \cdot 5}$ (equals $0 \cdot 082$ or $-10 \cdot 9$ dB)
(6.25) (a) 1 sone, (b) 52 sones
(6.26) (a) 40 phons, (b) $83 \cdot 2$ phons
(6.28) $117 \cdot 8$ dB, $20 \cdot 8$ dB
(6.29) $90 \cdot 4$ dB
(6.30) $94 \cdot 8$ dB, $95 \cdot 6$ dB

Chapter 7

(7.1) $112 \cdot 9$ mV, $1 \cdot 80$ mW
(7.2) $0 \cdot 273$ mA
(7.3) $0 \cdot 4$ A, $1 \cdot 6$ W
(7.4) $0 \cdot 72$ mA, $17 \cdot 3$ mW
(7.5)   306 kΩ
(7.6) $19 \cdot 3$ V
(7.7) $2 \cdot 073 \times 10^3$ radian s^{-1}
(7.8)   (a) $180°$, (b) $90°$, (c) $135°$, (d) $-90°$, (e) $60°$
(7.10) $5 \cdot 28$ cm, $0 \cdot 756$ radian
(7.12) $V_R = 4 \sin \omega t$, $V_X = 6 \sin(\omega t + \pi/2)$
        $Z = \sqrt{13} = 3 \cdot 61$ ohms, $\phi = \arctan(3/2) = 0 \cdot 983$ radian ($56 \cdot 3°$)
        peak voltage = $Z \times$ peak current = $2\sqrt{13} = 7 \cdot 21$ volts

(7.14) $3 - 4j$, $(2 - 5j)(2 + 5j) = 2^2 + 5^2 = 29$, $\sqrt{29} = 5\cdot385$

(7.15) $Z = 10/17 + 11j/17$; $|Z| = [\sqrt{(10^2 + 11^2)}]/17 = 0\cdot874$;

$\phi = \arctan[(11/17) \div (10/17)] = \arctan(11/10) = 0\cdot833$ radian $(47\cdot7°)$

(7.17) $4\cdot25$ μm, $0\cdot942$ ms^{-2}, 75 Hz, $-3\cdot5$ dB re 1 ms^{-2}

(7.18) $1\cdot04 \times 10^{-2}$ Nsm^{-1}, $1\cdot76 \times 10^{-7}$ N

(7.19) 2800 Nm^{-1}, $8\cdot9$ Nsm^{-1}

(7.21) 195 Hz, $0\cdot083$ ms^{-1}, $2\cdot72 \times 10^{-3}$ ms^{-1}

(7.22) $20\cdot4$

(7.23) (a) independent, (b) proportional

(7.24) 414 Nsm^{-3}

(7.25) $1\cdot41 \times 10^{-8}$ Wm^{-2}

(7.26) -10.9 dB

(7.27) $6\cdot28 \times 10^{-10}$ m^{3}s^{-1}, $9\cdot87 \times 10^{-5}$ m^{3}s^{-1}

(7.28) $1\cdot5 \times 10^{-11}$ m^3, $1\cdot15 \times 10^{-4}$ ms^{-1}

(7.29) $4\cdot92$ 10^{-12} m^{3}Pa^{-1}, $1\cdot47 \times 10^8$ Nsm^{-5}

(7.30) $27\cdot6\%$ increase

(7.31) $250\cdot1$ Hz

(7.33) $G = 1\cdot95$ mmho, $B = 1\cdot30$ mmho, $Y = 2\cdot34 \times 10^{-8}$ N^{-1}s^{-1}m^5

(7.34) $9\cdot990 \times 10^7$ Nsm^{-5}, $1\cdot001 \times 10^{-8}$ N^{-1}s^{-1}m^5,

$7\cdot049 \times 10^{-12}$ m^{3}Pa^{-1}

(7.35) $6\cdot8 \times 10^{-9}$ N^{-1}s^{-1}m^5

(7.36) $101\cdot9$ mm water

(7.39) $1\cdot60 \times 10^6$ Nsm^{-5}, springlike

Chapter 8

(8.2) $57\cdot6$ dB HL

(8.3) 6, 5, -3, 8 dB

(8.4) $6\cdot3\%$, limit is $5\cdot5\%$

(8.5) 891, 1122 Hz; 63 dB SPL

(8.6) $79\cdot3$ pe SPL, $84\cdot1$ dB peak SPL

(8.7) $70\cdot6$ dB SPL, $59\cdot6$ dB HL

(8.9) 70 dB SPL

Chapter 9

(9.3) T_c, T_b, T_a

(9.4) 68 dB SPL

(9.7) (a) $4\cdot45$ m^2, $1\cdot81$ s; (b) $6\cdot35$ m^2, $1\cdot27$ s

(9.8) 212 ms, 82.4 ms

Index